AF352801

Aero/Hydrodynamics and Symmetry

Aero/Hydrodynamics and Symmetry

Editor

Mostafa Safdari Shadloo

MDPI • Basel • Beijing • Wuhan • Barcelona • Belgrade • Manchester • Tokyo • Cluj • Tianjin

Editor
Mostafa Safdari Shadloo
Normandie University
France

Editorial Office
MDPI
St. Alban-Anlage 66
4052 Basel, Switzerland

This is a reprint of articles from the Special Issue published online in the open access journal *Symmetry* (ISSN 2073-8994) (available at: https://www.mdpi.com/journal/symmetry/special_issues/Aero_Hydrodynamics_Symmetry).

For citation purposes, cite each article independently as indicated on the article page online and as indicated below:

LastName, A.A.; LastName, B.B.; LastName, C.C. Article Title. *Journal Name* **Year**, *Article Number*, Page Range.

ISBN 978-3-03936-892-1 (Hbk)
ISBN 978-3-03936-893-8 (PDF)

Contents

About the Editor

Mostafa Safdari Shadloo has been actively engaged in the fields of (i) (aero-)hydrodynamics, turbulence and transitional boundary layers, as well as (ii) multiphase, multi-physics fluid flows and heat transfer, for the last 10 years. His expertise is mainly in theoretical and computational fluid dynamics (CFD), but he has also been active in developing validation strategies and guidelines for CFDist. He aims to develop a new generation high-order coupled algorithm for compressible/incompressible fluid flows with complex physical behaviors, in relation to industrial applications. In this framework, he uses high-performance computing (HPC), high-fidelity direct numerical simulations (DNS) and large-eddy simulations (LES) to decipher complex instabilities and flow behaviors caused mainly by multi-phase and/or turbulent flows, with heat transfer and compressibility effects.

In summary, at national, European and international levels, Dr. Shadloo has actively been the PI and main participant (MP) in numerous projects dealing with unsteady multi-physics flows, including multi-disciplinary modeling, simulation and validation, with an overall budget of more than €2 M (not all of them are listed above). The main outcomes of Dr. Shadloo's research have been published as 75 (+1 book chapter) original scientific articles in highly prestigious peer-reviewed journals and 24 proceedings presented in international peer-reviewed conferences. His citation number exceeds 2750, and to date, 2000 have an h-index of 30 and 27, respectively, based on Google Scholar and Scopus citation reports.

Preface to "Aero/Hydrodynamics and Symmetry"

The existence of symmetry and its tendency to break in aero/hydrodynamics applications are two of the most important aspects of many engineering fields, such as mechanical, aerospace, chemical and process engineering. For instance, the existence of symmetry breaking at a critical Reynolds number confirmed the existence of a bifurcation in expansion pipe flows. Such a symmetry breaking mechanism may cause the appearance of turbulence, which in return increases the mixing, as well as the required pumping power, for several process engineering design applications. Meanwhile, in aerospace applications, the receptivity of symmetric laminar flow to internal/external perturbations may cause flow transition and dramatic change in a local drag coefficient, and heat removal from the surface. The latter needs to be considered in the design step for choosing proper materials that can also bear the unbalanced thermodynamics loads.

The applications of symmetry and its breaking are usually inter-disciplinary, and prior knowledge of them is crucial for many real-life applications. Therefore, the current Special Issue, "Aero/Hydrodynamics and Symmetry", invites original and review works in the field for participation. The scope of this Special Issue includes, but is not limited to, the state of the art computational, theoretical and experimental works that deal with symmetry and its breaking, that are in line with aero-hydrodynamics applications. Recent advances in numerical, theoretical and experimental methodologies, as well as finding new physics, new methodological developments and their limitations, are within the scope of the current Special Issue. Potential topics which are deemed suitable for publication include, but are not limited to:

- Mathematical models, such as: symmetry method, homotropy perturbation method (HPM), homotropy analysis method (HAM), lie group, integral transform, etc.

- Equilibrium and out of equilibrium thermodynamics and fluid mechanics

- Hydrodynamics for symmetric exclusion

- Hydrodynamics with multiple higher-form symmetries

- Ideal order and dissipative fluids with q-form symmetry

- Partial and fractional order differential equations

- Finite difference (FDM), finite volume (FVM), and finite element (FEM), smoothed particle hydrodynamics (SPH), moving particle semi-implicit (MPS), lattice Boltzmann (LBM) methods, etc.

- Multiphysics phenomena, such as non-Newtonian flows, multiphase flows, phase change, nanofluidic, magnetohydrodynamics (MHD), electrohydrodynamics (EHD), etc.

- Symmetry and its breakdown in transitional and turbulent flows

Mostafa Safdari Shadloo
Editor

Article

Numerical Investigation of the Savonius Vertical Axis Wind Turbine and Evaluation of the Effect of the Overlap Parameter in Both Horizontal and Vertical Directions on Its Performance

Mohammad Ebrahimpour [1], Rouzbeh Shafaghat [1,*], Rezvan Alamian [1] and Mostafa Safdari Shadloo [2]

[1] Sea-Based Energy Research Group, Babol Noshirvani University of Technology, 47148 Babol, Iran; m.ebrahimpour.0123@gmail.com (M.E.); ralamian@nit.ac.ir (R.A.)

[2] CORIA Lab./CNRS, University and INSA of Rouen, 76000 Rouen, France; msshadloo@coria.fr

* Correspondence: rshafaghat@nit.ac.ir; Tel.: +98-(11)-3233-2071-1333; Fax: +98-(11)-3231-0968

Received: 28 May 2019; Accepted: 19 June 2019; Published: 21 June 2019

Abstract: Exploiting wind energy, which is a complex process in urban areas, requires turbines suitable for unfavorable weather conditions, in order to trap the wind from different directions; Savonius turbines are suitable for these conditions. In this paper, the effect of overlap ratios and the position of blades on a vertical axis wind turbine is comprehensively investigated and analyzed. For this purpose, two positive and negative overlap situations are first defined along the X-axis and examined at the different tip speed ratios of the blade, while maintaining the size of the external diameter of the rotor, to find the optimum point; then, the same procedure is done along the Y-axis. The finite volume method is used to solve the computational fluid dynamics. Two-dimensional numerical simulations are performed using URANS equations and the sliding mesh method. The turbulence model employed is a realizable K-ε model. According to the values of the dynamic torque and power coefficient, while investigating horizontal and vertical overlaps along the X- and Y-axis, the blades with overlap ratios of HOLR = +0.15 and VOLR = +0.1 show better performances when compared to other corresponding overlaps. Accordingly, the average C_m and C_p improvements are 16% and 7.5%, respectively, compared to the base with a zero overlap ratio.

Keywords: Savonius vertical axis wind turbine; horizontal overlap ratio; vertical overlap ratio; torque coefficient; power coefficient

---

1. Introduction

The increasing need for energy and the reduction of fossil fuel resources on one hand, and the strict laws on the environment and global warming on the other hand, draw governments' attention to renewable energy resources [1–3]. According to recent reports, the global use of energy by 2015 based on fossil, nuclear, and renewable energy resources were 78.4%, 2.3%, and 19.3%, respectively [4].

Renewable energy that includes wind, solar, geothermal, marine, biomass, and hydropower energy, seems to be the best alternative to humankind's exceedingly growing energy consumption and replacing of fossil sources [5]. Among others, wind energy is considered the least costly source of available renewable energy and is growing at a very fast pace. Since 1996, the capacity to generate energy from wind power has grown significantly as one of the most important renewable energy sources in the world today. The pioneers of this route are developed countries such as China, the United States, and Germany. The total wind energy capacity at the end of 2016 was about 487 gigawatts and is expected to reach 2000 gigawatts by 2030.

In the wind energy industry, there are two main types of wind turbines: horizontal axis wind turbines (HAWT) and vertical axis wind turbines (VAWT). In general, the efficiency of horizontal axis turbines is better than that of vertical axis turbines in wind power extraction (Figure 1). Therefore, most wind turbines in the commercial market today are horizontal axis turbines.

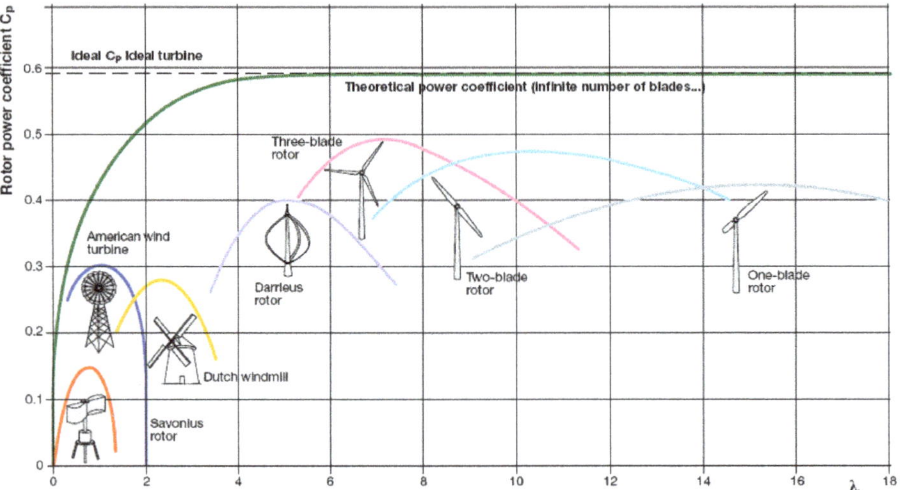

Figure 1. The variation curve of the rotor power coefficient (Cp) average to tip speed ratio (TSR) in different types of wind turbines [6].

Vertical axis wind turbines (VAWTs) represent a much less employed type of wind turbine. However, new trends in the use of VAWT technologies presented by researchers and manufacturers, as well as their benefits, have led to significant recent developments [7]. In some cases, these turbines have advantages over the horizontal ones, including a lack of dependence on the wind direction, easier maintenance, less visual impact, less noise pollution, and a better performance under skewed wind conditions. Urban winds include disordered, indirect, and transverse flows due to the existence of many obstacles (i.e., buildings). For this reason, VAWTs are more suitable than horizontal axis turbines for urban conditions [8]. VAWTs are made in various shapes. The two main types are lift-type turbines (Darrieus) and drag-type turbines (Savonius). Lift-type turbines are designed for high speeds and low torque and require an external or manual force to start working. Drag-type turbines are designed for low speeds and high torque. For conditions with turbulent flows and storms and whenever reliability and cost are more important than productivity, the latter turbines are the best option. This is because, unlike the Darrieus turbine, they do not need external forces to start working.

Numerical studies on renewable energy are very common [9–11]. The purpose of the present work is to conduct a numerical study of a particular type of drag-type wind turbine, called the conventional semi-circular Savonius rotor. In simulations, different turbulence models such as DNS, LES and RANS are used according to the required accuracy. The highest accuracy is expected from the DNS model [12,13], which has a high computational cost. However, the present article chose the RANS model due to the available facilities. The airflow around these turbines has a turbulent and transient nature, for which, in the present paper, the finite volume method is used for analysis. The general concept of drag-type turbines was established based on the developed principles of the Feltner model [14]. In recent years, various studies have been done to improve their performance. In 2015, Fredericks et al. [15] examined the impact of the number of blades on the efficiency of the Savonius turbine using empirical and numerical studies. They found that a four-blade turbine is more effective for a low tip speed ratio, while for a high tip speed ratio, a turbine with three blades is more efficient. Roy and Saha [16] experimentally compared a turbine with a novel geometry with four previous models, claiming an increase in its efficiency compared to the standard Savonius turbine. Tahani et al. [17] simulated five three-dimensional rotor models, including a Savonius with a simple circular blade, twisted Savonius with a simple circular blade, Savonius with a simple circular blade and variable cut plane, twisted Savonius with a simple circular blade and variable cut plane, and Savonius with two or three blades and a conical shaft. They examined the effect of different parameters such as rotor height and power coefficient to eventually find the optimal conditions for these turbines. Lee et al. [18] investigated the functional characteristics and shape of the helical Savonius turbine, with varying twist angles, and calculated the power and torque coefficients for different azimuth angles, both numerically and experimentally. Additionally, the highest power coefficient was obtained at a 45 degree twist angle. This value was calculated as 0.13. Roy et al. [19] investigated the height, diameter, and aspect ratio of the semicircular-bladed Savonius style wind turbine using a differential evolution-based inverse optimization methodology and performed optimization by reducing the space occupied by the rotor, and the overall dimensions were reduced by up to 9.8%. Wang and Zhan [20] compared three models of helical, semi-cylindrical, and semicircular Savonius turbines by investigating the effect of rotor height and twist angle, with respect to the parameter of the output dynamic torque, as well as the urban aesthetic theme. The simulations were carried out three-dimensionally by the sliding mesh method using the RANS equation and turbulence realizable k-ε model via the SIMPLE algorithm for the pressure-speed coupling. Müller et al. [21] experimentally tested the Persian or Sistan wind mill, which is the oldest wind energy device. The efficiency of this machine was assumed to be between 5 and 14%. A series of tests were conducted with a six-bladed model of a 0.6 m diameter and 0.5 m high runner. Two geometries were investigated: with an open downstream side and with a closed downstream side. The second geometry showed a better performance. It was found that a gap between the blades and axis of approximately 1/6 of the blade width is essential and with minimum torque applied, blade velocities can reach up to 2.5 times the wind speed. Roy and Saha [22] performed two-dimensional simulations using the k-ε model under the influence of the wall function and the SIMPLE algorithm in order to investigate the overlap ratios in conventional Savonius wind turbines. They acknowledged that these turbines would have a better efficiency at the overlap ratio of 0.2. Mohammad et al. [23] compared the results of a two-dimensional simulation for turbulent models SST k-ω, RSM, standard k-ε, and realizable k-ε to optimize the conventional Savonius turbines. They found that the realizable k-ε model exhibited the smallest error when compared to Hayashi's [24] experimental data by considering the uncertainty errors. This is an important part of experimental works. A detailed review of this topic was completed by Rizzo and Caracoglia [25], where wind-tunnel experimental errors, associated with the measurement of aeroelastic coefficients of bridge decks, was explored, and expressed no unexpected large irregularity, potentially linked to a systematic error. Tian et al. [26] used a BANKI wind rotor on the medians of the highway to recover energy from the wake of vehicles on both sides of the highway. To evaluate the performance of the rotor, 3D computational fluid dynamics simulations were performed. Five typical situations, including one car on the passing lane, one bus on the passing lane, two opposite moving cars on the passing lane, one car on the fast main lane, and one bus on the fast main lane, were considered and studied. The SST k-ω was used to model the turbulence terms of the RANS equations. The results showed that (1) the highest power coefficient of 0.00464 occurs from the wake of a bus on the passing lane, (2) the maximum power coefficient of two opposite moving cars on the passing lane is a little (7.5%) higher than the power coefficient of one car on the passing lane, (3) the rotor exerts negligible influences on the forces of the vehicles, and (4) the rotor cannot generate power from vehicles on the fast main lane because of the large distance between the rotor and the vehicle. Krzysztof Rogowski [27] analyzed the flow around a one-bladed Darrieus-type wind turbine numerically, by employing a laminar model and two SST k-ω and RNG k-ε turbulence models, and showed that the RNG k-ε turbulence model has a good precision in computing aerodynamic blade loads for the up- and downwind parts of the rotor. The laminar model and the SST k-ω turbulence model a bit more than the tangential aerodynamic blade loads at the downwind part of the rotor. Ferrari et al. [28] simulated the dynamic of a conventional Savonius wind turbine two- and three-dimensionally at different wind speeds and different angular velocities using Open FOAM software. They calculated the values of lift, drag, power, and torque coefficients, and compared the error values and differences between them. To do this, they evaluated three models of one and two equations of Spalart–Allmaras, realizable k-ε, and SST k-ω, among which the highest sensitivity was determined for SST k-ω; therefore, they applied this method. Jin et al. [29] examined the effect of different barrier plate parameters on the upstream flow, including the height, width, and distance of rotors, to evaluate the performance of vertical axis wind turbines. The simulation results were compared to experimental data with and without using the deflector, which showed a good fit. To do this, the SST k-ω and realizable k-ε models were compared with the experimental result and they stated that the SST k-ω model had positive characteristics of the k-ω method for the internal parts of the boundary layer. Simultaneously, this model operated in a free flow in the same way as the k-ε model and did not have the problems of the k-ω model. In sum, they stated that the results obtained by solving the SST k-ω method were more accurate than those of the k-ε model.

As can be seen from the literature, the change in geometry has a significant effect on wind turbine efficiency. However, the effects of overlap distance between the blades of Savonius-type turbines, in two directions of the rotor's cross section, which can be used to achieve the optimal placement of blades relative to each other, have not yet been considered to the authors' best knowledge. Therefore, in the present work, the main parameter that is used to optimize the performance of VAWT of the Savonius-type is the overlap distance between two blades. The work that has been done in relation to overlap ratios has presented different results, from 0.15 to 0.25 overlap ratios [22,30,31]. Fujisawa [30] carried out surveys by measuring the pressure distribution in the blade and monitoring the flow of fluid in and around the rotating and non-rotating rotors. The experiments were performed on four rotors with a half-diameter blade and an overlap ratio of 0 to 0.5. Increasing the overlap ratio, especially in the return mode, showed a better static torque recovery, and the maximum torque and rotor power were obtained at the overlap ratio of 0.15. Alom et al. [32] performed two-dimensional simulations using the SST k-ω model in order to investigate the overlap ratios in an elliptical-bladed Savonius wind turbine. They acknowledged that an elliptical-bladed rotor with a 0.15 overlap ratio exhibited the highest performance relative to the other overlap ratios, in the range of 0 to 0.3. Kumbernuss et al. [31] examined the overlap ratios of 0, 0.16, and 0.32, and the angular changes of phases 0, 15, 30, 45, and 60 degrees between two stages in the Savonius turbine. Experiments were carried out at various wind speeds. The best power coefficients were reported for the overlap ratios of 0 and 0.32 at a phase change angle equal to 60 degrees, and for the overlap ratio of 0.16, at a phase change angle equal to 30 degrees. In the present work, the geometric parameter of the overlap distance was also investigated. The difference between this and previous work is that it is employed to examine the overlap ratio as positive and negative in comparison to the base state without overlap. For this purpose, the overlaps were performed along both the X- and Y-axes, in both positive and negative directions. To illustrate the process, a guideline was selected on one of the blades and this point was denoted as the reference. In the optimal overlapping process, in the first step, the guide point is changed in the X-axis to reach an optimal point. Then, at the optimum point, the overlaps are checked along the Y-axis to eventually reach an optimal point.

2. Subject Theory

The overlap distance is the region between the two blades in the Savonius turbine. This distance is used to compensate for the differential pressures in the concave and convex sections in the leading blade. The overlap ratio in different papers is sometimes defined slightly differently, but, in general, it expresses a common concept that Roy and Saha described in their article as the relation (1) [33]:

$$Overlap\ Ratio = e/d \tag{1}$$

where *d* is the chord of the blade and *e* is the overlap distance between the two blades (Figure 2).

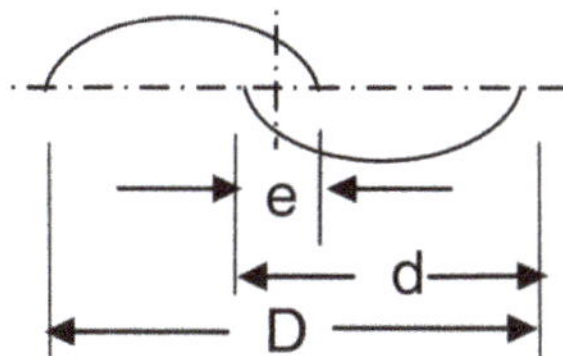

Figure 2. Schematic diagram of the overlap ratio according to Roy and Saha [33].

The difference between the current study and previous ones in the field of the overlap ratio is that the overlap ratio is defined as positive and negative values considering the base state with zero overlap. For this purpose, overlaps are performed along both the X- and Y-axes. To explain the process, a Guide Point (GP) is defined, which is located on the inner end of one of the blades marked with a cross sign in Figure 3. The process of achieving optimal overlap is as follows: in the first step, the GP is displaced on the X-axis, and the coordinates of these displaced points change as GP (x, 0). Here, x is the horizontal variable while keeping the vertical overlap zero. Then, at the optimal point obtained in the previous step, the overlaps in the direction of the Y-axis are checked and the GP moves on (C, y), where y is the vertical variable while keeping C as the optimum constant point obtained in the previous step. In this way, the optimum position of the blades in a region with a constant swept area is obtained for different dimensions and positions of the blade. The overlap ratios defined in the present work are as follows:

Horizontal overlap ratio:

$$HOLR = e_1/R \tag{2}$$

Vertical overlap ratio:

$$VOLR = e_2/R \tag{3}$$

Depending on the placement of the GP, each of the overlap ratios may get a positive or negative value. This means that if the GP is located in the negative region of the X-axis, the horizontal overlap is a negative amount, and if it is located in the positive region of the X-axis, the horizontal overlap is a positive amount. Similarly, if the GP is located in the negative region of the Y-axis, the vertical overlap is a negative amount, and if it is located in the positive region of the Y-axis, the vertical overlap is a positive amount.

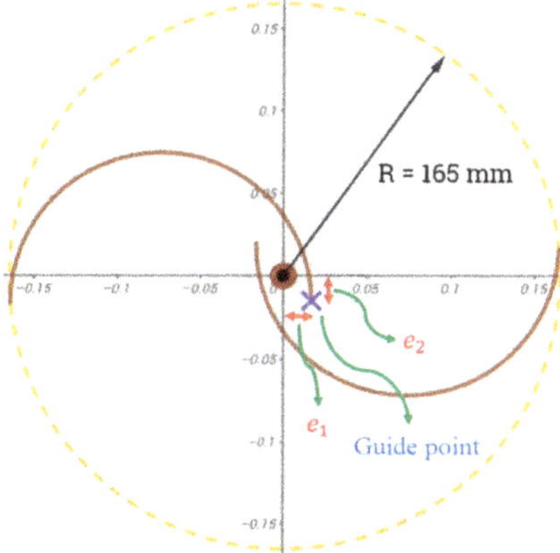

Figure 3. Schematic diagram. The overlap ratio is defined separately for the horizontal and vertical position by transferring the guide point in the Cartesian coordinate in two dimensions.

The wind speed suitable for the current simulations is considered based on the geographic atlas of wind speed in Iran (Figure 4). According to Figure 4, for the average altitude of 50 m in 2016, the appropriate and accessible wind speeds that can be reconstructed based on these simulation conditions are between 6 and 9 m/s. In the present work, the maximum of this value, i.e., 9 m/s, is selected and the simulations are carried out accordingly [34]. The swept area in this article is 0.33 m^2, the thickness of the blade is considered to be 2 mm according to the validation work, and the diameter of the mid-shaft is 15 mm [23]. Simulations begin by examining horizontal overlap ratios. The overlaps of 0, ±0.1, ±0.25, and ±0.4 are investigated. Then, according to the results and values, the overlap ratios of ±0.05, ±0.15, and +0.2 are also studied to obtain the best possible ratio (Figure 5).

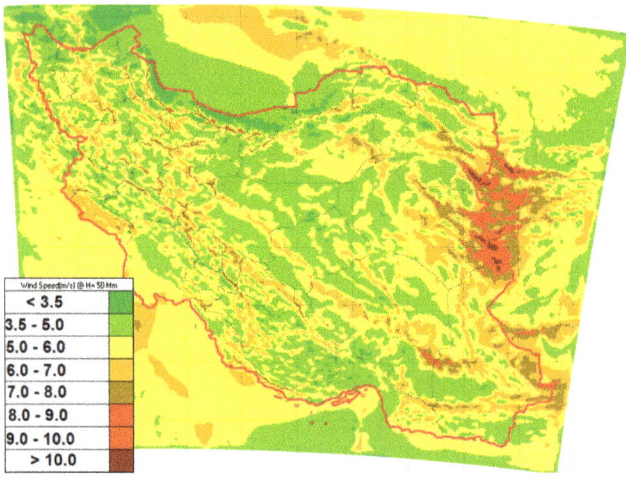

Figure 4. Wind Atlas of Iran for a height of 50 m, 2016, which is used to determine the acceptable wind speed for simulations [34].

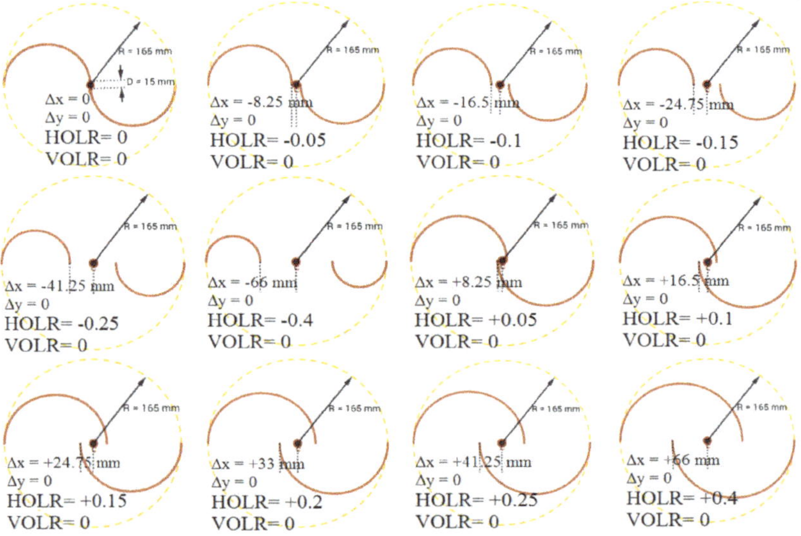

Figure 5. 2D schematic of rotors with investigated horizontal overlap ratios along the X-axis.

It may be mentioned that it is predictable that negative horizontal overlaps probably have a low efficiency, but they should be studied to check the amount of power difference in all cases. To understand whether the power reduction can be justified by a reduction in the amount of raw material consumption and production costs could be useful for future work.

2.1. Governing Equations and the Numerical Solution Method

The URANS equations are used for numerical solutions, in which the mass continuity and momentum conservation equations for the incompressible flow of Newtonian fluid are used.

A review of previous research revealed that in the examinations of wind turbines, the realizable K-ε turbulence model and the SST k-ω are preferred to the other turbulence models. This is mainly due to their satisfactory precision and speed of calculations and their highly precise solutions, respectively. In this work, the realizable K-ε turbulence model was selected with the following equations [22]:

$$\frac{\partial \rho}{\partial t}(\rho k) + \frac{\partial}{\partial x_j}(\rho k u_j) = \frac{\partial}{\partial x_j}\left[\left(\mu + \frac{\mu_t}{\sigma_k}\right)\frac{\partial k}{\partial x_j}\right] + G_k + G_b - \rho\varepsilon - Y_M + S_k \tag{4}$$

$$\frac{\partial \rho}{\partial t}(\rho\varepsilon) + \frac{\partial}{\partial x_j} = \frac{\partial}{\partial x_j}\left[\left(\mu + \frac{\mu_t}{\sigma_\varepsilon}\right)\frac{\partial \varepsilon}{\partial x_j}\right] + \rho C_1 - \rho C_2\frac{\varepsilon^2}{k + \sqrt{v\varepsilon}} + C_{1\varepsilon}\frac{\varepsilon}{k}C_{3\varepsilon}G_b + S_\varepsilon \tag{5}$$

Here, G_k is the turbulence kinetic energy generation due to the gradient of average velocity; G_b is the turbulent kinetic energy generation due to the gradient of average buoyancy; and σ_k, and σ_ε are, respectively, the turbulent Prandtl number for k and ε equations. $C_{1\varepsilon}$, $C_{2\varepsilon}$, $C_{3\varepsilon}$, and C_μ are constants, and S_k and S_ε are source terms. Y_M is the effect of changing the expansion in compressible turbulence to a total dissipation rate, which is defined as

$$Y_M = 2\rho\varepsilon\frac{k}{\gamma RT} \tag{6}$$

and C_1 is defined as

$$S = \sqrt{2S_{ij}S_{ij}} \quad , \quad \eta = S\frac{k}{\varepsilon} \quad , \quad C_1 = \max\left[0.43, \frac{\eta}{\eta + 5}\right] \tag{7}$$

In a wind turbine, the most important parameters in displaying the output efficiency of the system are the torque and output power. In the present work, with respect to these two parameters, the results are compared and analyzed. The non-dimensional results are expressed as torque and power coefficients. Their equations are as follows:

$$C_m = T/\left[(1/2)\rho ARU^2\right] \tag{8}$$

$$Cp = P/\left[(1/2)\rho AU^3\right] \tag{9}$$

T is the produced torque, A is the swept area in front of the wind stream, R is the rotor radius, U is the free stream velocity, and P is the output power.

Another dimensionless parameter which is used in wind turbine analysis is the tip speed ratio (TSR) and is defined as (12)

$$\lambda = \frac{R\omega}{U} \tag{10}$$

where ω is the rotational speed.

2.2. Mesh and Boundary Layers

The sliding mesh method is used to mesh the solution area. This consists of two fixed and rotating regions. Both areas are divided by triangle meshes (Figure 6). In order to eliminate the effects of walls and independence from the solution domain, according to Mohamed et al. [23], the size of the sides of the constant area needs to be 25 times larger than the diameter of the rotor. Additionally, the diameter of the rotating zone has to be 1.25 times larger than the diameter of the turbine rotor. The length of the sides of the constant area was 8250 mm and the diameter of the rotating zone was 412.5 mm. The element size varied in different parts of the blade, and this value increased at the edges and sharp angles of 0.4 and on flat surfaces up to about 1 mm. Around the rotor and shaft, an inflation mesh with 15 to 20 layers was used to consider the walls effects. As a result of this fine meshing, the Y^+ value on the rotor blades was always less than 2.5.

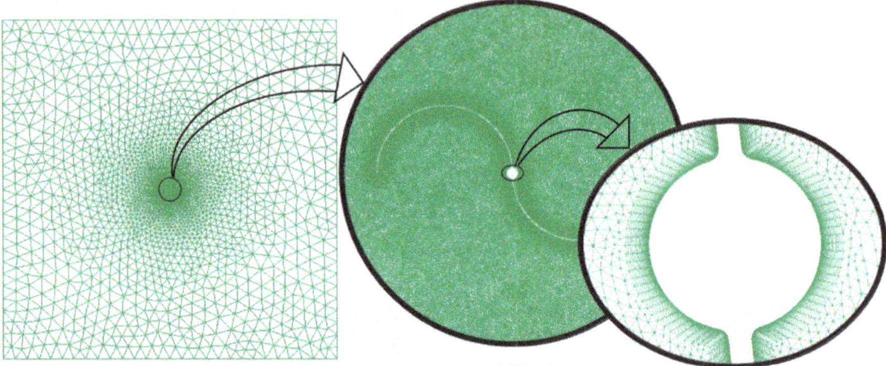

Figure 6. Triangular grid with 15 layers of inflation for a rotor with a horizontal overlap ratio (HOLR) of 0 and vertical overlap ratio (VOLR) of 0.

Given the physical conditions of the problem, the inlet boundary was assigned an inlet velocity, the outlet boundary was assigned an outlet pressure, the slide walls boundary was assigned a non-slip wall, and the rotor was assigned a non-slip wall.

3. Results and Discussion

In order to investigate the mesh independency, a criterion of overlap ratios is considered. For this purpose, the size of the elements on the blades and the interface between zones, the number of elements of cells was changed from 5400 to 730,000 (Figure 7). The mesh independence results showed that as the number of cells increases by more than 70,000, the gradient of the torque coefficient variation curve relative to the number of cells reaches approximately zero, so it was considered that a grid with 70,000 of cells would be appropriate. An examination of the results shows that when the number of cells is between 70,000 and 90,000 for different overlap ratios, acceptable responses will be achieved. It should be noted that the expression of a range for a mesh is due to the fact that for different geometries, the number of cells is slightly different. Meanwhile, considering the size of larger negative overlaps, the number of cells also decreases. Torque coefficients were recorded after two complete revolutions to ensure that the air flow was stable.

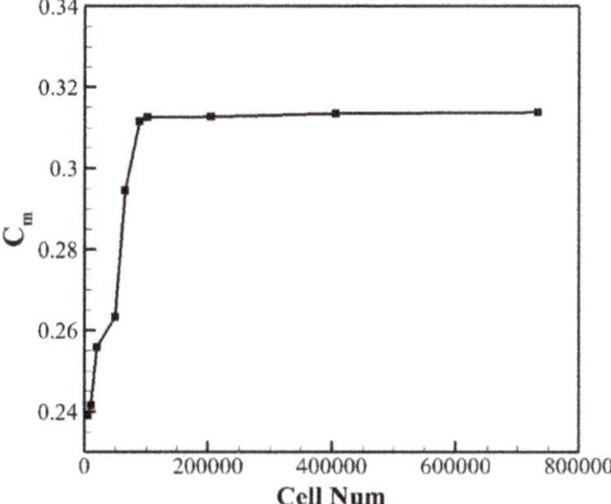

Figure 7. The mesh independency was investigated by changing the number of cells from 5400 to 730,000.

Validation was carried out according to the article by Mohamed et al. [23] at the wind speed of 10 m/s using the same turbulence models, where the validation diagrams related to the reference torque and the present work were compared. With a slight error, validation was achieved. The average error was found to be 3.73% (Figure 8). The difference in results could be because of the differences in meshing, the constant of the turbulence model, and uncertain environmental conditions, which are considered in the numerical simulations.

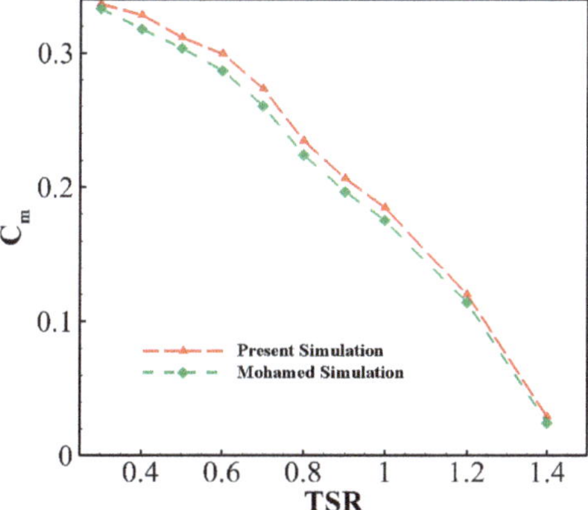

Figure 8. Validation of the numerical solution with Mohamed's results for U = 10 m/s.

After analyzing the horizontal overlap ratios, while the vertical overlap ratio was zero, the torque and power coefficients were obtained according to Figure 9. This shows that the horizontal overlap has a significant effect on the rotor performance. In horizontal–positive overlaps, the torque coefficients have higher values for lower TSRs. At the horizontal overlap ratio of +0.15, the maximum torque coefficient with the value of 0.4 is obtained at TSR = 0.25.

By increasing the tip speed ratios, in a range of TSR from 0.55 to 0.9, geometries with horizontal overlaps close to zero produce slightly higher torque coefficients compared to other horizontal overlaps. At values above a TSR of 0.9, the maximum torque coefficient is obtained at the horizontal overlap ratio of +0.15. The maximum produced power coefficient with an approximate value of 0.18 is related to zero overlap at the tip speed ratio of 0.7. However, the problem of the configuration with a slight overlap is a faster drop than the positive overlaps, due to the increase of the dimensionless TSR parameter.

This is what happens for the positive overlaps at slower rates. The reason for this is the presence of an overlap distance and steering return flow from the concave section of the backward blade to the convex section of the forward blade. This reduces the negative pressure produced by the Coanda flow, i.e., the tendency of a fluid jet to stay attached to a convex surface on the concave section of the forward blade. At negative overlap ratios, due to the absence of compensation for the negative pressure behind the forward blade, the torque coefficient and power coefficient show lower values. Accordingly, among horizontal overlaps, the value of +0.15 is suggested due to the higher efficiency for the wider range of tip speed ratios.

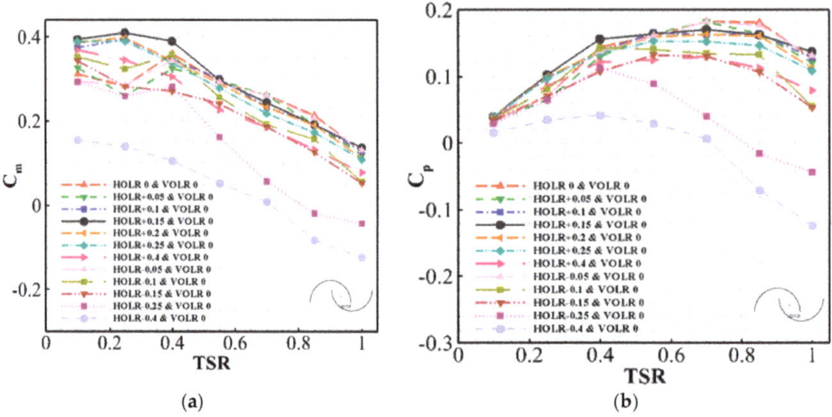

(a)   (b)

Figure 9. (a) Torque and (b) power coefficients obtained at different horizontal overlaps for a vertical overlap ratio (VOLR) of 0.

After choosing the optimal horizontal overlap ratio point, i.e., HOVR = +0.15, the variations of the vertical overlap ratio were investigated. For this, at first, the vertical overlap intervals ±0.1 and ±0.2 were checked and then, according to the resulting values, the overlaps ±0.05 and −0.15 were checked (Figure 10).

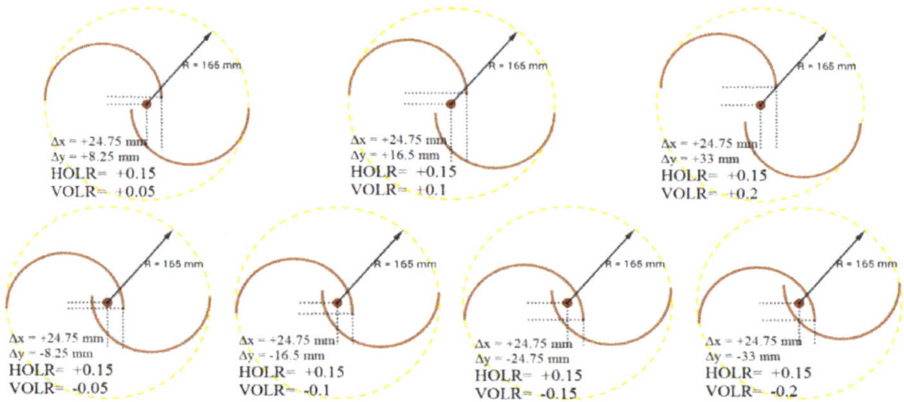

Figure 10. 2D schematic of rotors with investigated vertical overlap ratios along the Y-axis for a constant horizontal overlap ratio (HOLR) of +0.15.

The torque and power coefficients are shown in Figure 11. According to this figure, the vertical overlap ratio has a significant effect on the rotor performance. The torque coefficients have better performances for lower tip speed ratios in negative vertical overlaps, reaching their optimal point at the overlap ratio of −0.1. Additionally, the highest drops are related to negative horizontal overlaps that do not have the ability to use return flow. Furthermore, the shorter the cord length of the blade becomes, the more effective it can be. The lowest efficiency is related to HOLR = 0 and VOLR = −0.4.

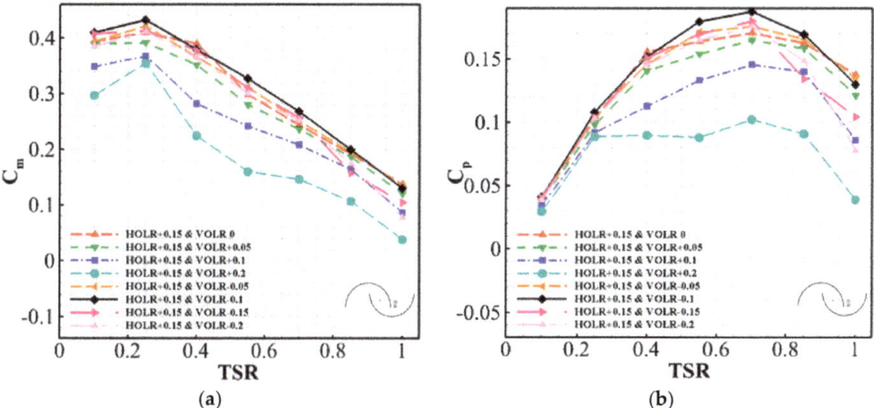

Figure 11. (**a**) Torque and (**b**) power coefficients obtained at different vertical overlap ratios along the Y-axis when the horizontal overlap ratio (HOLR) is +0.15.

The torque and power coefficient diagrams show that as the blades become farther away from each other in the vertical direction, the flow steering decreases sharply over the blade overlap, which is also observed when the blades are located at a very close distance to each other.

The power and torque coefficients of the base state with zero overlap, optimal overlap in the horizontal state, and the overall optimal overlap diagrams are shown in Figure 12. Only for a TSR of 0.85 is the zero overlap a little better. At the remaining TSRs, the rotor produces higher torque and power coefficients with HOLR = +0.15 and VOLR = −0.1 coordinates. For further consideration, a comparison of the average values is presented in Table 1.

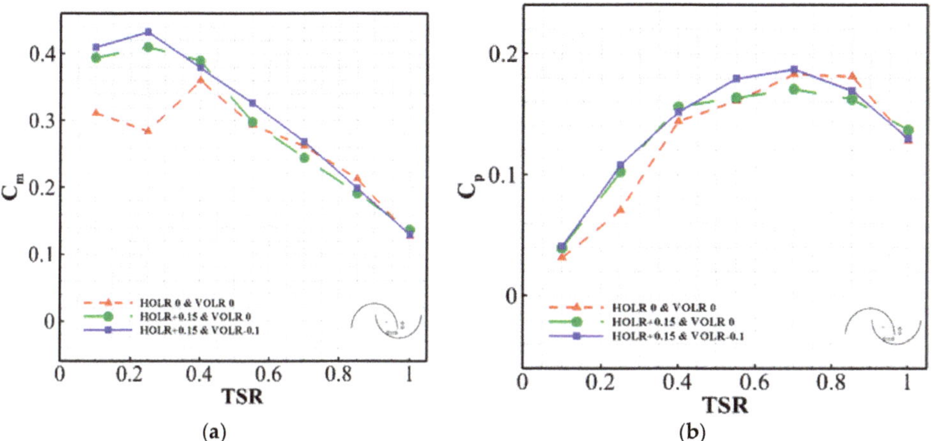

Figure 12. Variations trend and relative improvement of (**a**) torque and (**b**) power coefficients in three states of zero overlap ratio, optimal overlap ratio in horizontal investigation, and overall optimal overlap ratio.

Table 1. Comparison of the zero overlap ratio and optimal mode at horizontal and overall overlap ratios.

	HOLR = 0 VOLR = 0	HOLR = +0.15 VOLR = 0	HOLR = +0.15 VOLR = −0.1	% Increase with Regard to Horizontal Overlap Ratio Optimization	% with Regard to Overall Overlap Ratio Optimization
Average C_m	0.264	0.295	0.306	11.5	15.8
Average C_p	0.128	0.133	0.138	3.7	7.5

Regarding the optimization, the improvement percentage of the average C_m in the horizontal overlap state is 11.5%, and in the vertical overlap state, it is 16%, when compared to the base state with a zero overlap ratio. The improvement percentage in the average C_p, however, is less pronounced. Its increment in the most optimized horizontal overlap state is 3.7%, and in the best vertical overlap state, it is 7.5%, when compared to the same base state.

Given the similarity of the variations of the speed contours at different overlap ratios toward the TSR and the high number of contours, only the TSR of 0.4 is selected and presented in Figure 13 to observe the variation trend. The most obvious factor in these contours is the effect of overlaps on the utilization of the return flow from the concave section. As can be seen, steering the return flow reduces the pressure difference caused by the Coanda flow behind the forward blade. This, in return, results in an increase in the turbine efficiency.

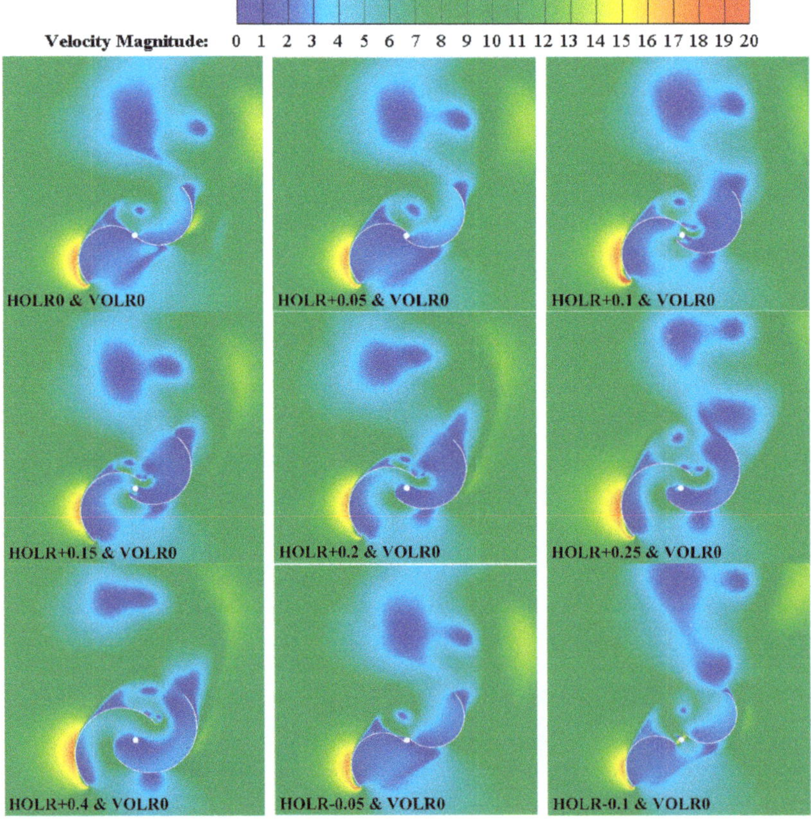

Figure 13. *Cont.*

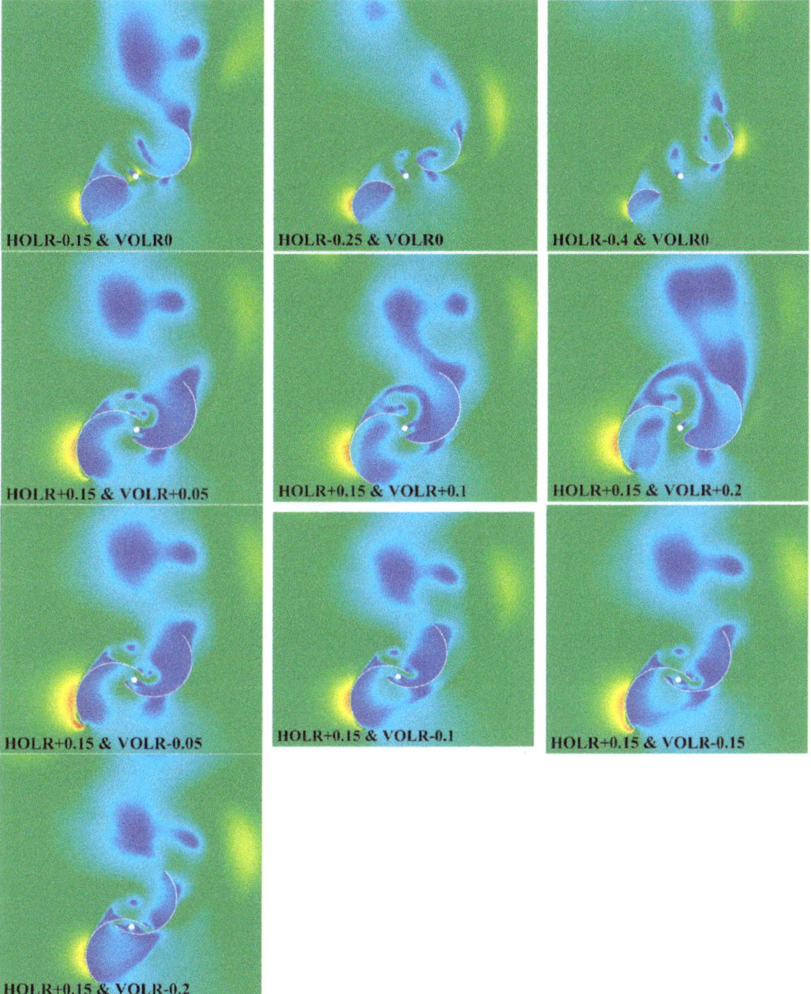

Figure 13. Velocity contours at different overlap ratios to display the gradient of velocity around the rotor for U = 9 m/s and the tip speed ratio (TSR) of 0.4.

The velocity contours show a growth of the vortex region independent of the tip speed ratio at the higher overlap ratios and near the positive overlap section. This indicates a deterioration of the aerodynamic efficiency at high tip speed ratios. The reason for this is due to the increase in the distance between the two blades, meaning that the return flow is scattered and does not properly hit the surface of the driving blade. Furthermore, the Torque coefficient variation graphs with time, during a cycle of rotor rotation, in the negative horizontal overlap state, indicate an extreme disturbance which results in system efficiency degradation. In fact, the system cannot achieve a trend with constant periodicity, which is one of the weak points of these geometries. With regard to the velocity contours, inside the overlap area, the return flow becomes turbulent by hitting the shaft, preventing the steered flow from fully colliding with the forward blade, thereby reducing the efficiency to a certain extent. By increasing the horizontal overlap angle, the effect of the shaft reduces, and this can have a positive effect on the system, but on the other hand, increasing the overlap distance reduces the efficiency. Therefore, to achieve the optimal state, the outcome of these two parameters must be considered. All things considered, the rotor with a horizontal overlap of +0.15 and vertical overlap of −0.1 is a suitable option, especially at low tip speed ratios. Of course, it should be kept in mind that the Savonius rotors often operate at low tip speed ratios, which also indicates the efficiency of the geometry chosen.

4. Conclusions

In this study, a typical Savonius turbine with a constant diameter of the rotor was studied, where the variation of the overlap ratios in both horizontal and vertical directions was considered. In this paper, simulations were carried out by using the computational fluid dynamics (CFD) and solving the URANS equations and realizable K-ε turbulence model. The results showed that at near zero and positive horizontal overlap ratios, a relatively higher efficiency than that of the other states could be achieved. In fact, for lower tip speed ratios, the rotor with an overlap ratio of +0.15 produces a higher torque ratio and is more efficient in a wider range of tip speed ratios. Therefore, it was selected as the optimal horizontal overlap. In the next step, at this optimal point, vertical overlaps were investigated. Among these overlaps, the value of −0.1 showed the best torque and power coefficients. As a result, a rotor with a horizontal overlap ratio of +0.15 and vertical overlap ratio of −0.1 was selected as the optimal model. The improvement percentage in the average C_m in the horizontal overlap state, was 11.5%, and in the vertical overlap state, was 16%, when compared to the base state with zero overlap. Additionally, the improvement percentage in the average C_p in the horizontal overlap state was 3.7%, and in the vertical overlap state, was 7.5%, when compared to the same base state.

This article is a part of an ongoing project. In the next step, this optimal blade application will three-dimensionally examine an oscillating water columns (OWC) system. Finally, the effect of tensile forces on the rotor's structure will be investigated by using FSI simulation so that the information obtained can be used to select the raw materials and manufacture the rotor.

Author Contributions: Conceptualization, M.E., R.S., and R.A.; methodology, M.E., R.S., R.A., and M.S.S.; software, M.E.; validation, M.E.; formal analysis, M.E., R.S., R.A., and M.S.S.; investigation, M.E.; resources, R.S. and R.A.; data curation, M.E., R.S., and R.A.; writing—original draft preparation, M.E., writing—review and editing M.E., R.S., R.A., and M.S.S.; visualization, M.E., R.S., and R.A.; supervision, R.S., R.A., and M.S.S.; project administration, R.S. and R.A.; funding acquisition, R.S.

Funding: This research received no external funding.

Acknowledgments: The Sea-Based Energy Research Group at Babol Noshirvani University of Technology is acknowledged for providing technical and administrative assistance.

Conflicts of Interest: The authors declare no conflict of interest.

Nomenclature

TSR (λ) [-]	Tip speed ratio
VAWT	Vertical axis wind turbine
CFD	Computational fluid dynamics
U [m/s]	Free stream velocity
A [m^2]	Rotor swept area (A = DH)
C_P	Power coefficient
C_m	Torque coefficient
HOLR	Horizontal overlap ratio
VOLR	Vertical overlap ratio
D	Rotor diameter
RANS	Reynolds Averaged Navier–Stokes

References

1. Alamian, R.; Shafaghat, R.; Miri, S.J.; Yazdanshenas, N.; Shakeri, M. Evaluation of technologies for harvesting wave energy in Caspian Sea. *Renew. Sustain. Energy Rev.* **2014**, *32*, 468–476. [CrossRef]

2. Ghasemian, M.; Ashrafi, Z.N.; Sedaghat, A. A review on computational fluid dynamic simulation techniques for Darrieus vertical axis wind turbines. *Energy Convers. Manag.* **2017**, *149*, 87–100. [CrossRef]
3. Roy, S.; Ducoin, A. Unsteady analysis on the instantaneous forces and moment arms acting on a novel Savonius-style wind turbine. *Energy Convers. Manag.* **2016**, *121*, 281–296. [CrossRef]
4. Sawin, J.L.; Sverrisson, F.; Seyboth, K.; Adib, R.; Murdock, H.E.; Lins, C.; Appavou, F.; Brown, A.; Chernyakhovskiy, I.; Epp, B.; et al. *Renewables 2017 Global Status Report*; REN21: Paris, France, 2013.
5. Mohammadi, M.; Lakestani, M.; Mohamed, M. Intelligent parameter optimization of Savonius rotor using artificial neural network and genetic algorithm. *Energy* **2018**, *143*, 56–68. [CrossRef]
6. Ajayi, O. Application of Automotive Alternators in Small Wind Turbines. Master's Thesis, Delft University of Technology, Delft, Netherlands, 2012.
7. Svorcan, J.; Stupar, S.; Komarov, D.; Peković, O.; Kostić, I. Aerodynamic design and analysis of a small-scale vertical axis wind turbine. *J. Mech. Sci. Technol.* **2013**, *27*, 2367–2373. [CrossRef]
8. Chong, W.-T.; Muzammil, W.K.; Wong, K.-H.; Wang, C.-T.; Gwani, M.; Chu, Y.-J.; Poh, S.-C. Cross axis wind turbine: Pushing the limit of wind turbine technology with complementary design. *Appl. Energy* **2017**, *207*, 78–95. [CrossRef]
9. Alamian, R.; Shafaghat, R.; Safaei, M.R. Multi-objective optimization of a pitch point absorber wave energy converter. *Water* **2019**, *11*, 969. [CrossRef]
10. Esmaeelpour, K.; Shafaghat, R.; Alamian, R.; Bayani, R. Numerical study of various geometries of breakwaters for the installation of floating wind turbines. *J. Nav. Archit. Mar. Eng.* **2016**, *13*, 27–37. [CrossRef]
11. Alamian, R.; Shafaghat, R.; Ketabdari, M.J. Wave simulation in a numerical wave tank, using BEM. *AIP Conf. Proc.* **2015**, *648*, 770008. [CrossRef]
12. Méndez, M.; Shadloo, M.; Hadjadj, A.; Ducoin, A. Boundary layer transition over a concave surface caused by centrifugal instabilities. *Comput. Fluids* **2018**, *171*, 135–153. [CrossRef]
13. Ducoin, A.; Shadloo, M.; Roy, S. Direct numerical simulation of flow instabilities over Savonius style wind turbine blades. *Renew. Energy* **2017**, *105*, 374–385. [CrossRef]
14. Shikha; Bhatti, T.; Kothari, D. Early development of modern vertical and horizontal axis wind turbines: A review. *Wind Eng.* **2005**, *29*, 287–299. [CrossRef]
15. Wenehenubun, F.; Saputra, A.; Sutanto, H. An experimental study on the performance of Savonius wind turbines related with the number of blades. *Energy Procedia* **2015**, *68*, 297–304. [CrossRef]
16. Roy, S.; Saha, U.K. Wind tunnel experiments of a newly developed two-bladed Savonius-style wind turbine. *Appl. Energy* **2015**, *137*, 117–125. [CrossRef]
17. Tahani, M.; Rabbani, A.; Kasaeian, A.; Mehrpooya, M.; Mirhosseini, M. Design and numerical investigation of Savonius wind turbine with discharge flow directing capability. *Energy* **2017**, *130*, 327–338. [CrossRef]
18. Lee, J.-H.; Lee, Y.-T.; Lim, H.-C. Effect of twist angle on the performance of Savonius wind turbine. *Renew. Energy* **2016**, *89*, 231–244. [CrossRef]
19. Roy, S.; Das, R.; Saha, U.K. An inverse method for optimization of geometric parameters of a Savonius-style wind turbine. *Energy Convers. Manag.* **2018**, *155*, 116–127. [CrossRef]
20. Wang, Y.-F.; Zhan, M.-S. 3-Dimensional CFD simulation and analysis on performance of a micro-wind turbine resembling lotus in shape. *Energy Build.* **2013**, *65*, 66–74. [CrossRef]
21. Müller, G.; Chavushoglu, M.; Kerri, M.; Tsuzaki, T. A resistance type vertical axis wind turbine for building integration. *Renew. Energy* **2017**, *111*, 803–814. [CrossRef]
22. Roy, S.; Saha, U.K. Computational study to assess the influence of overlap ratio on static torque characteristics of a vertical axis wind turbine. *Procedia Eng.* **2013**, *51*, 694–702. [CrossRef]
23. Mohamed, M.; Janiga, G.; Pap, E.; Thévenin, D. Optimal blade shape of a modified Savonius turbine using an obstacle shielding the returning blade. *Energy Convers. Manag.* **2011**, *52*, 236–242. [CrossRef]
24. Hayashi, T.; Li, Y.; Hara, Y. Wind tunnel tests on a different phase three-stage Savonius rotor. *JSME Int. J. Ser. B Fluids Therm. Eng.* **2005**, *48*, 9–16. [CrossRef]
25. Rizzo, F.; Caracoglia, L. Examining wind tunnel errors in Scanlan derivatives and flutter speed of a closed-box. *J. Wind Struct.* **2018**, *26*, 231–251.
26. Tian, W.; Mao, Z.; An, X.; Zhang, B.; Wen, H. Numerical study of energy recovery from the wakes of moving vehicles on highways by using a vertical axis wind turbine. *Energy* **2017**, *141*, 715–728. [CrossRef]
27. Rogowski, K. Numerical studies on two turbulence models and a laminar model for aerodynamics of a vertical-axis wind turbine. *J. Mech. Sci. Technol.* **2018**, *32*, 2079–2088. [CrossRef]

28. Ferrari, G.; Federici, D.; Schito, P.; Inzoli, F.; Mereu, R. CFD study of Savonius wind turbine: 3D model validation and parametric analysis. *Renew. Energy* **2017**, *105*, 722–734. [CrossRef]
29. Jin, X.; Wang, Y.; Ju, W.; He, J.; Xie, S. Investigation into parameter influence of upstream deflector on vertical axis wind turbines output power via three-dimensional CFD simulation. *Renew. Energy* **2018**, *115*, 41–53. [CrossRef]
30. Fujisawa, N. On the torque mechanism of Savonius rotors. *J. Wind Eng. Ind. Aerodyn.* **1992**, *40*, 277–292. [CrossRef]
31. Kumbernuss, J.; Chen, J.; Yang, H.; Lu, L. Investigation into the relationship of the overlap ratio and shift angle of double stage three bladed vertical axis wind turbine (VAWT). *J. Wind Eng. Ind. Aerodyn.* **2012**, *107*, 57–75. [CrossRef]
32. Arriving at the optimum overlap ratio for an elliptical-bladed Savonius rotor. In *ASME Turbo Expo 2017: Turbomachinery Technical Conference and Exposition*; Alom, N.; Saha, U.K., Eds.; American Society of Mechanical Engineers: New York, NY, USA, 2017.
33. Roy, S.; Saha, U.K. Review of experimental investigations into the design, performance and optimization of the Savonius rotor. Proceedings of the Institution of mechanical engineers, part A. *J. Power Energy* **2013**, *227*, 528–542. [CrossRef]
34. Atlas Iw. Available online: http://www.satba.gov.ir/fa/regions/windatlas (accessed on 17 June 2018).

Article

Investigation of Magneto Hydro-Dynamics Effects on a Polymer Chain Transfer in Micro-Channel Using Dissipative Particle Dynamics Method

Ramin Zakeri [1], Moslem Sabouri [1], Akbar Maleki [1] and Zahra Abdelmalek [2,3,*]

[1]   Faculty of Mechanical Engineering, Shahrood University of Technology, P.O. Box 3619995161, Shahrood, Iran; r_zakeri@shahroodut.ac.ir (R.Z.); sabouri@shahroodut.ac.ir (M.S.); a_maleki@shahroodut.ac.ir (A.M.)
[2]   Institute of Research and Development, Duy Tan University, Da Nang 550000, Vietnam
[3]   Faculty of Medicine, Duy Tan University, Da Nang 550000, Vietnam
*   Correspondence: zahraabdelmalek@duytan.edu.vn

Received: 26 January 2020; Accepted: 27 February 2020; Published: 4 March 2020

Abstract: In this paper, the effect of Magneto Hydro-Dynamics (MHD) on a polymer chain in the micro channel is studied by employing the Dissipative Particle Dynamics simulation (DPD) method. First, in a simple symmetric micro-channel, the results are evaluated and validated for different values of Hartmann (Ha) Number. The difference between the simulation and analytical solution is below 10%. Then, two types of polymer chain including short and long polymer chain are examined in the channel and the effective parameters such as Ha number, the harmony bond coefficient or spring constant (K), and the length of the polymer chain (N) are studied in the MHD flow. It is shown that by increasing harmony bond constant to 10 times with $Ha = 20$, the reduction of about 80% in radius of gyration squared, and half in polymer length compared to $Ha = 1$ would occur for both test cases. For short and long length of polymer, proper transfer of a polymer chain through MHD particles flow is observed with less perturbations (80%) and faster polymer transfer in the symmetric micro-channel.

Keywords: magneto hydro-dynamics (MHD); dissipative particle dynamics (DPD); Hartmann number (Ha-value); harmony bond coefficient or spring constant (K)

---

1. Introduction

Understanding the behavior of polymer chains in micro/nano-scale flows is an important issue, as it is the gateway to different scientific and technological research activities in different fields of study such as biology, genetics, etc. The translocation of polymers through nano/micro-scale passages is encountered in many biological processes in living cells or chemical processes such as DNA (Deoxyribonucleic acid) motion through narrow pores, protein translocation through cell membranes, and penetration of viruses into the cell nucleus. Knowledge of such processes can be beneficial in developing some technical analysis procedures concerning genomic partitioning and rapid DNA sequencing [1–3]. There are several ways to transfer polymer through narrow pores or micro channel including electroosmotic micro pump, magneto hydrodynamic method, and pressure driven flow [3]. Due to high controllability on fluid flow using electrical field or magnetic field, they are proper methods for fluid pumping [3,4].

The difficulties and costs associated with the experimental studies promote the researchers to use the computational simulation methods as the preliminary design and analysis tools to narrow down the design parameters' envelopes before getting into the actual manufacturing process. Regarding the small scale of simulated systems, molecular simulation methods would be the best choices. Such simulation methods have been successfully applied to different nano-scale flow problems. Researches have used different numerical simulation methods such as computational fluid dynamics [5–7], molecular dynamics [8], Langevin dynamics [9,10], and Brownian dynamics [11] to study prediction of polymer chains behavior in fluid flows. Based on the Lagrangian methods such as Molecular Dynamics [12–15], lattice Boltzmann method [16,17] or smoothed particle hydrodynamics method [18–21], the dissipative particle dynamics (DPD) method is a mesoscopic method [12,13,22] which has been vastly used in micro/nano-scale simulations. It benefits from the lower computational cost compared with the molecular dynamics method via using the clusters of molecules, known as beads, instead of considering all actual molecules.

There are several studies in relation of DPD and polymer chain motion in nano/micro flows. Zhang and Manke [23] used the DPD method to study the motion of polymers and polymer solutions rheology in the spherical particles with adsorbed polymers. They found that Newtonian behavior is governing on polymer solutions or polymer in sphere suspensions at low shear rates, but shear-thinning behavior is formed at higher shear rates. Willemsen et al. [24] used the DPD method to investigate the motion of a polymer within a square capillary and the effect of polymers on melting process in a shear flow. Pastorino et al. [25] compared the dynamics of Langevin and DPD as a thermostat term in non-equilibrium simulations of polymeric systems. They studied polymer brushes in different systems including the relative sliding motion, Poiseuille and Couette flows of polymeric liquids, and brush-melt interfaces to compare these two different thermostats. Based on the DPD method, Duong-Hong et al. [26] introduced an electrophoresis model for DNA, which they simulated the coupled DNA electro-osmotic and electrophoretic motion in micro/nano-scale passages. Using this model, they were able to capture the free-draining mobility of DNA while avoiding the expensive electrostatic interactions in the molecular simulations. They also computed DNA mobilities in realistic geometries with a good accuracy. Their results indicated that the Ti-channel has a better separating performance than the Tp-channel. Pan et al. [27] used the DPD method to study the DNA separation in a micro-device using an entropic trapping mechanism. They showed that longer DNA strands have a higher speed than shorter ones. They concluded that the entropic trapping is the consequence of delayed entrance. Moreover, they concluded that corner trapping does not contribute to DNA separation. Masoud and Alexeev [28] used the DPD method to design nano-structured surfaces capable of selective regulation of collision between microchannel walls and polymer, which is suspended in fluid though the microchannel. By utilizing different geometries for nanoscopic posts attached to the internal channel surfaces, they could attract the suspended nanoparticles and polymeric chains to the walls or repel them. Guo et al. [29] studied the translocation of polymers in fluid through a microchannel using the DPD method. They predicted the relation between shear stress and length of polymer chain for the average translocation time. Moreover, they observed two different mechanisms for translocation including single-file and double-folded translocation. They also mentioned the possibility of clogging at the entrance of the channel for polymers longer than a critical length.

All of the mentioned studies used pressure driven flow for transfer of polymer through microchannel. However, lack of studies in the area of different body forces are observed. Yang et al. [30] studied the motion of a polymer chain through a hole using DPD method. They considered two different driving forces, namely the uniform hydrostatic force implemented to whole solvent particles and polymer chains, and also uniform electrostatic force which is employed as a body force, applied to selected charged particles in the chain and some ions in the solvent which were charged oppositely. They found that the power-law correlations should be used for coil-like chains and it is not proper for globular chains. Ranjith et al. [31] used the DPD method to investigate the effect of finite slip at hydrophobic microchannel walls on the hydrodynamics and the dynamics of the DNA chain. They showed that an asymmetric velocity profile caused by hydrophobic and hydrophilic walls can affect the location of the DNA molecules. They used this effect to propose a simple arrangement for separation of short and long DNA chains. Zakeri [32] used the DPD method to simulate the performance of a soft polymer micro-actuator in electro-osmotic flow in a simple micro-channel and a convergent–divergent one. The results indicated that the amplitude of reciprocating motion of polymer increases as the electric field is enhanced, the number of beads is decreased, the spring constant is increased, or more length of a polymer chain is exposed against the fluid flow motion.

A thorough review on the literature shows that there were studies in which the electrical field is used as the transport driving force for polymer chains, e.g., [26,32,33]. Base on the Zakeri [32,34], electroosmotic is a proper external force to move DPD particles in a micro channel, However, the transport of polymer chains using the magnetic forces is not regarded frequently. Magnetic force [35,36] is also another body force capable of inducing fluid flow. This driving force plays an important role in micro pumps (e.g., see [37–42]). Kefayati [43] employed LBM method to simulate the effect of MHD flow in a lid-driven cavity problem for various Hartman numbers. Ghahderijani [44] used LBM method to simulate MHD flow in simple micro channels. Javaherdeh and Najjarnezami [45] used LBM to investigate natural convection in a porous cavity with sinusoidally heated walls considering the effects of magnetic field. They investigated the effects of Hartmann number, porosity and Darcy number on the fluid flow, and heat transfer. Chaabane and Jemni [46] used the LBM to study the convection heat transfer in a 2D enclosure containing a conductive fluid. They examined the effects of Hartmann number, Rayleigh number, Prandtl number on the flow, and temperature fields. Although LBM is a proper simulation method in microscale, the freedom degree of DPD method is higher and it treats more realistically based on the real physics of particles interaction [17,47]. Based on the several references [32,34], the effect of polymer chain from electroosmotic flow was presented, but polymer transfer from in MHD flow requires more investigation.

In this paper, the DPD simulation method is employed to simulate MHD flow in simple symmetric micro channel, also the motion of a polymer chain through micro channel is studied to investigate the various physical properties of polymer chain influenced from MHD flow.

2. Numerical Simulation

To present the simulation of MHD in micro channels which affects the polymer chain transfer through fluid particles, we discuss in the three main subjects in the Sections 2.1–2.3, including the magneto-hydrodynamics equations, the details of DPD method, and the molecular model of polymer chain.

2.1. Magneto-Hydrodynamics

Magneto-Hydro-Dynamics (MHD) equations consist of equations of electromagnetics and hydrodynamics. Electromagnetics equation relates the current density J to the magnetic field strength H [48–50].

$$curl\ \vec{H} = \vec{J} \tag{1}$$

$$div\ \vec{J} = 0 \tag{2}$$

In the case that the substance such as fluid has a velocity, J is calculated as follows:

$$\vec{J} = \sigma(\vec{E} + \vec{v} \times \mu\vec{H}) \tag{3}$$

where σ is the electrical conductivity, v is the velocity of flow, μ is the magnetic permeability, and E is the electric field intensity. The body force or external force $\vec{F_e}$ which will be used in DPD equations, Section 2.2 (as hydrodynamic equations) is calculated as follows:

$$\vec{F_e} = \vec{J} \times \mu\vec{H} \tag{4}$$

2.1.1. Analytical Solution for MHD in Simple Channel

By using the source term F_{ex} calculated in Equation (4) in the steady equation of fluid motion, applying the assumption of uniform conductivity of liquid between two parallel walls (simple channel), implementing H_0 perpendicular to the walls in z-direction while implementing E in y-direction, and ignoring the gravity effect, a fluid motion should occur in x-direction satisfying the following momentum equation [47,51].

$$0 = \frac{-\partial p}{\partial x} + \mu J_y H_0 + \eta \frac{\partial^2 v}{\partial z^2} \tag{5}$$

where p is pressure term, η is the dynamic viscosity, and $J_y = \sigma(E - v\mu H_0)$.

Substituting the J_y in Equation (5) and assuming that velocity is zero at the walls and shear stress is zero at the middle of channel, would result in the following solution of MHD flow in the simple channel.

$$v = V_v(1 - \frac{\cos h(\frac{Ha\,z}{L})}{\cos h(Ha)}) \tag{6}$$

where, $V_v = (\frac{E}{\mu H_0} + \frac{\frac{-\partial p}{\partial x}}{\sigma \mu^2 H_0^2})$ and $Ha = \mu H_0 L (\frac{\sigma}{\mu})^{0.5}$.

If the parameter H_0 varies while the other parameters are kept constant, the value of Ha (or Hartmann number) will change linearly. Hartmann number or Ha is the ratio of magnetic force to viscous force.

If we assume that $V_v = 1$ then $E = (1 - \frac{\frac{-\partial p}{\partial x}}{\sigma \mu^2 H_0^2})\mu H_0$ and $J_y = \sigma((1 - \frac{\frac{-\partial p}{\partial x}}{\sigma \mu^2 H_0^2})\mu H_0 - \mu v H_0)$.

2.2. Dissipative Particle Dynamics Method

The DPD method treats the simulated system as a cloud of beads each having the mass, m_i, position vector $\vec{r}_i$, and velocity vector $\vec{v}_i$. The evolution of beads' velocity vectors follows the basic kinematic and dynamic laws of motion [47].

$$\vec{v}_i = \frac{d\vec{r}_i}{dt} \tag{7}$$

$$m_i \frac{d\vec{v}_i}{dt} = \vec{F}_i \tag{8}$$

The net force exerted on bead i and $\vec{F}_i$, can be decomposed in two parts, namely the force exerted by an external force-field $\vec{F}_e$ like gravitational, electrical, or magnetic fields, and the intermolecular forces exerted by polymer $\vec{F}_{p,i}$ and fluid particles $\sum_{j\neq i} \vec{f}_{ij}$:

$$\vec{F}_i = \vec{F}_{p,i} + \sum_{j\neq i} \vec{f}_{ij} + \vec{F}_e \tag{9}$$

In DPD method, it is assumed that the intermolecular force exerted by bead j on bead i and $\vec{f}_{ij}$, consists of three parts, namely the conservative force, the dissipative force, and the random force [52].

$$\vec{f}_{ij} = \vec{f}_{ij}^{\,C} + \vec{f}_{ij}^{\,D} + \vec{f}_{ij}^{\,R} \tag{10}$$

Of course, $\vec{f}_{ij}$ would be equal to , as the third law of motion indicates. $\vec{F}_{ext}$ is the external force such a electro-osmotic force or magnetic force [32].

The conservative force can be described using the following force filed:

$$\vec{f}_{ij}^{\,C} = \begin{cases} a_{ij}(1 - r_{ij}/r_c)\hat{r}_{ij} & r_{ij} < r_c \\ 0 & r_{ij} \geq r_c \end{cases} \tag{11}$$

where a_{ij} is the maximum repulsion force between beads i and j, r_{ij} is the distance between those beads, i.e., $r_{ij} = \left|\vec{r}_{ij}\right|$, and $\hat{r}_{ij} = \vec{r}_{ij}/\left|\vec{r}_{ij}\right|$ is the unit vector pointing from bead j to bead i, and r_c is the cut-off radius beyond which, the intermolecular forces are assumed to diminish effectively.

The dissipative intermolecular force is calculated as follows:

$$\vec{f}_{ij}^{\,D} = -\gamma\omega^D r_{ij}(\hat{r}_{ij}.\vec{v}_{ij})\hat{r}_{ij} \tag{12}$$

where γ is a constant that determines the strength of dissipative force. The weighting function ω^D is calculated as follows:

$$\omega^D(r_{ij}) = \begin{cases} (1 - r_{ij}/r_c)^2 & r < r_{ij} \\ 0 & r \geq r_{ij} \end{cases} \tag{13}$$

On the other hand, the random force is calculated as follows:

$$\vec{f}_{ij}^{\,R} = -\sigma^R\omega^R r_{ij}\theta_{ij}\hat{r}_{ij} \tag{14}$$

in which, the constant σ^R determines the strength of random force. $\vec{v}_{ij}$ is the relative velocity vector between beads i and j. θ_{ij} is a random number chosen from a symmetric Gaussian distribution having the zero mean and unit variance. σ^R and ω^R relate to γ and ω^D as follows:

$$\omega^R = \sqrt{\omega^D} \tag{15}$$

$$\sigma^R = \sqrt{2\gamma k_B T} \tag{16}$$

where k_B is the Boltzmann constant and T is the temperature.

In this study, the modified velocity-Verlet algorithm is used for time integration of position and velocity of each particle is calculated explicitly. The Ref. [53] is suggested for more information. Implementation of wall boundary condition has been follow based on the ref [48,54–56].

2.3. Polymer Chain

Polymer chain model consists of a number of masses and springs which are connected together. The mechanism of polymer chain motion influenced from interaction of fluid particles and polymer beads is depicted in Figure 1. According to number of different types of particles, we have different interactions. In this case, we have three types of interactions, including fluid to fluid particles, polymer and fluid, and polymer and polymer beads interactions. Each mass has its own repulsion force to fluid particles or other masses. Thus, the conservative force F_c should be modified for this type of simulation. In this paper, the harmonic spring force, $\vec{f}_{i,i\pm1}^{\,S}$, or $\vec{F}_{p,i}$ is used between beads which is added to F_c in DPD formulation. For the beads consisting a polymer chain, the following conservative harmonic force presents the intermolecular bonds [32,34,53].

$$\vec{f}_{i,i\pm1}^{\,S} = -K(r_{i,i\pm1} - r_{i,i\pm1}^{eq})\hat{r}_{i,i\pm1} = \vec{F}_{p,i} \tag{17}$$

where $r_{i,i\pm1}^{eq}$ is the equilibrium (zero-force) distance between two beads in the polymer chain and K is the harmonic bond constant.

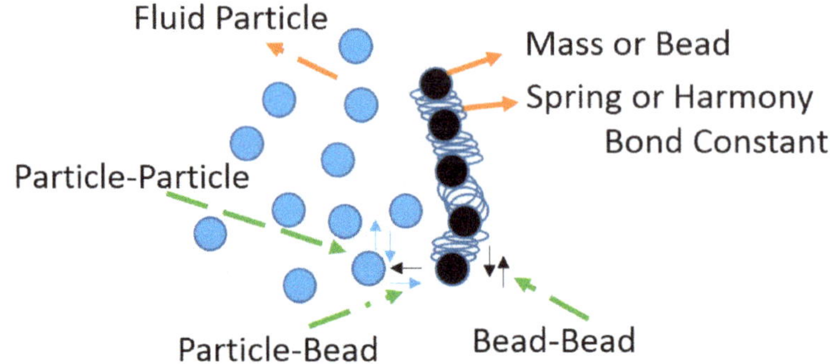

Figure 1. Mechanism of polymer and fluid particles interaction and motion of polymer chain.

Due to difficulty of studying the properties of polymer chain in motion, some physical properties such as mean square radius of gyration and velocity of mass center are employed for finding effect of MHD on polymer chain motion.

In equilibrium, the radius of gyration is defined as [30,51]:

$$\langle R_G^2 \rangle = \frac{1}{2N^2}\langle r_{ij}^2 \rangle \tag{18}$$

where, $r_{ij} = \left| \vec{r}_i - \vec{r}_j \right|$ and N is the number of beads of polymer chain. The end to end distance (Ee) is defined as $\langle R_G^2 \rangle = \langle r_{N1}^2 \rangle$. Also, the center of mass velocity of system is calculated as follows:

$$v_{\text{cm}} = \frac{\sum_1^N m_i v_i}{\sum_1^N m_i} \tag{19}$$

where m_i is the mass of bead i. Also, the average kinetic temperature is calculated as follows [51].

$$\langle k_B T \rangle = \frac{m}{3n-3}\langle \sum_1^n v_i^2 \rangle \tag{20}$$

where n is the number of particles.

3. Results

In this section, the results of numerical simulations of MHD flow in micro channel are presented regarding the polymer chain motion under the influence of magnetic field (MHD flow) using DPD method. We developed our DPD code for current simulation. First, the DPD results for the simple channel flow with the results of the analytical solution are compared to evaluate the accuracy of suggested method for this type of simulation. Then, the study by embedding the polymer chain in a simple channel is extended to study the effect of MHD flow on polymer chain and the related motion characteristics. Different conditions have been considered including changing the magnetic field or *Ha*-value, spring constant (harmonic bond constant of polymer), or K parameter and the number of polymer beads in the polymer chain.

3.1. Validation of MHD-DPD Results with Analytical Solution

In order to evaluate and validate the results, the simple microchannel flow under external force of magnetic field is simulated. The simulation parameters are presented in Tables 1–3. The analytical velocity profiles are also calculated in accordance with Equation (6) in Section 2.1.1 with the assumption of $V_v = 1$. As can be seen in Figure 2 (left), by increasing the value of parameter Ha Number, dimensionless velocity profiles not only increase to a certain extent in terms of value, but they also flatten more in plug like shape and tend to remain constant thereafter. For example, there is a dramatic change of about 80% for maximum value of dimensionless velocity when the value *Ha* Number is increased from 1.0 to 2.0. However, the changes are much less evident from *Ha* = 10 to *Ha* = 20. For a better study of variation of MHD flow, the dimensionless average velocity of particles in channel considering different values of *Ha* is compared with the analytical calculation in Figure 2 (right). As it is expected, increasing the value of *Ha*, the differences would decrease remarkably and would approach the constant value of unity. Also, the results show a proper agreement with the analytical results. The range of discrepancy between DPD solution and analytical solution is between 3% to 7%. The level of accuracy depends on time steps and number of iterations. Therefore, the parameter of *Ha* number is a suitable criterion for studying the velocity profile behavior in micro-channels and DPD method is a proper method for this type of fluid simulation.

Table 1. Setting parameters for particles of DPD.

Variables	a_{ij} Different Particles	a_{jj} Same Particles	Number of Particles	Simulation Box (Channel Size)	Time Step	σ	γ	Cut Off Radious	Periodic Boundary Condition
Value	3	10	4000	20 (length) × 50 (height)	0.001	3	4.5	1	*x*-direction

Table 2. Setting parameters for polymer chain.

Variables	Spring Constant	Number of Beads
Value	500, 5000	20, 50

Table 3. Setting parameters for magnetic parameter.

Variables	L	$\mu L(\frac{\sigma}{\rho v})^{0.5}$	*Ha* Number
Value	25	1	1, 2, 7, 10, 20

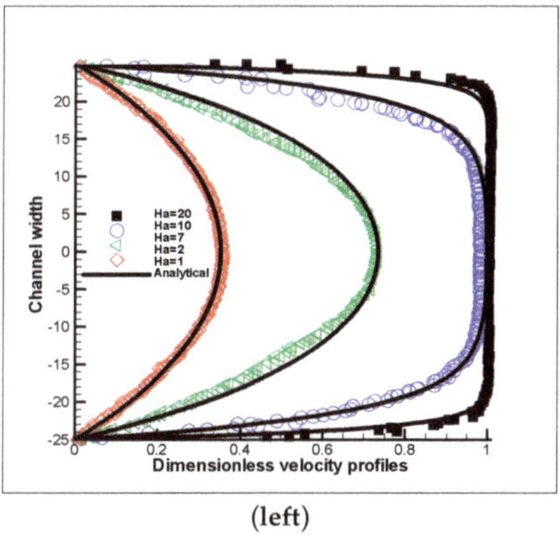

(left)

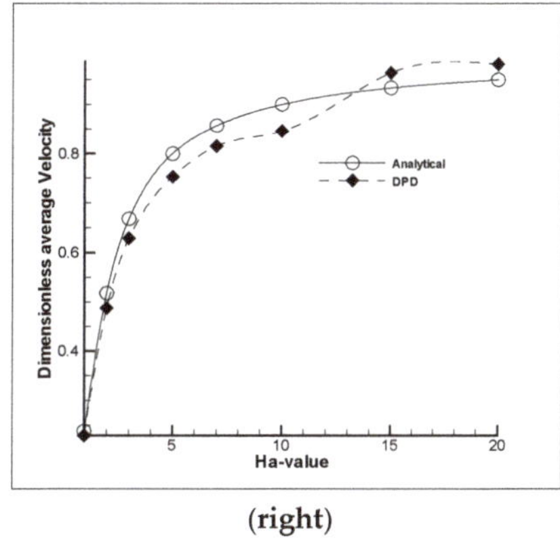

(right)

Figure 2. Comparison of dimensionless velocity profiles of DPD particles (**left**) and dimensionless average velocities (**right**) with analytical results under influence of MHD in simple channel by changing values of *Ha*.

3.2. Short Polymer Chain Transfer in MHD Flow

In order to extend the results to the case of polymer translocation in microchannel, the movement of a short no charge polymer chain consisting of 20 beads in a microchannel has been analyzed under influence of MHD flow. Figure 3 illustrates the movement of the polymer chain under different magnetic field strengths (or *Ha*-values) as well as different polymer harmonic bond coefficients (or spring constants), considering periodic boundary condition at inlet/outlet boundaries. As can be observed, by increasing the *Ha* values from 1 to *Ha* = 20, the relative movement of the short polymer would be more with respect to the hardness coefficient (spring constant) of 500. It has been simulated that conspicuous differences in *x*-direction for the cases of *Ha* = 1 and 20 with the harmonic bond constant of *K* = 500 (red lines with circular symbol and blue lines with diamond symbol) is occurred compared to the cases of *K* = 5000 (green lines with triangle symbols and black lines with gradient symbols). By increasing the harmonic bond constant from 500 to 5000 (10 times), the length of polymers is decreased almost to the half. Results show that choosing the higher *Ha* values and higher harmonic bond constants provides proper polymer chain transfer for low length of polymer cases.

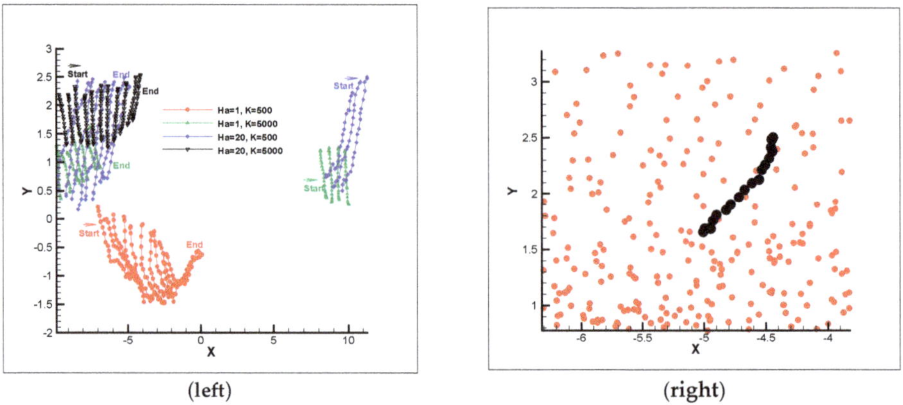

(left)   (right)

Figure 3. Motion of polymer chain with different values for *Ha* parameter and harmonic bond constant (*K*) from number of time steps 1 to 15000 for 20 beads (**left**) and a chain polymer through DPD particles at number of time step 14000 with *K* = 5000 and *Ha* = 20 (**right**).

In the following, the effects of magnetic field and hardness of polymers on the properties of polymer chain are investigated. The variation of dimensionless velocity of mass center of the polymer during the time is depicted in the Figure 4 (left). As can be seen, the oscillations decrease over time and, in the case of more severe magnetic field and higher spring coefficient value, the oscillation is greatly reduced. Therefore, in the condition of *Ha* = 20 and *K* = 5000 causes the polymer to move with less velocity changes. As is evident from Figure 4 (right), temporal evolutions of the average kinetic temperature for all cases reach to unity and fluid condition approaches the equilibrium. Also, by examining the radius of gyration squared for the polymer chain in Figure 5 (left), it is observed that the amount of distortion and perturbation is greatly reduced around the 80% by increasing the harmonic bond constant from 500 to 5000. Changing the magnetic field has little effect on the polymer chain perturbations. Such behavior can be expected by studying the end-to-end distance of the polymer chain as shown in Figure 5 (right). Results indicate that in the non-equilibrium situation, the amount of perturbation for a short polymer chain is high for a short period of time, and in the case of higher *Ha* and *K* values, low oscillation is resulted for those mentioned parameters. Therefore, again, having a chain with a higher harmonic bond constant has about 80% lower oscillation in this study.

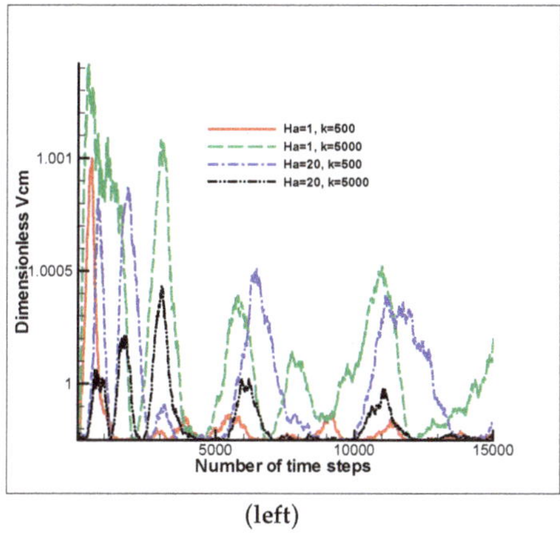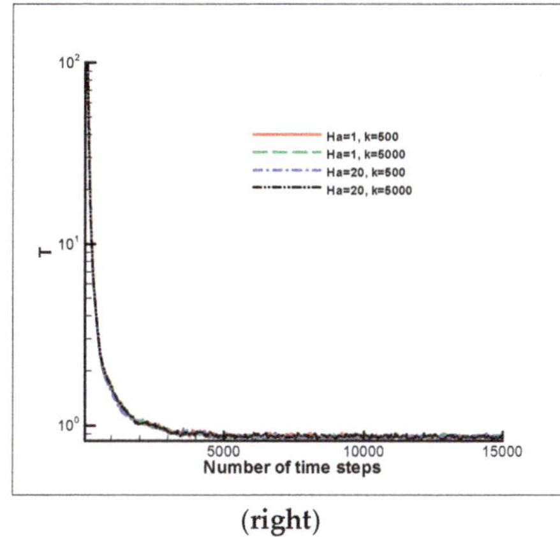

(left)   (right)

Figure 4. Dimensionless velocity of polymer mass center for different *Ha* and *K* values for a polymer chain consisting of 20 beads (**left**) and temporal variation of average kinetic temperature (**right**).

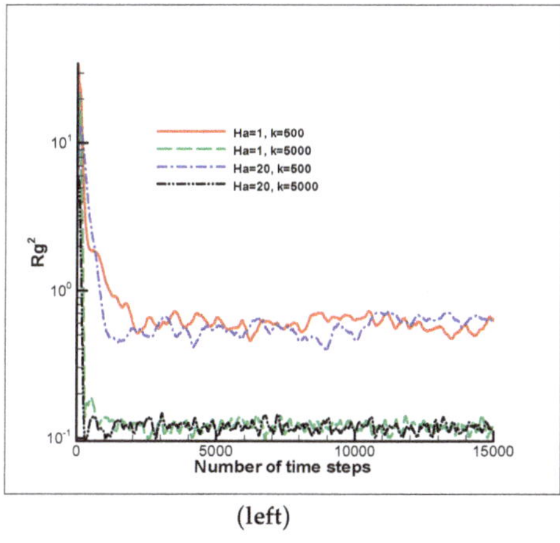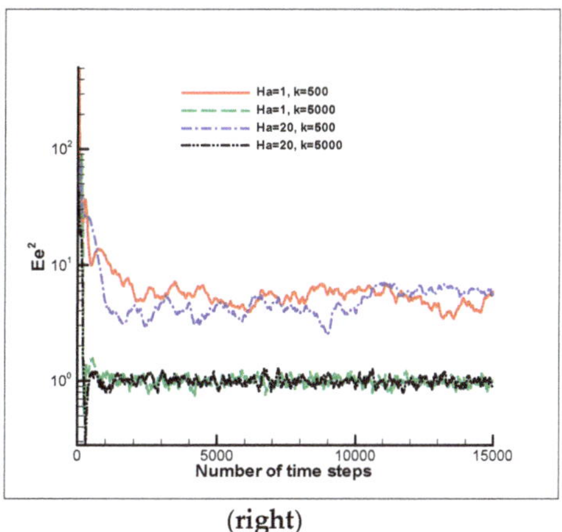

(left)   (right)

Figure 5. Temporal evolution of radius of gyration squared (**left**) and end-end distance (**right**) for different *Ha* and *K* values for a polymer chain consisting of 20 beads.

3.3. Long Polymer Chain Transfer in MHD Flow

One of the significant factors in polymer chain transfer is the length of polymer chain. In this study, the effect of MHD is investigated on no charge polymer chain motion and consequently, the length of chain would be influenced from magnetic field. By consideration of previous results, the motion of a polymer chain consisting of 50 beads having different harmonic bond constant values is depicted in Figure 6. As can be expected, higher magnetic field has more effect on the motion in the case of the higher spring constant or $K = 5000$ compared to $K = 500$. Also, by increasing the harmonic bond constant, the polymer prefers to collapse and in higher magnetic field, compression is enhanced compared to other conditions. In the maximum circumstance, 40% compression is observed in the case of $Ha = 20$ and $K = 5000$. It can be concluded that the proper selection of parameters for transfer of polymer chain are the higher *Ha* and *K* values.

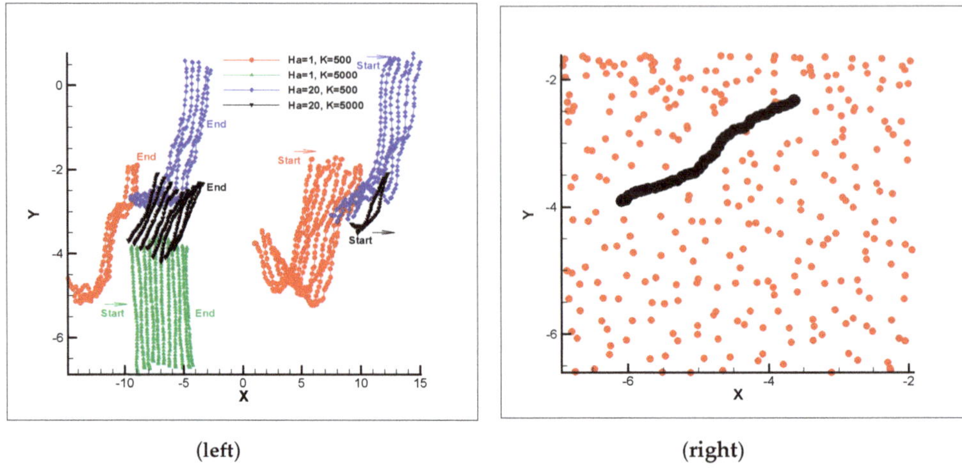

(left)   (right)

Figure 6. Motion of polymer chain with different *Ha* parameter and harmonic bond constant (*K*) from time step 1 to time step 15000 for a 50-beads polymer chain (**left**) and a snapshot of polymer chain among DPD particles at time step14000 for *K* = 5000 and *Ha* = 50 (**right**).

It can also be concluded from the study of the dimensionless velocity of mass center the rate of velocity variation would greatly decrease as the harmonic bond constant and chain length increase. As Figure 7 indicates, more reduction would be resulted for the magnetic field with *Ha* = 20 and *K* = 5000. Similar to the previous results, the average kinetic temperature would converge to unity over the time and fluid would move towards the equilibrium condition. In these cases, for high length of polymer chain, the higher magnetic field and the higher polymer hardness has a proper result.

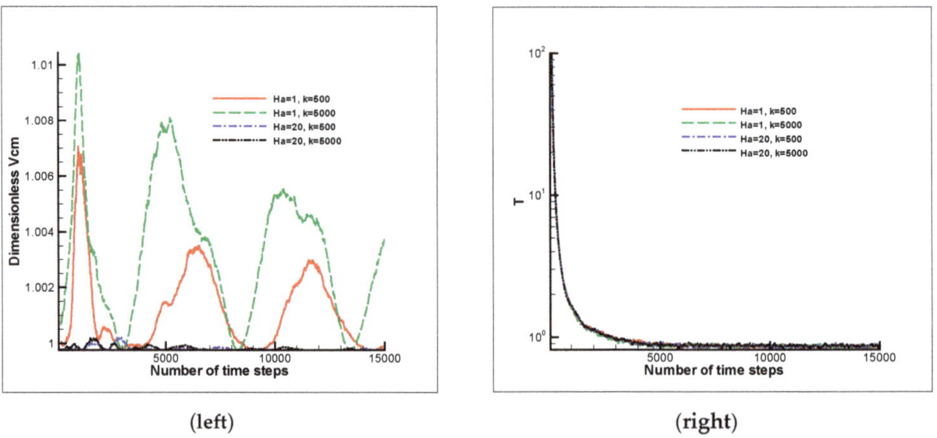

(left)   (right)

Figure 7. Dimensionless velocity of polymer mass center for different *Ha* and *K* values for a 50-beads polymer chain (**left**) and the temporal evolution of average kinetic temperature (**right**).

By examining the radius of gyration squared and the end-to-end distance, it can be concluded from Figure 8 that a significant decrease of about 75% would occur by increasing the harmonic bond constant from 500 to 5000. It should be noted that higher length of polymer chain delays the equilibrium condition. In this case, higher *Ha* value along with a higher harmonic bond constant present proper polymer chain transfer. In Figure 9, these transfers for both cases of short and long polymer chain with *Ha* = 20 and *K* = 5000 as a high *Ha* and *K* values are shown as a proper polymer chain transfer in this study. For more information from quantity aspect, average properties of radius of gyration squared, dimensionless velocity of polymer mass center, temporal evolution of average kinetic temperature, and the length of polymer chain by consideration of Figure 2 to 8 and different input DPD variable are presented in Table 4. As can be seen again, test case of *Ha* = 20 and *K*=5000 show less perturbation and less strength in polymer chain through transfer in the micro channel.

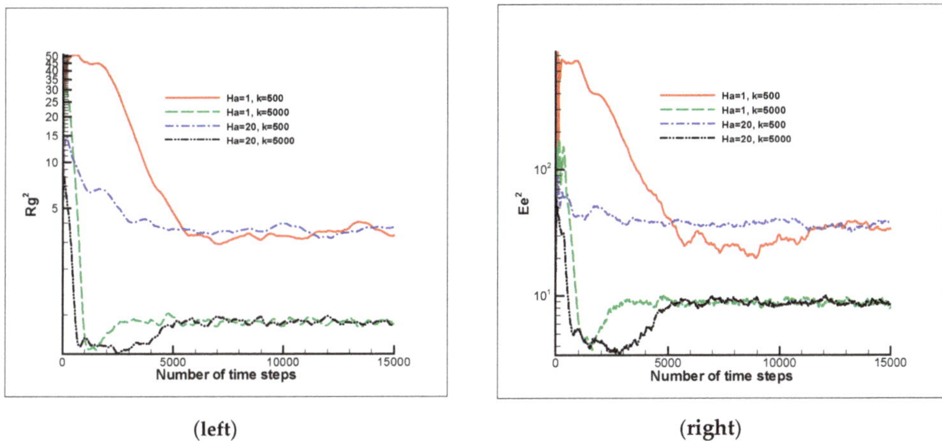

(left)　　　　　　　　　　　　**(right)**

Figure 8. Radius of gyration squared (**left**) and end-end distance (**right**) for different *Ha* and *K* values for a 50-beads polymer chain.

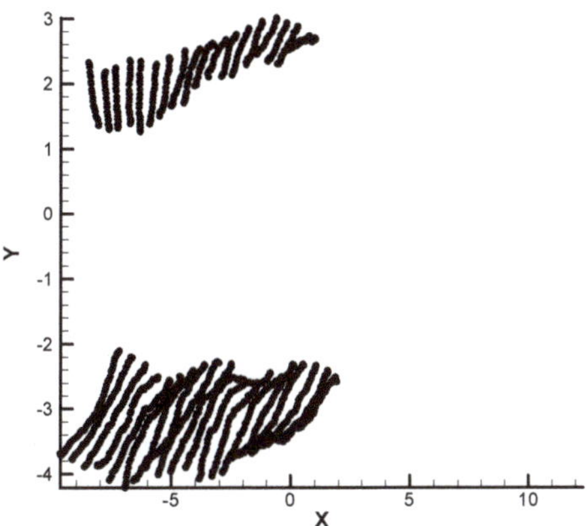

Figure 9. Short (*N* = 20) and long (*N* = 50) polymer chain transfer in microchannel for *Ha* = 20 and *K* = 5000 from number of time steps from 1 to 20000.

Table 4. Average values of test cases calculation from number of time steps from 1 to 15000 with different effective input parameters.

Spring Constant	Number of Beads	*Ha* Number	$\overline{Rg2}$	$\overline{Vcm}$	$\overline{T}$	$\overline{N}$
500	20	1	0.54919	0.99985	0.87459	1.5659
5000	20	1	0.11650	1.00033	0.86930	0.7998
500	20	20	0.53191	1.00027	0.86012	1.96924
5000	20	20	0.11838	0.99982	0.87195	0.78627
500	50	1	6.7853	1.00161	0.860313	2.9855
5000	50	1	0.90651	1.00477	0.86932	3.3996
500	50	20	4.3087	0.99984	0.86096	2.7783
5000	50	20	0.85215	0.99985	0.87459	1.8455

4. Conclusions

In this paper, the DPD simulation of a polymer chain through micro channel was investigated by consideration of Magneto Hydro-Dynamics (MHD) body force. The validation of results with analytical solution was presented and accuracy of simulation was in proper range (max. discrepancy was below 10%). Various physical properties of polymer chain in transition were studied, such as radius of gyration squared, velocity of mass center, and average kinetic temperature for short (20 beads) and long polymer chain (50 beads). For all cases, enhancing the *Ha*-value and K parameters provide less perturbations for both short and long length of polymer chains, in the case of $Ha = 20$ and $K = 5000$ around the 80% reduction in radius of gyration squared was resulted in this paper.

Author Contributions: R.Z. conducted modeling; the writing and revising of the article was carried out by R.Z. and M.S.; R.Z., A.M. and Z.A. supervised the research and edited it. All authors have read and agree to the published version of the manuscript.

Funding: This research received no external funding.

Conflicts of Interest: The authors declare no conflicts of interest.

References

1. Li, X.; Pivkin, I.V.; Liang, H. Hydrodynamic effects on flow-induced polymer translocation through a microfluidic channel. *Polymer* **2013**, *54*, 4309–4317. [CrossRef]
2. Xu, Z.; Yang, Y.; Zhu, G.; Chen, P.; Huang, Z.; Dai, X.; Hou, C.; Yan, L.T. Simulating Transport of Soft Matter in Micro/Nano Channel Flows with Dissipative Particle Dynamics. *Adv. Theory Simul.* **2019**, *2*, 1800160. [CrossRef]
3. Li, P.C. *Microfluidic Lab-on-A-Chip for Chemical and Biological Analysis and Discovery*; CRC Press: Boca Raton, FL, USA, 2005.
4. Darbandi, M.; Zakeri, R.; Schneider, G.E. Simulation of Polymer Chain Driven by DPD Solvent Particles in Nanoscale Flows. In Proceedings of the ASME 2010 8th International Conference on Nanochannels, Microchannels, and Minichannels Collocated with 3rd Joint US-European Fluids Engineering Summer Meeting, Montreal, QC, Canada, 1–5 August 2010; pp. 1035–1040.
5. Chen, C.-W.; Jiang, Y. Computational Fluid Dynamics Study of Magnus Force on an Axis-Symmetric, Disk-Type AUV with Symmetric Propulsion. *Symmetry* **2019**, *11*, 397. [CrossRef]
6. Irandoost Shahrestani, M.; Maleki, A.; Safdari Shadloo, M.; Tlili, I. Numerical Investigation of Forced Convective Heat Transfer and Performance Evaluation Criterion of Al_2O_3/Water Nanofluid Flow inside an Axisymmetric Microchannel. *Symmetry* **2020**, *12*, 120. [CrossRef]
7. Maleki, A.; Elahi, M.; Assad, M.E.H.; Nazari, M.A.; Shadloo, M.S.; Nabipour, N. Thermal conductivity modeling of nanofluids with ZnO particles by using approaches based on artificial neural network and MARS. *J. Therm. Anal. Calorim.* **2020**. [CrossRef]

8. Guillouzic, S.; Slater, G.W. Polymer translocation in the presence of excluded volume and explicit hydrodynamic interactions. *Phys. Lett. A* **2006**, *359*, 261–264. [CrossRef]

9. Muthukumar, M.; Kong, C. Simulation of polymer translocation through protein channels. *Proc. Natl. Acad. Sci. USA* **2006**, *103*, 5273–5278. [CrossRef]

10. Liu, S.; Ban, X.; Wang, B.; Wang, X. A Symmetric Particle-Based Simulation Scheme towards Large Scale Diffuse Fluids. *Symmetry* **2018**, *10*, 86. [CrossRef]

11. Ikonen, T.; Bhattacharya, A.; Ala-Nissila, T.; Sung, W. Unifying model of driven polymer translocation. *Phys. Rev. E* **2012**, *85*, 051803. [CrossRef]

12. Jin, H.; Chen, B.; Zhao, X.; Cao, C. Molecular dynamic simulation of hydrogen production by catalytic gasification of key intermediates of biomass in supercritical water. *J. Energy Resour. Technol.* **2018**, *140*, 041801. [CrossRef]

13. Xu, B.; Jin, H.; Li, H.; Guo, Y.; Fan, J. Investigation on the evolution of the coal macromolecule in the process of combustion with Molecular dynamics method. *J. Energy Resour. Technol.* **2020**, *142*. [CrossRef]

14. Hoogerbrugge, P.; Koelman, J. Simulating microscopic hydrodynamic phenomena with dissipative particle dynamics. *EPL (Europhys. Lett.)* **1992**, *19*, 155. [CrossRef]

15. Zakeri, R.; Lee, E.S. Similar Region in Electroosmotic Flow Rate for Newtonian and non-Newtonian Fluids using dissipative particle dynamics (DPD). In Proceedings of the ASME 2014 International Mechanical Engineering Congress and Exposition, Montreal, QC, Canada, 14–20 November 2014.

16. Safdari Shadloo, M. Numerical simulation of compressible flows by lattice Boltzmann method. *Numer. Heat Transf. Part A Appl.* **2019**, *75*, 167–182. [CrossRef]

17. Vasheghani Farahani, M.; Foroughi, S.; Norouzi, S.; Jamshidi, S. Mechanistic Study of Fines Migration in Porous Media Using Lattice Boltzmann Method Coupled With Rigid Body Physics Engine. *J. Energy Resour. Technol.* **2019**, *141*. [CrossRef]

18. Almasi, F.; Shadloo, M.; Hadjadj, A.; Ozbulut, M.; Tofighi, N.; Yildiz, M. Numerical simulations of multi-phase electro-hydrodynamics flows using a simple incompressible smoothed particle hydrodynamics method. *Comput. Math. Appl.* **2019**. [CrossRef]

19. Fatehi, R.; Rahmat, A.; Tofighi, N.; Yildiz, M.; Shadloo, M.S. Density-based smoothed particle hydrodynamics methods for incompressible flows. *Comput. Fluids* **2019**, *185*, 22–33. [CrossRef]

20. Hopp-Hirschler, M.; Shadloo, M.S.; Nieken, U. A smoothed particle hydrodynamics approach for thermo-capillary flows. *Comput. Fluids* **2018**, *176*, 1–19. [CrossRef]

21. Shadloo, M.S.; Oger, G.; Le Touzé, D. Smoothed particle hydrodynamics method for fluid flows, towards industrial applications: Motivations, current state, and challenges. *Comput. Fluids* **2016**, *136*, 11–34. [CrossRef]

22. Hopp-Hirschler, M.; Shadloo, M.S.; Nieken, U. Viscous fingering phenomena in the early stage of polymer membrane formation. *J. Fluid Mech.* **2019**, *864*, 97–140. [CrossRef]

23. Zhang, K.; Manke, C.W. Simulation of polymer solutions by dissipative particle dynamics. *Mol. Simul.* **2000**, *25*, 157–166. [CrossRef]

24. Willemsen, S.; Hoefsloot, H.; Iedema, P. Mesoscopic simulation of polymers in fluid dynamics problems. *J. Stat. Phys.* **2002**, *107*, 53–65. [CrossRef]

25. Pastorino, C.; Kreer, T.; Müller, M.; Binder, K. Comparison of dissipative particle dynamics and Langevin thermostats for out-of-equilibrium simulations of polymeric systems. *Phys. Rev. E* **2007**, *76*, 026706. [CrossRef] [PubMed]

26. Duong-Hong, D.; Han, J.; Wang, J.S.; Hadjiconstantinou, N.G.; Chen, Y.Z.; Liu, G.R. Realistic simulations of combined DNA electrophoretic flow and EOF in nano-fluidic devices. *Electrophoresis* **2008**, *29*, 4880–4886. [CrossRef] [PubMed]

27. Pan, H.; Ng, T.; Li, H.; Moeendarbary, E. Dissipative particle dynamics simulation of entropic trapping for DNA separation. *Sens. Actuators A Phys.* **2010**, *157*, 328–335. [CrossRef]

28. Masoud, H.; Alexeev, A. Selective control of surface properties using hydrodynamic interactions. *Chem. Commun.* **2011**, *47*, 472–474. [CrossRef]

29. Guo, J.; Li, X.; Liu, Y.; Liang, H. Flow-induced translocation of polymers through a fluidic channel: A dissipative particle dynamics simulation study. *J. Chem. Phys.* **2011**, *134*, 134906. [CrossRef] [PubMed]

30. Yang, K.; Vishnyakov, A.; Neimark, A.V. Polymer translocation through a nanopore: DPD study. *J. Phys. Chem. B* **2013**, *117*, 3648–3658. [CrossRef]

31. Ranjith, S.K.; Patnaik, B.; Vedantam, S. Transport of DNA in hydrophobic microchannels: A dissipative particle dynamics simulation. *Soft Matter* **2014**, *10*, 4184–4191. [CrossRef]

32. Zakeri, R. Dissipative particle dynamics simulation of the soft micro actuator using polymer chain displacement in electro-osmotic flow. *Mol. Simul.* **2019**, *45*, 1488–1497. [CrossRef]

33. Mao, J.; Yao, Y.; Zhou, Z.; Hu, G. Polymer translocation through nanopore under external electric field: Dissipative particle dynamics study. *Appl. Math. Mech.* **2015**, *36*, 1581–1592. [CrossRef]

34. Zakeri, R.; Lee, E.S. Simulation of nano polymer chain sensor in electroosmotic flow using dissipative particle dynamics (DPD) method. In Proceedings of the ASME 2014 International Mechanical Engineering Congress and Exposition, Montreal, QC, Canada, 14–20 November 2014.

35. Alarifi, I.M.; Abokhalil, A.G.; Osman, M.; Lund, L.A.; Ayed, M.B.; Belmabrouk, H.; Tlili, I. MHD flow and heat transfer over vertical stretching sheet with heat sink or source effect. *Symmetry* **2019**, *11*, 297. [CrossRef]

36. Khan, I.; Alqahtani, A.M. MHD Nanofluids in a Permeable Channel with Porosity. *Symmetry* **2019**, *11*, 378. [CrossRef]

37. Laser, D.J.; Santiago, J.G. A review of micropumps. *J. Micromech. Microeng.* **2004**, *14*, R35. [CrossRef]

38. Lim, S.; Choi, B. A study on the MHD (magnetohydrodynamic) micropump with side-walled electrodes. *J. Mech. Sci. Technol.* **2009**, *23*, 739–749. [CrossRef]

39. Kang, H.-J.; Choi, B. Development of the MHD micropump with mixing function. *Sens. Actuators A Phys.* **2011**, *165*, 439–445. [CrossRef]

40. Ito, K.; Takahashi, T.; Fujino, T.; Ishikawa, M. Influences of channel size and operating conditions on fluid behavior in a MHD micro pump for micro total analysis system. *J. Int. Counc. Electr. Eng.* **2014**, *4*, 220–226. [CrossRef]

41. Khan, M.A.; Hristovski, I.R.; Marinaro, G.; Kosel, J. Magnetic Composite Hydrodynamic Pump With Laser-Induced Graphene Electrodes. *IEEE Trans. Magn.* **2017**, *53*, 1–4. [CrossRef]

42. Zhou, X.; Gao, M.; Gui, L. A Liquid-Metal Based Spiral Magnetohydrodynamic Micropump. *Micromachines* **2017**, *8*, 365. [CrossRef]

43. Kefayati, G.R.; Gorji-Bandpy, M.; Sajjadi, H.; Ganji, D. Lattice Boltzmann simulation of MHD mixed convection in a lid-driven square cavity with linearly heated wall. *Sci. Iran.* **2012**, *19*, 1053–1065. [CrossRef]

44. Ghahderijani, M.J.; Esmaeili, M.; Afrand, M.; Karimipour, A. Numerical simulation of MHD fluid flow inside constricted channels using lattice Boltzmann method. *J. Appl. Fluid Mech.* **2017**, *10*, 1639–1648. [CrossRef]

45. Javaherdeh, K.; Najjarnezami, A. Lattice Boltzmann simulation of MHD natural convection in a cavity with porous media and sinusoidal temperature distribution. *Appl. Math. Mech.* **2018**, *39*, 1187–1200. [CrossRef]

46. Chaabane, R.; Jemni, A. Lattice Boltzmann approach for MagnetoHydroDynamic convective heat transfer. *Energy Procedia* **2019**, *162*, 181–190. [CrossRef]

47. Jafari, S.; Zakeri, R.; Darbandi, M. DPD simulation of non-Newtonian electroosmotic fluid flow in nanochannel. *Mol. Simul.* **2018**, *44*, 1444–1453. [CrossRef]

48. Elmars, B.; Yu, M.; Ozols, R. *Heat and Mass Transfer in MHD Flows*; World Scientific: Singapore, 1987; Volume 3.

49. Gold, R.R. Magnetohydrodynamic pipe flow. Part 1. *J. Fluid Mech.* **1962**, *13*, 505–512. [CrossRef]

50. Asma, M.; Othman, W.; Muhammad, T.; Mallawi, F.; Wong, B. Numerical Study for Magnetohydrodynamic Flow of Nanofluid Due to a Rotating Disk with Binary Chemical Reaction and Arrhenius Activation Energy. *Symmetry* **2019**, *11*, 1282. [CrossRef]

51. Karniadakis, G.; Beskok, A.; Aluru, N. *Microflows and Nanoflows: Fundamentals and Simulation*; Springer Science & Business Media: Berlin, Germany, 2006; Volume 29.

52. Jehser, M.; Zifferer, G.; Likos, C.N. Scaling and Interactions of Linear and Ring Polymer Brushes via DPD Simulations. *Polymers* **2019**, *11*, 541. [CrossRef]

53. Nikunen, P.; Karttunen, M.; Vattulainen, I. How would you integrate the equations of motion in dissipative particle dynamics simulations? *Comput. Phys. Commun.* **2003**, *153*, 407–423. [CrossRef]

54. Duong-Hong, D.; Phan-Thien, N.; Fan, X. An implementation of no-slip boundary conditions in DPD. *Comput. Mech.* **2004**, *35*, 24–29. [CrossRef]

55. Sheikholeslami, M.; Shehzad, S. Magnetohydrodynamic nanofluid convection in a porous enclosure considering heat flux boundary condition. *Int. J. Heat Mass Transf.* **2017**, *106*, 1261–1269. [CrossRef]
56. Yapici, M.K.; Al Nabulsi, A.; Rizk, N.; Boularaoui, S.M.; Christoforou, N.; Lee, S. Alternating magnetic field plate for enhanced magnetofection of iron oxide nanoparticle conjugated nucleic acids. *J. Magn. Magn. Mater.* **2019**, *469*, 598–605. [CrossRef]

Article

Impact of Nonlinear Thermal Radiation on the Time-Dependent Flow of Non-Newtonian Nanoliquid over a Permeable Shrinking Surface

A. Zaib [1], Umair Khan [2], Ilyas Khan [3],*, El-Sayed M. Sherif [4,5], Kottakkaran Sooppy Nisar [6] and Asiful H. Seikh [4]

[1] Department of Mathematical Sciences, Federal Urdu University of Arts, Science & Technology, Gulshan-e-Iqbal, Karachi 75300, Pakistan; aurangzaib@fuuast.edu.pk

[2] Department of Mathematics and Social Sciences, Sukkur IBA University, Sukkur 65200, Sindh, Pakistan; umairkhan@iba-suk.edu.pk

[3] Faculty of Mathematics and Statistics, Ton Duc Thang University, Ho Chi Minh City 72915, Vietnam

[4] Center of Excellence for Research in Engineering Materials (CEREM), King Saud University, P.O. Box 800, Al-Riyadh 11421, Saudi Arabia; esherif@ksu.edu.sa (E.-S.M.S.); aseikh@ksu.edu.sa (A.H.S.)

[5] Electrochemistry and Corrosion Laboratory, Department of Physical Chemistry, National Research Centre, El-Behoth St. 33, Dokki, Cairo 12622, Egypt

[6] Department of Mathematics, College of Arts and Sciences, Prince Sattam bin Abdulaziz University, Wadi, Al-Dawaser 11991, Saudi Arabia; drnisarks1@gmail.com or n.sooppy@psau.edu.sa

* Correspondence: ilyaskhan@tdtu.edu.vn

Received: 5 December 2019; Accepted: 9 January 2020; Published: 28 January 2020

Abstract: Symmetry and fluid dynamics either advances the state-of-the-art of mathematical methods and extends the limitations of existing methodologies to new contributions in fluid. Physical scenario is modelled in terms of differential equations as mathematical models in fluid mechanics to address current challenges. In this work a physical problem to examine the unsteady flow of a third-grade non-Newtonian liquid induced through a permeable shrinking surface containing nanoliquid is considered. The model of Buongiorno is utilized comprising the thermophoresis and Brownian effects through nonlinear thermal radiation and convective condition. Based on the flow symmetry, suitable similarity transformations are employed to alter the partial differential equations into nonlinear ordinary differential equations and then these ordinary differential equations are numerically executed via three-stage Lobatto IIIa formula. The flow symmetry is discussed for interesting physical parameters and thus this work is concluded. More exactly, the impacts of pertinent constraints on the concentration, temperature and velocity profiles along together drag force, Sherwood and Nusselt numbers are explained through the aid of the tables and plots. The outcomes reveal that the dual nature of solutions is gained for a specific amount of suction and flow in the decelerating form $A < 0$. However, the unique result is obtained for flow in accelerating form $A \geq 0$. In addition, the non-linear parameter declines the liquid velocity and augments the concentration and temperature fields in the first result, whereas the contrary behavior is scrutinized in the second result.

Keywords: Buongiorno model; unsteady flow; nanoliquid; special third-grade liquid; non-linear thermal radiation

---

1. Introduction

In recent times, non-Newtonian liquids play an imperative role in industrial and engineering processes due to their numerous applications. Here, blood, paints, clay coatings, molten plastics, certain oils, artificial fibers, ketchup, etc., are a few examples. These liquids defy Newton's law of viscosity due to their vital elastic properties. These kinds of liquids are found in an ample variety of realistic problems, enclosing essential significance in polymer depolarization, composite processing, boiling, bubble absorption, etc. Non-Newtonian liquids are scrutinized via three foremost classes, for instance, the integral type, differential type, and rate type. A third-grade liquid is a subclass liquid of the differential kind of model which can envisage the normal stresses along with the phenomena of shear thickening and thinning. Keçebaş and Yürüsoy [1] discussed the time-dependent flow of a special type of third-grade liquid and obtained the solution through similarity variables. Ellahi and Riaz [2] obtained an analytical result of the magnetohydrodynamics (MHD) flow of a third-grade liquid through erratic viscosity analysis. Sahoo and Do [3] explored the impact of the slip factor on a magnetic field comprising a third-grade liquid induced through a stretched sheet. The slip impact on a third-grade liquid from a stretched surface moving in exponential form was discussed by Sahoo and Poncet [4]. Abbasbandy and Hayat [5] discussed the special third-grade liquid through a permeable moving surface. Rahman et al. [6] inspected the mixed convective flow of a third-grade liquid close to a stagnation position from an exponentially stretching surface. Hussain et al. [7] presented solar radiation model for the magnetic field along with third-grade nanoliquid from a convectively heated stretched surface. Naganthran et al. [8] scrutinized the time-dependent flow of a special third-grade liquid through a porous stretched/shrinking surface. They observed the dual nature of results which remained true for the shrinking and stretched surfaces, and they also executed stability analysis. Recently, Reddy et al. [9] have inspected the time-dependent flow of a third-grade liquid through a cylinder. They found that a significant impact of third-grade fluid can observed in the flow field as compared to Newtonian fluid.

In recent times, the study of nanoliquids has gathered considerable curiosity, due to the field's assorted realistic applications, including use in pharmaceutical medicines, thermal systems, electronics, nuclear reactors, chemical industry, etc. Nanoliquids are solid liquid particles which hold nanofiber or nanometer sized particles, having sizes of 1–100 nm, which are scattered in liquids such as water, lubricants, ethylene glycol, bio-fluids, oil, and polymer solutions. In recent times, it has been shown that nanoliquids demonstrate enhanced properties, including an enhanced thermophysical behavior, a modified viscosity, and enhanced thermal diffusivity, density, and thermal conductivity. A novel category of liquid has been described by Choi [10], which has been acknowledged as a nanoliquid. The phenomenon of the augmentation the thermal conductivity through dispersion of the nanomaterial in the liquid was shown by Masuda et al. [11]. Later, Buongiorno [12] observed that the motion of Brownian and thermophoretic diffusion of nanomaterials offers immense potential improvement in liquid thermal conductivity. As a result of these impacts, he recommended variations in situations of convective transport. This plays a significant role in several applications in various industries, like in the collection of aerosols, the safety of nuclear reactors, and eradicating tiny particles from gas streams. Khan and Pop [13] scrutinized flow involving a nanoliquid from a stretched surface using Buongiorno's model. Then, Rana and Bhargava [14] expanded this problem through the use of a different geometry, namely, a non-linear stretched surface. The impact of the heat transport containing nanoliquid induced through a heated stretching surface was inspected through Makinde and Aziz [15]. The results revealed that the preserved thermal features were significantly distorted by escalating the impacts of thermophoresis and Brownian movement. Recently, several researchers [16–20] have discussed the importance of thermophoresis along with Brownian motion with different physical aspects.

The impact of thermal radiation through convective boundary circumstances is involved in numerous industrial and engineering processes, including gas turbines, the storage of thermal energy, die forging, nuclear turbines, and chemical reactions. Aziz [21] looked at the flow of a flat surface by employing a convective condition. Makinde and Aziz [22] scrutinized the magnetic impact of free and forced convective flow on the characteristic of heat transfer in an erect plate in a convective condition in a porous medium. The boundary layer flow provoked through a permeable stretched surface through a convective condition was scrutinized by Ishak [23]. Yao et al. [24] investigated flow with a permeable convectively heated shrinking and stretching wall and found an exact result. Rahman et al. [25] utilized the convective condition to find the solution numerically for mixed convective flow through an upright flat plate in a convective condition. Mustafa et al. [26] analyzed Maxwell liquid and an exponential stretched surface with a convective condition engrossed in a nanoliquid. Ibrahim and Haq [27] explored electrically conducting flow involving a nanoliquid from a stretched sheet in a convective condition. Makinde et al. [28] looked into the impact of MHD on flow comprising a nanoliquid through a connectivity-heated stretched sheet through radiation and slip impacts. The influence of convective condition on MHD flow with nonlinear radiation involving Carreau liquid induced by a stretched sheet was examined by Khan et al. [29]. Recently, Mabood and Khan [30] considered the impact of time-dependent flow with magnetic field from a stretched surface with convective condition.

However, time-dependent flow in the presence of a precise third-grade nanoliquid induced by a shrinking surface through nonlinear thermal radiation and a convective condition has not yet been explored. Although, the literature review has revealed that just a small amount of research has been carried out in a two-phase model comprising nanofluids in non-Newtonian fluid. Second, most researchers have found only one solution regarding the flow field. In addition, with a shrinking surface, a boundary layer flow through a shrinking surface is not achievable because the vorticity produced in this case is not confined inside the boundary layer. To maintain the structure of the boundary layer, the flow requires a certain value of exterior suction at the permeable sheet. Thus, we inspect the impact of nonlinear radiation on time-dependent flow through a shrinking surface involving a special third-grade nanofluid and obtain the dual solutions. The significant technique bvp4c (which is actually a finite difference technique which employs the 3-stage Lobatto IIIa formula) is utilized to solve transmuted ordinary differential equations (ODE's). It is estimated that this study may be helpful for the examination of bundle and shrink wrappings, etc., and is a procedure through which manufactured goods or a group of goods may be wrapped in a movable cover or wrapper of plastic film, leading to the application of heat shrinkage and firmly conforming to the shape of the enclosed contents. The important aspect of this procedure is shrinking film. Owing to its real advantages, shrinking film has seen utilization in numerous industries at several stages of processes relating to packaging.

2. Formulation of the Problem

An unsteady nonlinear radiative flow, together with mass and heat transport, containing a special third-grade liquid induced through a permeable shrinking surface crammed by nanoliquids was developed. It is supposed that the x and y axes denote alongside the sheet and normal to it, respectively. The nanoliquid velocity of shrinking sheet is $ax/(1-ct) + u_w(x,t) = 0$ with $a > 0$ and c showing the unsteadiness of the problem. In addition, the sheet was convectively heated by temperature T_f, which suggests a heat transport coefficient h_f. The considerations of the physical flow problem for third-grade liquids were considered as was considered by Fosdick and Rajagopal [31].

$$T_1 + pI = \mu A_1 + \alpha_1 A_2 + \alpha_2 A_1^2 + \beta\left(\mathrm{tr}A_1^2\right)A_1 \tag{1}$$

where μ, T_1, p, and I represent the viscosity, Cauchy stress tensor, pressure, and identity tensor, respectively, and α_1, α_2, and β represent the material moduli. Following Fosdick and Rajagopal [31], we imposed the following conditions:

$$0 \leq \mu,\ 0 \leq \beta,\ 0 \leq \alpha_1,\ (24\mu\beta)^{0.5} \geq |\alpha_1 + \alpha_2| \tag{2}$$

Through these statements, the leading equations that oversee the time-dependent flow are presented as [4,8,26]:

$$\frac{\partial v}{\partial y} = -\frac{\partial u}{\partial x} \tag{3}$$

$$\frac{\partial u}{\partial t} + v\frac{\partial u}{\partial y} + u\frac{\partial u}{\partial x} - v\frac{\partial^2 u}{\partial y^2} - 6\kappa\left(\frac{\partial u}{\partial y}\right)^2\frac{\partial^2 u}{\partial y^2} = 0 \tag{4}$$

$$\frac{\partial T}{\partial t} + v\frac{\partial T}{\partial y} + u\frac{\partial T}{\partial x} - \alpha\frac{\partial^2 T}{\partial y^2} = \tau\left[\left(\frac{D_T}{T_\infty}\right)\left(\frac{\partial T}{\partial y}\right)^2 + D_B\frac{\partial C}{\partial y}\frac{\partial T}{\partial y}\right] - \frac{1}{\left(\rho c_p\right)_f}\frac{\partial q_r}{\partial y}$$

(5)

$$\frac{\partial C}{\partial t} + v\frac{\partial C}{\partial y} + u\frac{\partial C}{\partial x} - D_B\frac{\partial^2 C}{\partial y^2} = \left(\frac{D_T}{T_\infty}\right)\left(\frac{\partial T^2}{\partial y^2}\right)$$

(6)

The boundary conditions are:

$$t < 0: \quad C = C_\infty,\ v = 0,\ u = 0,\ T = T_\infty \text{ for all } x, y,$$
$$t \geq 0: \quad C = C_w,\ v + v_w(t) = 0,\ u - u_w(x,t) = 0,\ -k\frac{\partial T}{\partial y} = h_f(t)\left(T_f - T\right) \text{ at } y = 0,$$
$$u \to 0,\ C \to C_\infty,\ T \to T_\infty \text{ as } y \to \infty.$$

(7)

Here, (v, u) denotes the components of velocity in the $y-, x-$ directions, respectively, α, v and ρ denote the thermal diffusivity, kinematic viscosity, and density, respectively, C, D_B, D_T and κ denote the concentration of nanoparticles, the Brownian motion, thermophoresis diffusion coefficient, and the non-Newtonian parameter, respectively, and $\tau = \left(\rho c_p\right)_p/\left(\rho c_p\right)_f$ is the ratio of capacity.

In Equation (3), q_r (radiative heat-flux) is presented as was obtained by Khan et al. [29]:

$$q_r = -\frac{\partial T^4}{\partial y}\frac{4\sigma^*}{3k^*} = -T^3\frac{\partial T}{\partial y}\frac{16\sigma^*}{3k^*},$$

(8)

where (k^*, σ^*) denotes the mean-absorption and the Stefan–Boltzmann constants, respectively. Utilizing Equation (8), Equation (5) became:

$$\frac{\partial T}{\partial t} + v\frac{\partial T}{\partial y} + u\frac{\partial T}{\partial x} - \tau\left[D_B\frac{\partial C}{\partial y}\frac{\partial T}{\partial y} + \left(\frac{D_T}{T_\infty}\right)\left(\frac{\partial T}{\partial y}\right)^2\right] = \frac{\partial}{\partial y}\left[\left(\alpha + \frac{16\sigma^* T^3}{3(\rho c_p)_f k^*}\right)\frac{\partial T}{\partial y}\right]$$

(9)

Currently, we establish the similarity variables:

$$\eta = y\sqrt{\frac{a}{v(1-ct)}},\quad \psi = \sqrt{\frac{av}{(1-ct)}}xf(\eta),\quad \phi(\eta) = \frac{C - C_\infty}{C_w - C_\infty},\quad \theta(\eta) = \frac{T - T_\infty}{T_f - T_\infty}.$$

(10)

Here η and ψ denote the similarity variable and the stream function, respectively, and we get $T = [T_\infty + T_\infty(\theta_w - 1)\theta]$ through $\theta_w > 1$, where $\theta_w = T_f/T_\infty$ is the ratio of temperature. For the purpose of the similarity result, we presume that $\kappa = \kappa_0(1 - ct)^3/x^2$, with $\kappa_0 > 0$ being a constant (see Naganthran et al. in [8]), $v_w(t) = v_0/\sqrt{1 - ct}$ is the suction with $v_0 > 0$, and $h_f = d(1 - ct)^{-0.5}$ with $d > 0$ (see Mahapatra and Nandy in [32]).

Employing Equation (10), Equations (4)–(9) are transmuted into the ODE's:

$$\left(1 + Kf''^2\right)f''' + ff'' - A\left(f' + \frac{1}{2}\eta f''\right) - f'^2 = 0$$

(11)

$$\theta'' + \Pr f\theta' + \frac{4}{3R_d}\frac{d}{d\eta}\left[\{1 + (\theta_w - 1)\theta\}^3\theta'\right] - \frac{1}{2}\Pr A\eta\theta' + \Pr\left[Nt(\theta')^2 + Nb\theta'\phi'\right] = 0$$

(12)

$$\phi'' + Lef\phi' + \frac{Nt}{Nb}\theta'' - \frac{1}{2}\Pr LeA\eta\phi' = 0$$

(13)

Along with

$$f'(0) = -1,\ f(0) = S,\ \theta'(0) = -\gamma(1 - \theta(0)),\ \phi(0) = 1,$$
$$\phi(\infty) \to 0,\ f'(\infty) \to 0,\ \theta(\infty) \to 0,$$

(14)

$$K = 6\kappa_0 a^3/v^2$$

$$\mathrm{Pr} = v/\alpha$$

$$A = -c/a$$

$$Nt = \tau D_T\left(T_f - T_\infty\right)/T_\infty v$$

$$\gamma = d\sqrt{v/a}/k$$

$$Nb = \tau D_B(C_w - C_\infty)/v$$

$$S = v_0/\sqrt{av} > 0$$

$$R_d = kk^*/4\sigma^* T_\infty^3$$

$$Le = v/D_B.$$

In above, K is the dimensionless non-Newtonian constraint, Pr is the Prandtl number, A is the unsteady constant, Nt is the thermophoresis, γ is the convective parameter, Nb is the Brownian motion, S is the suction, Rd id the thermal radiation and Le is the Lewis parameter.

The imperative engineering quantities of interest are the friction factor, heat, and mass transfer, which are presented as follows:

$$C_{fx} = \frac{\tau_w}{\rho u_w^2}, \ Nu_x = -\frac{xq_w}{k(T_f - T_w)}, \ Sh_x = \frac{xm_w}{D_B(C_w - C_\infty)}, \tag{15}$$

where q_w, m_w, and τ_w denote the shear stress, mass flux and heat flux, respectively, and are shown as follows:

$$\tau_w = \mu\left(\frac{\partial u}{\partial y}\right)_{y=0}, \ m_w = -D_B\left(\frac{\partial C}{\partial y}\right)_{y=0}, \ q_w = -k\left(\frac{\partial T}{\partial y}\right)_w + (q_r)_w, \tag{16}$$

Utilizing (10), we obtain:

$$\begin{aligned}
\mathrm{Re}_x^{0.5}C_f &= f''(0), \ \mathrm{Re}_x^{-0.5}Sh_x = -\phi'(0), \\
\mathrm{Re}_x^{-0.5}Nu_x &= -\left[1 + \tfrac{4}{3R_d}\{1 + (\theta_w - 1)\theta(0)\}^3\right]\theta'(0).
\end{aligned} \tag{17}$$

where $Re_x = \hat{x}\hat{u}_w(\hat{x})/v$ is the Reynolds number.

3. Numerical Procedure

The nonlinear ordinary differential Equations (11)–(13) with the boundary restriction shown in Equation (11) are tackled by bvp4c, based on the finite difference technique, which implements a 3-stage Lobatto formula. This formula is also known as a collocation technique with fourth order accuracy. Here, we re-write the ODE's shown in Equations (11)–(13) by translating them as an initial value problem (IVP). Further, it is supportive to give a fixed value to $\eta \to \infty$, known as η_∞. The above ODE's are converted into a system of first order as follows:

$$f' = y_1, \tag{18}$$

$$y_1{}' = y_2, \tag{19}$$

$$y_2{}' = \frac{1}{\left(1 + Ky_2^2\right)}\left[y_1^2 - fy_2 + A\left(y_1 + \frac{1}{2}\eta y_2\right)\right], \tag{20}$$

$$\theta' = z, \tag{21}$$

$$z' = \frac{1}{\left[1 + \tfrac{4}{3R_d}\{1 + (\theta_w - 1)\theta\}^3\right]}\left[\begin{array}{l} -\mathrm{Pr}fz - \tfrac{4}{R_d}\{1 + (\theta_w - 1)\theta\}^2(\theta_w - 1)z^2 + \tfrac{1}{2}\mathrm{Pr}A\eta z \\ -\mathrm{Pr}\left(Nt\, z^2 + Nb\, zp\right) \end{array}\right], \tag{22}$$

$$\phi' = p, \tag{23}$$

$$p' = -Lefp + \frac{Nt}{Nb}z' + \frac{1}{2}PrA\eta p, \tag{24}$$

where

$$f(0) = S, \; y_1(0) = -1, \; z(0) = -\gamma[1 - \theta(0)], \; \phi(0) = 1,$$
$$\phi(\eta_\infty) = 0, \; y_1(\eta_\infty) = 0, \; \theta(\eta_\infty) = 0. \tag{25}$$

We have now a set of 'partial' initial conditions:

$$f(0) = S, \; y_1(0) = -1, \; y_2(0) = \beta_1, \; \theta(0) = \beta_2,$$
$$z(0) = -\gamma[1 - \beta_2], \; \phi(0) = 1, \; p(0) = \beta_3. \tag{26}$$

Holding the system of Equations (18)–(26) as an IVP numerically, we need the value for $y_2(0)$, $\theta(0)$, and $p(0)$, i.e., β_1, β_2, and β_3, respectively. However, these values are not mentioned in Equation (25). The initial estimated values for $y_2(0)$, $\theta(0)$, and $p(0)$ are speculated and bvp4c was obtained via the MATLAB software to achieve accurate results. It is noted that this was carried out by setting different initial guesses to obtain multiple solutions. Afterward, the considered values of $f'(\eta)$, $\theta(\eta)$ and $\phi(\eta)$ at η_∞ (=10) were appraised through the boundary conditions $f'(\eta_\infty) = 0$, $\theta(\eta_\infty) = 0$ and $\phi(\eta_\infty) = 0$, in which the estimated values of $f''(0)$, $\theta(0)$ and $\phi'(0)$ were controlled by the Secant technique to get a better conjecture for the results. The step-size was taken as $\Delta\eta = 0.01$. The process was repeated iteratively until required results with an acceptable level of accuracy (i.e., up to 10^{-5}) were acquired, in order to fulfill the convergence criterion.

4. Results and Discussion

The distorted nonlinear ODE's shown in Equations (11)–(13), together with the boundary conditions shown in Equation (14), were worked out via the bvp4c solver. The numerically-obtained outcomes for the Nusselt number, skin factor, and Sherwood number, along with the liquid velocity, temperature distribution, and concentration were plotted for distinct values of the pertinent parameters encountered in the problem.

Figures 1–3 show the deviation of time-dependent constraint A on the fluid velocity, temperature distribution and concentration profile. From Figure 1, it can be seen that the velocity profile increases when augmenting A in the first result, and, thus, the thickness of the velocity boundary layer reduces, whilst for the second solution, a tendency to repel can be observed. Also, it can be seen through the figure that the velocity of liquid initially increasing when growing η in the first result and then it starting to decrease after $\eta = 1$. The unsteady impact is significant in the second solution. Physically, the impact of the unsteadiness parameter resists the motion of liquid flow, and, consequently, reduces the liquid velocity. Figures 2 and 3 explain the temperature and concentration fields, respectively, which increase due to the increase in A in both results. Therefore, the thicknesses of the temperature and concentration boundary layers rise in both results. Physically, an increase in the unsteadiness parameter enhances the heat and mass of the nanoliquid, leading to an increase in the temperature as well as the concentration and its related boundary layer thicknesses. In addition, these sketches gratify the border conditions asymptotically and the survival of the dual nature of the results, sustaining the justification of our achieved numerical solutions. Figures 4–6 have been prepared to inspect the influence of non-Newtonian constraint K on the fluid velocity, temperature distribution, and concentration profile. Figure 4 shows that the fluid velocity declines due to K in the first result and enhances in the second result. Consequently, the thickness of the velocity boundary layer expands in the first result and shrinks in the second solution. $K = 0$ signifies a Newtonian fluid. A rise in K implies an increase in the behavior of non-Newtonian behavior. Thus, the non-Newtonian parameter widens the thickness of the boundary layer. Alternatively, the temperature and concentration profile enhance when introducing K in the first result, as portrayed in Figures 5 and 6, whereas, the contrary behavior is scrutinized in the second result. These figures also suggest that the profiles are superior for special third-grade liquid, in contrast to the Newtonian liquid. The impact of the Brownian parameter Nb on the temperature and concentration fields is depicted in Figures 7 and 8, respectively. Figure 7 shows that the temperature distribution increases due to Nb in both solutions. In contrast, the concentration of the nanoliquid moves in the opposite direction to Nb in both solutions, as depicted in Figure 8. Therefore, the thickness of concentration boundary layer shrinks and the thickness of the thermal boundary layer expands. Physically, the nanoparticles of kinetic energy increase with the chaotic motion strength, increasing the temperature of liquid. Also, the Brownian motion at both the molecular and nanoscale levels is an imperative mechanism which governs the thermal behavior. In systems with nanoliquids, the Brownian motion occurs due to the nanoparticle size, which can alter the heat transport properties. As the material size moves toward a nanosized level, Brownian activity and its effect on the liquids is significantly responsibility for the features of heat transport. Figures 9 and 10 show the impact of Nt on the liquid temperature and nanoliquid concentration. These outcomes suggest that the profiles of the temperature and concentration increase when increasing Nt in both solutions. Physically, diffusion produces a profound effect in the liquid due to the increasing Nt, which grounds the widening of the thermal and concentration boundary layers. Figures 11–14 have been created to depict the impacts of the convective radiation, γ, and thermal radiation, R_d, on the temperature of the liquid and concentration. Figure 11 shows that the temperature gradient of sheet rises when increasing γ. This influences deeper thermal piercing into the sluggish liquid. Thus, the thickness and temperature rise with rising values of γ in both results. It is claimed that the constant wall temperature $\theta(0) = 1$ will be able to recover by capturing a sufficiently large amount of γ. In addition, $\gamma = 0$ communicates an insulated surface. Figure 12 explains the concentration, along with the boundary thickness, with a large amount of γ in both results. Figures 13 and 14 confirm that the temperature distribution and nanofluid concentration decline as R_d increases in both results. Consequently, the thicknesses of the concentration and thermal boundary layers become slimmer in both solution forms. Physically, a huge amount of R_d denotes the supremacy of conduction, and, as a result, the thicknesses of the concentration and the thermal boundary layers shrink. As expected, the influence of radiation is more prominent on the temperature in contrast to concentration. The impact of the time-dependent constraint A on the friction factor, $C_f \mathrm{Re}_x^{1/2}$, the heat transfer rate, $Nu_x \mathrm{Re}_x^{-1/2}$, and the mass transfer rate, $Sh_x \mathrm{Re}_x^{-1/2}$, against S is demonstrated in Figures 15–17 and in Table 1. Figure 15 shows that the friction factor increases when increasing A in both forms of the solution. However, the heat and mass transport rates diminish when increasing A in both solutions, as illustrated in Figures 16 and 17. Also, the trend of these solutions can be inspected through Table 1. Multiple results are attained when $S_c \leq S$ and no result is obtained for $S_c > S$. Here, S_c is identified as a critical value. Moreover, the influence of A towards the critical values is depicted in Table 2. The superior features of the time-dependent constraint reduce critical point value. Therefore, the time-dependent parameter influences boundary layer separation. Finally, the behavior of the friction factor and heat and mass transfer for distinct values of K against A are portrayed in Figures 18–20. These results suggest that the dual nature of the solutions of the ODE's shown in Equations (11)–(13), with respect to the boundary condition shown in Equation (14), are available only with decelerating ($A \geq 0$) flow (see Mahapatra and Nandy in [32]), whilst the solution is single for accelerating ($A < 0$) flow. The multiple results are attained when $A \leq A_c$ and no result is found when $A > A_c$, where A_c signifies the critical value. Here, it is interesting to observe that for Newtonian liquid ($K = 0$), the point of critical appears far, thus, we terminated calculations at $A = 10$. The result is consistent with the existing literature [33,34]. However, the multiple outcomes survive for $K = 1$ in the range of A, i.e., $A \leq 4.2080$, and no result survives when $A > 4.2080$. It is noteworthy that an additional quantity of K results in a notable decline of $(|A_c|)$ in the domain of the solution. For $K = 2$, the result survives when $A \leq 2.6890$ and no result exists for $A > 2.6890$. So, the critical values, $(|A_c|)$, reduce when the non-Newtonian constraint increases, which interrupts separation. Moreover, the friction factor and heat and mass transfer rate shrink in the first solution and increase in the second solution due to the rising K, as illustrated in Figures 18–20. It should be observed that the calculations have been executed awaiting the point where the result does not converge, and the computations were stopped at that position. According to the previous studies [35–38], the first solution (upper branch solution) is stable and physically appropriate, while the second solution (lower branch solution) is unstable and physically not relevant. However, these results are denied of physical meaning, where these solutions are still of interest as far as the differential equations are concerned. Similar results may occur in other conditions where the consequent solutions have more practical meaning [39].

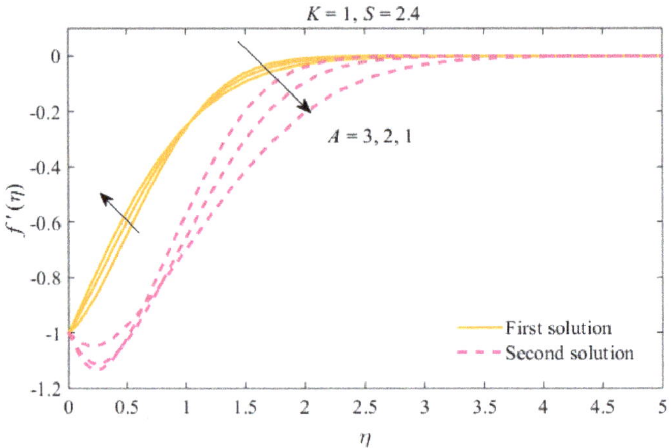

Figure 1. The velocity profiles for different values of A.

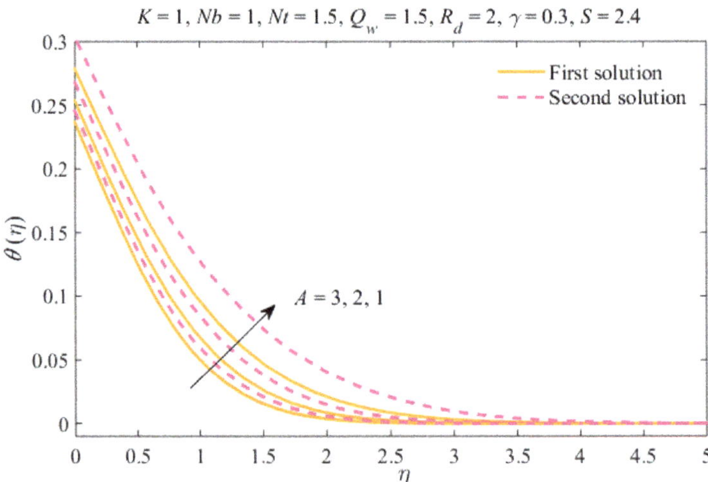

Figure 2. The temperature profiles for different values of A.

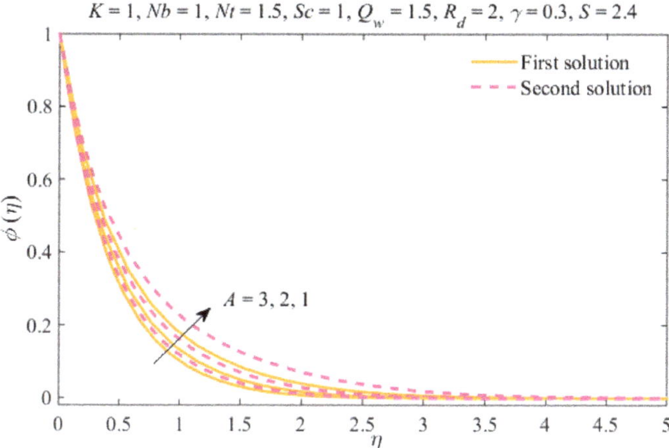

Figure 3. The concentration profiles for different values of *A*.

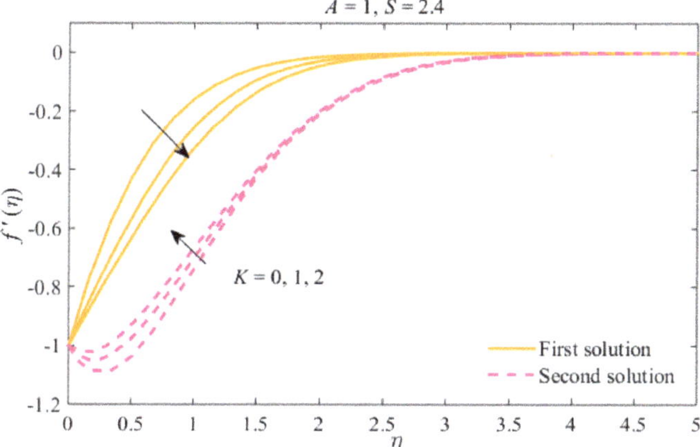

Figure 4. The velocity profiles for different values of *K*.

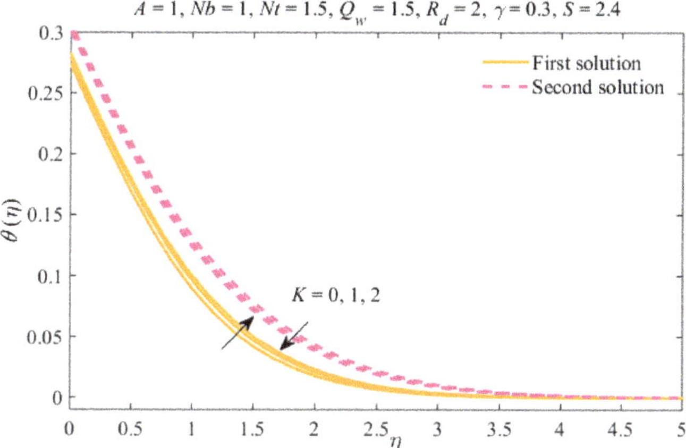

Figure 5. The temperature profiles for different values of *K*.

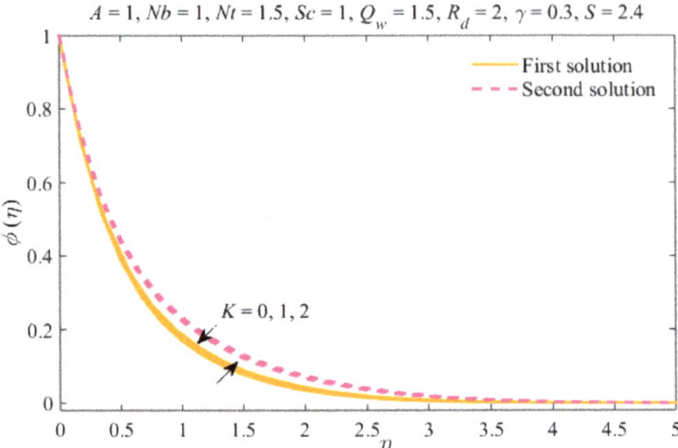

Figure 6. The concentration profiles for different values of *K*.

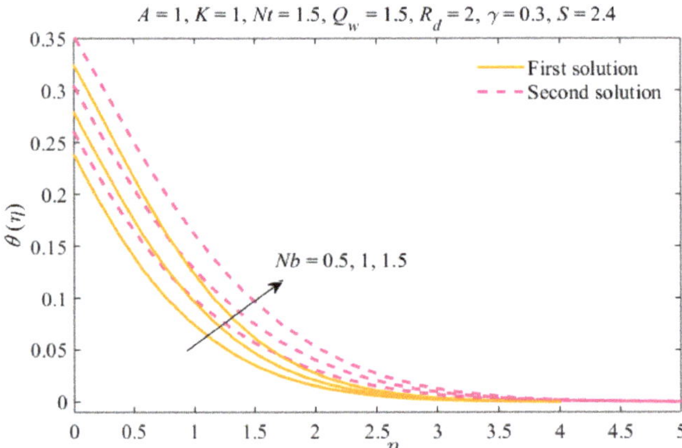

Figure 7. The temperature profiles for different values of *Nb*.

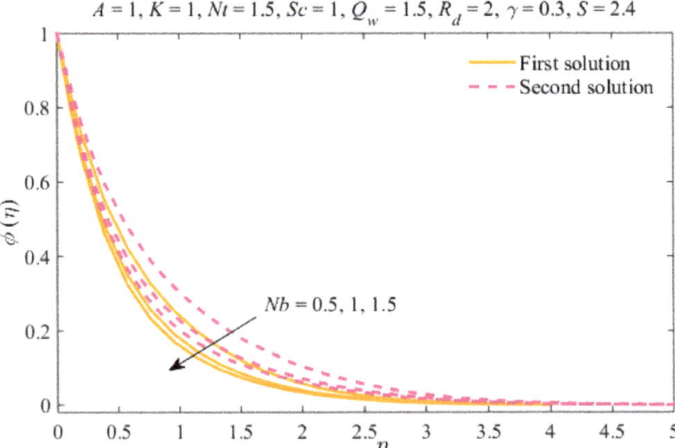

Figure 8. The concentration profiles for different values of *Nb*.

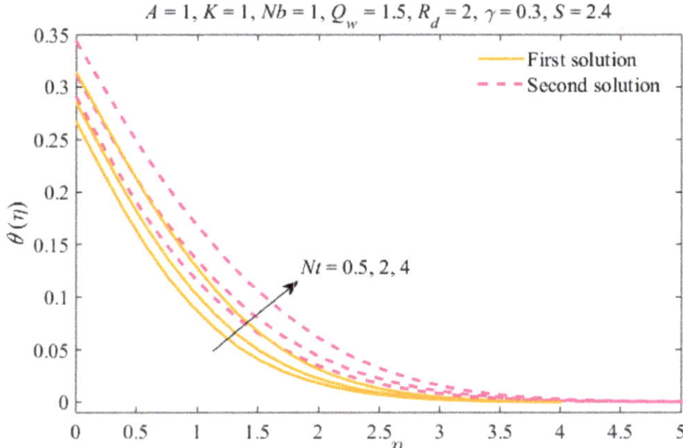

Figure 9. The temperature profiles for different values of Nt.

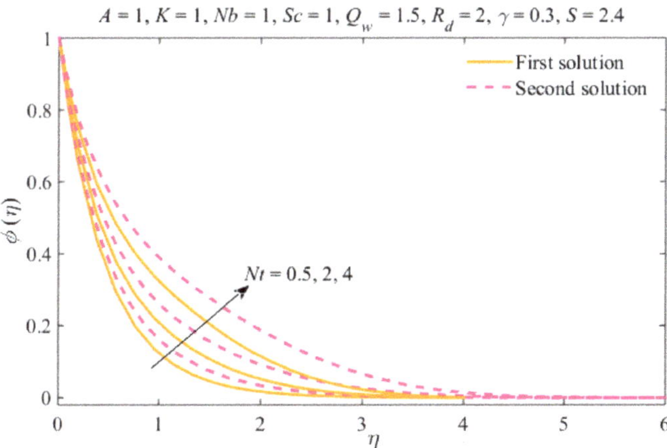

Figure 10. The concentration profiles for different values of Nt.

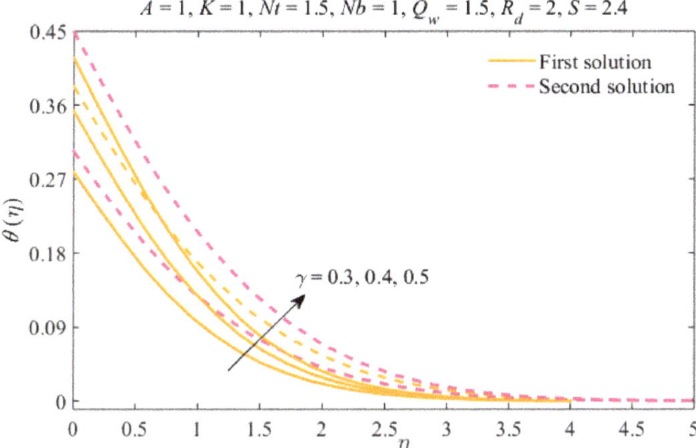

Figure 11. The temperature profiles for different values of γ.

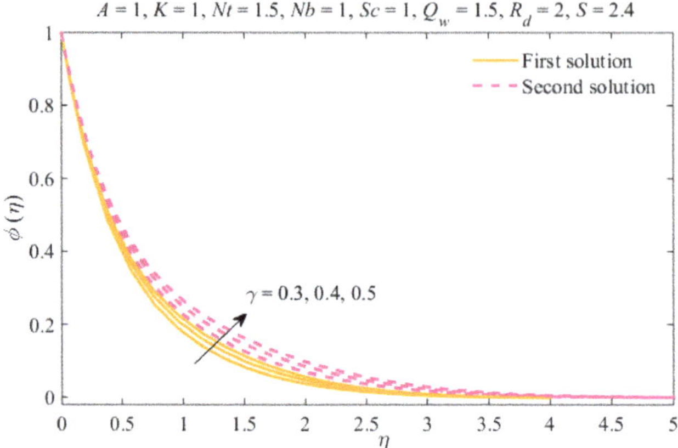

Figure 12. The concentration profiles for different values of γ.

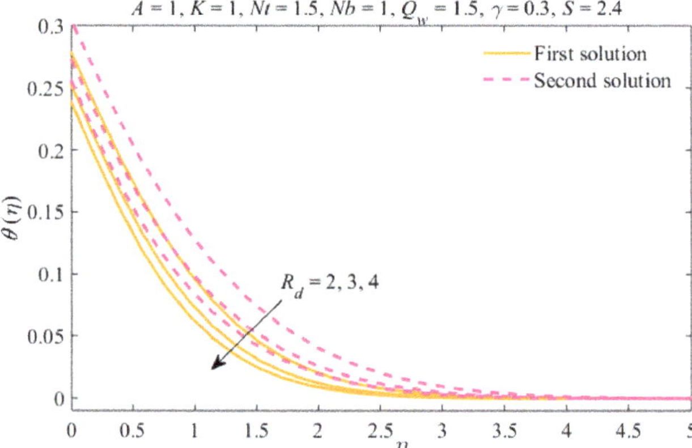

Figure 13. The temperature profiles for different values of R_d.

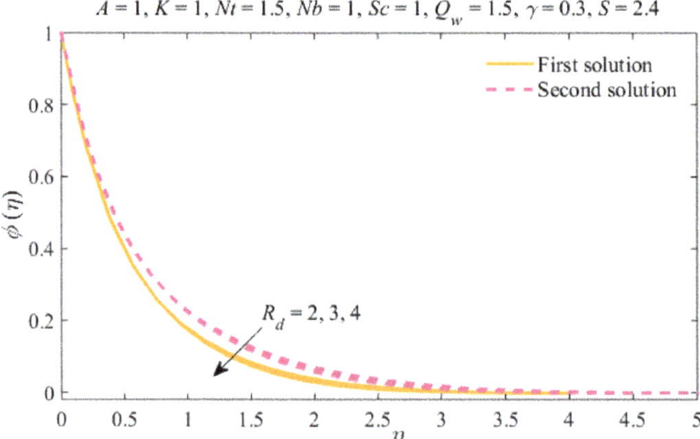

Figure 14. The concentration profiles for different values of R_d.

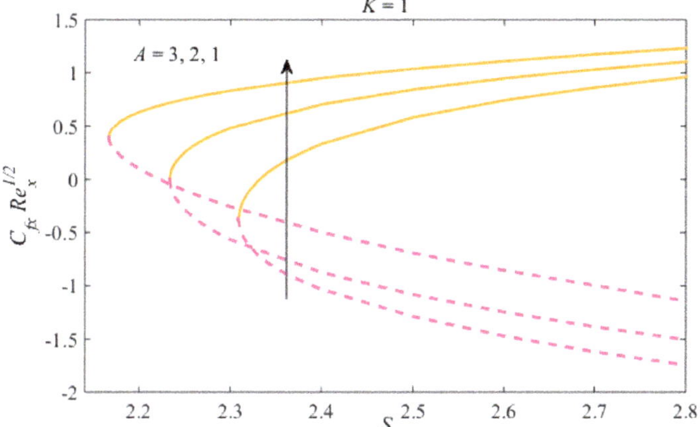

Figure 15. The skin friction $C_f \mathrm{Re}_x^{1/2}$ versus S for different values of A.

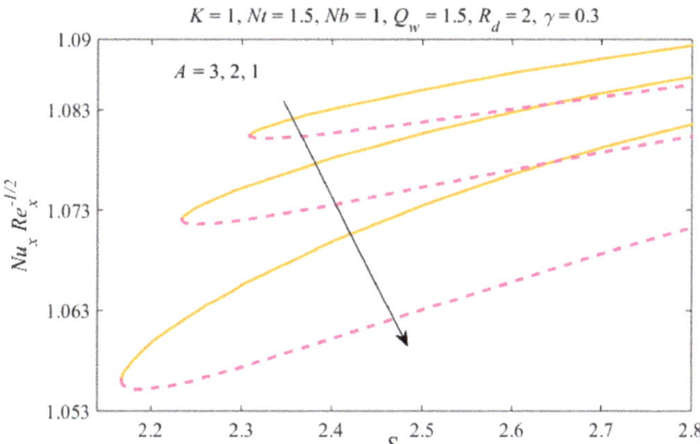

Figure 16. The Nusselt number $Nu_x \mathrm{Re}_x^{-1/2}$ versus S for different values of A.

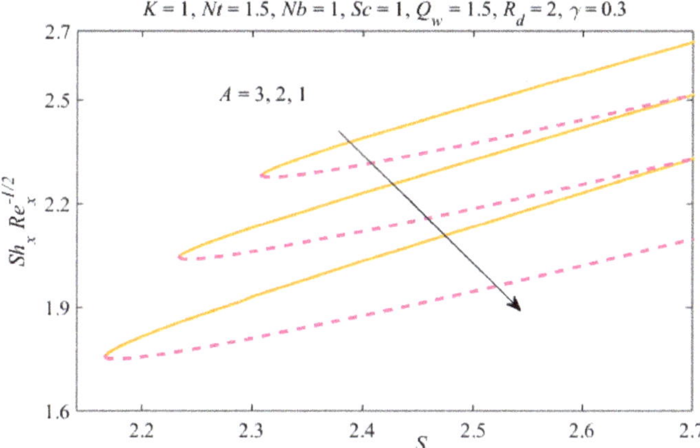

Figure 17. The Sherwood number $Sh_x \mathrm{Re}_x^{-1/2}$ versus S for different values of A.

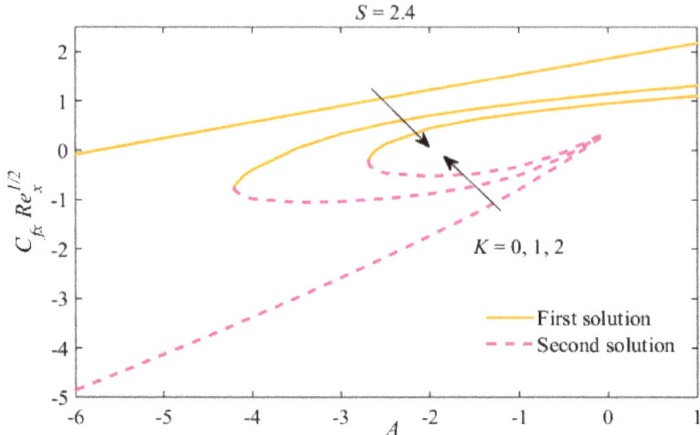

Figure 18. The skin friction $C_f \mathrm{Re}_x^{1/2}$ versus A for different values of K.

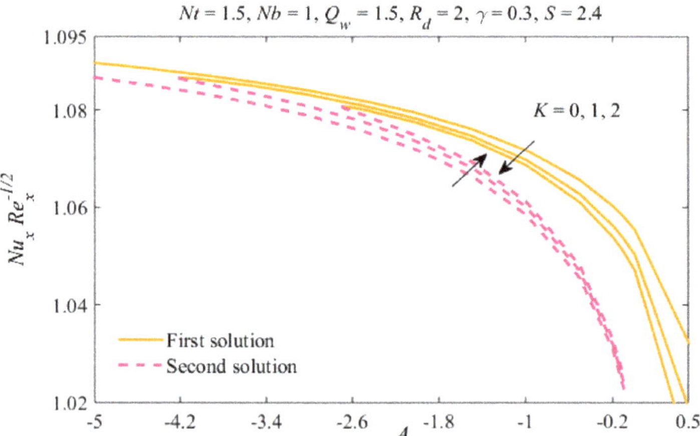

Figure 19. The Nusselt number $Nu_x \mathrm{Re}_x^{-1/2}$ versus A for different values of K.

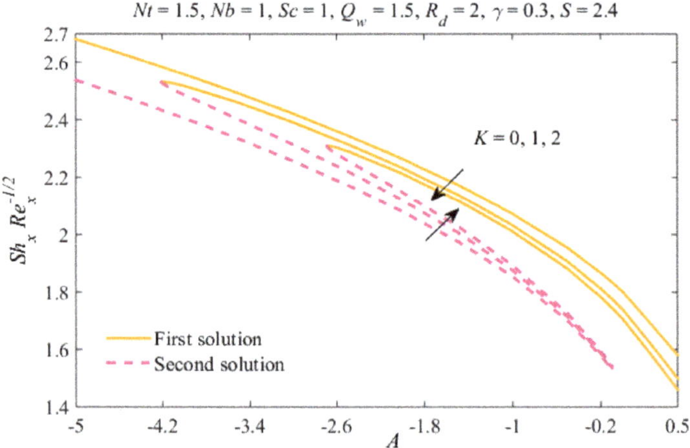

Figure 20. The Sherwood number, $Sh_x \mathrm{Re}_x^{-1/2}$, versus A for different values of K.

Table 1. Values of skin friction, Nusselt number, and Sherwood number versus S for different values of A when $K = 1$, $Nb = 1$, $Nt = 1.5$, $Le = 1$, $Q_w = 1.5$, $R_d = 2$, $\gamma = 0.3$ are fixed.

S	A	$C_f\mathrm{Re}_x^{1/2}$		$Nu_x\mathrm{Re}_x^{-1/2}$		$Sh_x\mathrm{Re}_x^{-1/2}$	
		First Solution	**Second Solution**	**First Solution**	**Second Solution**	**First Solution**	**Second Solution**
	−3	0.9627	−1.7424	1.0894	1.0855	2.7589	2.5893
2.8	−2	1.1099	−1.5024	1.0863	1.0804	2.6085	2.4075
	−1	1.2370	−1.1420	1.0816	1.0713	2.1800	2.0495
	−3	0.7457	−1.4673	1.0867	1.0831	2.5777	2.4429
2.6	−2	0.9505	−1.2430	1.0828	1.0770	2.4220	2.2570
	−1	1.1154	−0.8527	1.0766	1.0659	2.2332	2.0209
	−3	0.3372	−1.0314	1.0831	1.0807	2.3910	2.3136
2.4	−2	0.7068	−0.8738	1.0782	1.0735	2.2324	2.1203
	−1	0.9524	−0.4982	1.0699	1.0602	2.0347	1.8766

Table 2. Critical values of S_c for different values of A when $K = 1$, $Nb = 1$, $Nt = 1.5$, $Le = 1$, $Q_w = 1.5$, $R_d = 2$, $\gamma = 0.3$ are fixed.

A	S_c
−3	2.3079
−2	2.2338
−1	2.1665

5. Conclusions

In the current study we have discussed the time-dependent flow of a nanoliquid involving a special third-grade fluid induced through a heated shrinking surface through non-linear radiation. The transmuted ODE's were numerically calculated via a bvp4c solver for distinct values of the substantial constraints that appeared in the problem. This investigation reveals the following significant finding:

- Multiple results have been obtained for decelerating flow only, and for precise values of S.
- The liquid velocity declines in the first result and increases in the second due to K. The concentration and temperature fields rise in the first result and diminish in the second result.
- The thicknesses of the concentration and thermal boundary layers increase due to γ in both results.
- The liquid temperature and nanomaterial concentration decrease due to thermal radiation in both results.
- The nanoparticles distribution can be controlled through the mechanism of Brownian motion and the thermophoresis effect.
- The time-dependent and non-Newtonian parameters delay the separation of the boundary.

Author Contributions: A.Z. and U.K.; Writing—original draft, I.K.; Conceptualization, E.-S.M.S.; Methodology, I.K.; Software, K.S.N.; Validation, A.H.S.; Formal analysis, A.Z.; Writing—review and editing, Investigation, U.K. and I.K.; Resources, A.H.S.; Data curation, E.-S.M.S.; Visualization and supervision, E.-S.M.S.; Project administration and funding. All authors have read and agree to the published version of the manuscript.

Funding: The authors would like to extend their sincere appreciation to the Deanship of Scientific Research at King Saud University for the funding of this research. Research group project number RGP-160.

Conflicts of Interest: The authors declare no conflict of interest concerning this manuscript.

References

1. Keçebaş, A.; Yürüsoy, M. Similarity solutions of unsteady boundary layer equations of a special third grade fluid. *Int. J. Eng. Sci.* **2006**, *44*, 721–729. [CrossRef]

2. Ellahi, R.; Riaz, A. Analytical solutions for MHD flow in a third-grade fluid with variable viscosity. *Math. Comput. Model.* **2010**, *52*, 1783–1793. [CrossRef]

3. Sahoo, B.; Do, Y. Effects of slip on sheet-driven flow and heat transfer of a third grade fluid past a stretching sheet. *Int. Comm. Heat Mass Transf.* **2010**, *37*, 1064–1071. [CrossRef]

4. Sahoo, B.; Poncet, S. Flow and heat transfer of a third grade fluid past an exponentially stretching sheet with partial slip boundary condition. *Int. J. Heat Mass Transf.* **2011**, *54*, 5010–5019. [CrossRef]

5. Abbasbandy, S.; Hayat, T. On series solution for unsteady boundary layer equations in a special third grade fluid. *Commun. Nonlinear Sci. Numer. Simul.* **2011**, *16*, 3140–3146. [CrossRef]

6. Rehman, A.; Nadeem, S.; Malik, M.Y. Boundary layer stagnation-point flow of a third grade fluid over an exponentially stretching sheet. *Braz. J. Chem. Eng.* **2013**, *30*, 611–618. [CrossRef]

7. Hussain, T.; Hayat, T.; Shehzad, S.A.; Alsaedi, A.; Chen, B. A model of solar radiation and Joule heating in flow of third grade nanofluid. *Z. Nat. A* **2015**, *70*, 177–184. [CrossRef]

8. Naganthran, K.; Nazar, R.; Pop, I. Unsteady stagnation-point flow and heat transfer of a special third grade fluid past a permeable stretching/shrinking sheet. *Sci. Rep.* **2016**, *6*, 24632. [CrossRef]

9. Reddy, G.J.; Hiremath, A.; Kumar, M. Computational modeling of unsteady third-grade fluid flow over a vertical cylinder: A study of heat transfer visualization. *Results Phys.* **2018**, *8*, 671–682. [CrossRef]

10. Choi, S.U.S.; Eastman, J.A. Enhancing thermal conductivity of fluids with nanoparticles. In Proceedings of the 1995 International Mechanical Engineering Congress and Exhibition, San Francisco, CA, USA, 12–17 November 1995.

11. Masuda, H.; Ebata, A.; Teramae, K.; Hishinuma, N. Alteration of thermal conductivity and viscosity of liquid by dispersing ultra-fine particles. *Netsu Bussei* **1993**, *7*, 227–233. [CrossRef]

12. Buongiorno, J. Convective transport in nanofluids. *ASME J. Heat Transf.* **2006**, *128*, 240–250. [CrossRef]

13. Khan, W.A.; Pop, I. Boundary-layer flow of a nanofluid past a stretching sheet. *Int. J. Heat Mass Transf.* **2010**, *53*, 2477–2483. [CrossRef]

14. Rana, P.; Bhargava, R. Flow and heat transfer of a nanofluid over a nonlinearly stretching sheet: A numerical study. *Commun. Nonlinear Sci. Numer. Simul.* **2012**, *17*, 212–226. [CrossRef]

15. Makinde, O.D.; Aziz, A. Boundary layer flow of a nanofluid past a stretching sheet with a convective boundary condition. *Int. J. Therm. Sci.* **2011**, *50*, 1326–1332. [CrossRef]

16. Irfan, M.; Khan, M.; Khan, W.A.; Ayaz, M. Modern development on the features of magnetic field and heat sink/source in Maxwell nanofluid subject to convective heat transport. *Phys. Lett. A* **2018**, *382*, 1992–2002. [CrossRef]

17. Khan, M.; Irfan, M.; Khan, W.A. Impact of heat source/sink on radiative heat transfer to Maxwell nanofluid subject to revised mass flux condition. *Results Phys.* **2018**, *9*, 851–857. [CrossRef]

18. Zaib, A.; Abelman, S.; Chamkha, A.J.; Rashidi, M.M. Entropy generation in a Williamson nanofluid near a stagnation point over a moving plate with binary chemical reaction and activation energy. *Heat Transf. Res.* **2018**, *49*, 1131–1149. [CrossRef]

19. Bibi, M.; Rehman, K.-U.; Malik, M.Y.; Tahir, M. Numerical study of unsteady Williamson fluid flow and heat transfer in the presence of MHD through a permeable stretching surface. *Eur. Phys. J. Plus* **2018**, *133*, 154. [CrossRef]

20. Hayat, T.; Khan, S.A.; Khan, M.I.; Alsaedi, A. Theoretical investigation of Ree-Eyring nanofluid flow with entropy optimization and Arrhenius activation energy between two rotating disks. *Comp. Methods Prog. Biomed.* **2019**, *177*, 57–68. [CrossRef]

21. Aziz, A. A similarity solution for laminar thermal boundary layer over a flat plate with a convective surface boundary condition. *Commun. Nonlinear Sci. Numer. Simul.* **2009**, *14*, 1064–1068. [CrossRef]

22. Makinde, O.D.; Aziz, A. MHD mixed convection from a vertical plate embedded in a porous medium with a convective boundary condition. *Int. J. Therm. Sci.* **2010**, *49*, 1813–1820. [CrossRef]

23. Ishak, A. Similarity solutions for flow and heat transfer over a permeable surface with convective boundary condition. *Appl. Math. Comput.* **2014**, *217*, 837–842. [CrossRef]

24. Yao, S.; Fang, T.; Zhong, Y. Heat transfer of a generalized stretching/shrinking wall problem with convective boundary conditions. *Commun. Nonlinear Sci. Numer. Simul.* **2011**, *16*, 752–760. [CrossRef]

25. Rahman, M.M.; Merkin, J.H.; Pop, I. Mixed convection boundary-layer flow past a vertical flat plate with a convective boundary condition. *Acta Mech.* **2015**, *226*, 2441–2460. [CrossRef]

26. Mustafa, M.; Khan, J.A.; Hayat, T.; Alsaedi, A. Simulations for Maxwell fluid flow past a convectively heated exponentially stretching sheet with nanoparticles. *AIP Adv.* **2015**, *5*. [CrossRef]

27. Ibrahim, W.; Haq, R.U. Magnetohydrodynamic (MHD) stagnation point flow of nanofluid past a stretching sheet with convective boundary condition. *J. Braz. Soc. Mech. Sci. Eng.* **2016**, *38*, 1155–1164. [CrossRef]

28. Makinde, O.D.; Khan, W.A.; Khan, Z.H. Stagnation point flow of MHD chemically reacting nanofluid over a stretching convective surface with slip and radiative heat. *Proc. Inst. Mech. Eng. Part E J. Process Mech. Eng.* **2017**, *231*, 695–703. [CrossRef]

29. Khan, M.; Hashim; Hussain, M.; Azam, M. Magnetohydrodynamic flow of Carreau fluid over a convectively heated surface in the presence of non-linear radiation. *J. Magn. Magn. Mater.* **2016**, *412*, 63–68. [CrossRef]

30. Mabood, F.; Khan, W.A. Analytical study for unsteady nanofluid MHD flow impinging on heated stretching sheet. *J. Mol. Liq.* **2016**, *219*, 216–223. [CrossRef]

31. Fosdick, R.L.; Rajagopal, K.R. Thermodynamics and stability of fluids of third grade. *Proc. R. Sot. Lond. A* **1980**, *339*, 351. [CrossRef]

32. Mahapatra, T.R.; Nandy, S.K. Slip effects on unsteady stagnation-point flow and heat transfer over a shrinking sheet. *Meccanica* **2013**, *48*, 1599–1606. [CrossRef]

33. Ali, F.M.; Nazar, R.; Arifin, N.M.; Pop, I. Unsteady flow and heat transfer past an axisymmetric permeable shrinking sheet with radiation effect. *Int. J. Numer. Meth. Fluids* **2011**, *67*, 1310–1320. [CrossRef]

34. Rohni, A.M. Flow over an unsteady shrinking sheet with suction in a nanofluid. *Int. Conf. Math. Comput. Biol.* **2012**, *9*, 511–519. [CrossRef]

35. Merkin, J.H. On dual solutions occurring in mixed convection in a porous medium. *J. Eng. Math.* **1986**, *20*, 171–179. [CrossRef]

36. Weidman, P.D.; Kubitschek, D.G.; Davis, A.M.J. The effect of transpiration on self-similar boundary layer flow over moving surfaces. *Int. J. Eng. Sci.* **2006**, *44*, 730–737. [CrossRef]

37. Roşca, A.V.; Pop, I. Flow and heat transfer over a vertical permeable stretching/shrinking sheet with a second order slip. *Int. J. Heat Mass Transf.* **2013**, *60*, 355–364. [CrossRef]

38. Roşca, N.C.; Pop, I. Mixed convection stagnation point flow past a vertical flat plate with a second order slip: Heat flux case. *Int. J. Heat Mass Transf.* **2013**, *65*, 102–109. [CrossRef]

39. Ridha, A. Aiding flows non-unique similarity solutions of mixed-convection boundary-layer equations. *Z. Angew. Math. Phys.* **1996**, *47*, 341–352. [CrossRef]

Article

Analytical Solution of Heat Conduction in a Symmetrical Cylinder Using the Solution Structure Theorem and Superposition Technique

Rasool Kalbasi [1], Seyed Mohammadhadi Alaeddin [2], Mohammad Akbari [1] and Masoud Afrand [3,4,*]

[1] Department of Mechanical Engineering, Najafabad Branch, Islamic Azad University, Najafabad, Iran; rasool.kalbasi@gmail.com (R.K.); m.akbari.g80@gmail.com (M.A.)
[2] Department of Mechanical Engineering, University of Isfahan, Isfahan, Iran; alaedin.hadi.ma@gmail.com
[3] Laboratory of Magnetism and Magnetic Materials, Advanced Institute of Materials Science, Ton Duc Thang University, Ho Chi Minh City, Vietnam
[4] Faculty of Applied Sciences, Ton Duc Thang University, Ho Chi Minh City, Vietnam
* Correspondence: masoud.afrand@tdtu.edu.vn

Received: 13 November 2019; Accepted: 9 December 2019; Published: 16 December 2019

Abstract: In this paper, non-Fourier heat conduction in a cylinder with non-homogeneous boundary conditions is analytically studied. A superposition approach combining with the solution structure theorems is used to get a solution for equation of hyperbolic heat conduction. In this solution, a complex origin problem is divided into, different, easier subproblems which can actually be integrated to take the solution of the first problem. The first problem is split into three sub-problems by setting the term of heat generation, the initial conditions, and the boundary condition with specified value in each sub-problem. This method provides a precise and convenient solution to the equation of non-Fourier heat conduction. The results show that at low times ($t = 0.1$) up to about $r = 0.4$, the contribution of T_1 and T_3 dominate compared to T_2 contributing little to the overall temperature. But at $r > 0.4$, all three temperature components will have the same role and less impact on the overall temperature (T).

Keywords: heat conduction; non-fourier; solution structure theorems; superposition approach

---

1. Introduction

In recent years, some studies have focused on the deviation from the classical Fourier heat transfer equation. In the classical theory of conduction heat transfer based on Fourier's law, the thermal heat flux is a linear relationship with the gradient of temperature and the heat wave propagation speed is assumed to be unlimited (Equation (1)).

$$q^* = -k\nabla T^* \tag{1}$$

where k is the coefficient of thermal conductivity and T^* is temperature. When this equation is combined into the balance equation of energy,

$$\nabla \cdot q^* = -\rho c_p \frac{\partial T^*}{\partial t^*} \tag{2}$$

The classical parabolic heat conduction equation is derived,

$$\frac{\partial T^*}{\partial t^*} = \alpha \nabla^2 T^* \tag{3}$$

where $\alpha = \frac{k}{\rho c_p}$, ρ, and c_p are thermal diffusivity, density, and specific heat, respectively. Also, the effect of any thermal disturbance on the medium is instantaneously sensed through the entire molecular network. In the majority of engineering applications such as material processing (welding, cutting, forming, etc.), applying high power laser radiation, cryogenic applications, and materials which experience the high heat transfer rates [1–4], this equation is very useful. However, at very low times and very high thermal fluxes and very low temperatures close to absolute zero, the Fourier law has poor accuracy, and considering the effects of non-Fourier in describing the heat dissipation process and prediction of temperature distribution, non-Fourier are more reliable in these situations. Fourier's failure to exactly predict the temperature field in sufficiently high heat flux and low temperature engineering usages is because it assumes that thermal energy transport is occurring at an infinite propagation speed [2,5]. As a result, a more advanced method, in the situation of the thermal waves finite propagation speed, is required to analyze the high temperature gradients. Usually, when the infinite propagation speed was assumed, the temperature was calculated more than its actual values and it causes some errors in temperature prediction.

A modified equation of non-Fourier heat flux has been developed by Cattaneo [6] and Vernotte [7] in the present form,

$$q^* + \tau_0 \frac{\partial q^*}{\partial t^*} = -k\nabla T^* \tag{4}$$

where τ_0 is known thermal relaxation time. If the relaxation time ignores, $\tau_0 = 0$, the law of the non-Fourier model is converted to the Fourier law. The energy equation is derived as follow,

$$-\frac{\partial \vec{q}^*}{\partial r^*} + \dot{g}''' = \rho c_p \frac{\partial T^*}{\partial t^*} \tag{5}$$

In the Equation (5), $\dot{g}'''$ expresses the rate of internal generation of energy. Inserting Equation (4) into Equation (5), the equation of hyperbolic heat transfer, containing source the term, derived,

$$\alpha\nabla^2 T^* = \frac{\partial T^*}{\partial t^*} + \tau_0 \frac{\partial^2 T^*}{\partial t^{*2}} + Q(r,t) \tag{6}$$

where the source term is $Q(r,t)$. Different solution procedures for Equation (6) with different boundary and initial conditions for finite media can be found in literature.

The analytical, numerical, and experimental methods were used in many researches for analysis and calculating the rate of heat transfer in applied physics problems [8–10]. Using the heat sources such as lasers and microwaves at very small times of applying the heat or high frequencies has a great deal of application in analytical physics, applied sciences, and engineering. In fast and short processes and rapid and concentrated conduction heat transfer, the order of space and time is very small. So, the law of Fourier equation which assumed that heat propagates at an infinite velocity, cannot be used [11].

Ozisik and Vick [12] studied the heat propagation in a semi-infinite body with a volumetric source of energy by solving the thermal wave equation. They found that the classical Fourier's law was no longer suitable in obtaining the temperature field at short times. Jiang [13] applied the method of Laplace transform to study the hyperbolic conduction heat transfer in a hollow sphere whose boundaries are affected by a sudden change in temperature. Moosaie [14] solved, analytically, the equation of non-Fourier heat conduction for a finite body with an arbitrary initial condition and insulated boundaries. His results showed that the time needed for reaching steady state situation is enhanced with rising the relaxation time τ_0. Moosaie [15] investigated a finite body subjected to an arbitrary non-periodic surface disturbance. Their obtained solution is such that for a given non-periodic disturbance, analytically if possible, but in general numerically is a straight forward computational task. Ahmadikia and Riesmanian [16] presented an analytical method for solving the hyperbolic heat transfer in a blade under periodic boundary conditions applying the Laplace transform approach. The findings showed that in small blades under rapid phenomena, temperature behavior is the non-Fourier wave form. Bamdad et al. [17] studied the non-fluoride effects at extended surfaces. Their results showed that for all fins at the start times, the point of discontinuity is time, relaxation time, and cross-section of the dependent fin. In addition, the cross-sectional effects on the amplitude of the heat wave reflecting from the tip of the fin are such that there is no reflected heat wave in the fins with concave shape. Lam and Fang [18] presented an analytical method to investigate the heat conduction in a slab applied by various boundary conditions. They indicated that the solution accuracy depends on the terms number which were applied in the Fourier series expansion process.

Liu et al. [19] surveyed the non-Fourier heat conduction characteristics in the oil/water emulsions experimentally. Their results showed that in the ratio of time lag less than one, no thermal waves exist for oil/water emulsions.

The analysis of non-Fourier heat conduction in infinite hollow cylinders subjected to a heat source, which is a function of time, was investigated by Daneshjou et al. [20]. They used the Laplace transform method and demonstrated that their approach is valuable in correctness and exactness. Ma et al. [21] studied the C-V wave model for a plate which is irradiated by a non-Gaussian laser pulse. The method of mode superposition was applied for solving the equation. They discussed the dependence of the wave velocity on relaxation time. The influence of scanning and wave velocity on the temperature field was also presented. Wankhade et al. [22] investigated the heat transfer response of wet fins using the models of Fourier and non-Fourier. The method of separation of variables was applied, and the results showed a considerable deviation in temperature response using the non-Fourier heat conduction compared to the Fourier model. Also, the effect of fin surface conditions was studied. Han and Peddieson [23] investigated the Non-Fourier one-dimensional unsteady equation in a body for medium speeds less than (sub-critical), equal to (critical), and greater than (super-critical) the thermal wave speed. Liu et al [24] studied a hyperbolic lattice Boltzmann method (HLBM) compare to the parabolic lattice Boltzmann method (PLBM) to survey the non-Fourier effects. The results show that the electron temperatures simulated by the two-step HLBM/PLBM and two-temperature models are not much different from each other and both of them coincide with the experimental data.

Review of this research indicated that various solutions are studied in different works. In each study, different features of non-Fourier heat conduction were investigated and, therefore, various results were established which could be useful in its position. However, the shortage of a general problem with nonzero initial conditions and boundary conditions with internal heat generation in these studies is observed. In this study, an exact solution is presented to non-Fourier heat conduction in a cylinder with nonzero initial and boundary conditions. As it will be mentioned later, analytical solutions of this problem were obtained by applying the theory of solution structure combined with the superposition method. This approach can be used for obtaining the heat transfer in many physical applications which solved by different methods [25–31].

2. Formulation

2.1. Hyperbolic Heat Conduction

According to the Figure 1, we assume a cylinder composed of a homogenous heat conducting material with different boundary conditions at both sides: A symmetry boundary condition in the cylinder's central line and convection in the cylinder surface ($r = R$) with ambient.

The heat is conducted through the cylinder in r-direction, where one dimensional heat conduction dominates. This problem was solved for L/r > 10 and a one-dimensional assumption for this problem is reasonable. For simplifying the solution, by using the following parameters, we can non-dimensionalize the governing equations,

$$r = \frac{cr^*}{2\alpha}, \ t = \frac{c^2 t^*}{2\alpha}, \ T = \frac{kcT^*}{\alpha f_r}, \ T_\infty = \frac{kcT_\infty^*}{\alpha f_r}, \ q = \frac{q^*}{f_r}, \ g = \frac{4\alpha \dot{g}'''}{c f_r} \tag{7}$$

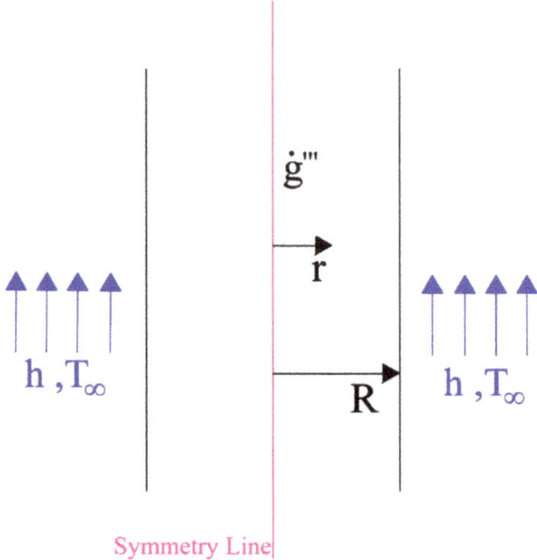

Figure 1. A schematic illustration of the problem.

Combining the Equations (4)–(6) with Equation (7), we can derive the non-dimensional form of non-Fourier heat conduction equations as follows,

$$\frac{\partial q}{\partial t} + \frac{\partial T}{\partial r} = -2q \tag{8}$$

$$\frac{\partial T}{\partial t} + \frac{\partial q}{\partial r} = \frac{g}{2} \tag{9}$$

$$\frac{\partial^2 T}{\partial t^2} + 2\frac{\partial T}{\partial t} = \frac{\partial^2 T}{\partial r^2} + f(r,t) \tag{10}$$

where $f(r,t)$ as a total internal heat generation in system can be determined as follows,

$$f(r,t) = \frac{1}{2}\frac{\partial g}{\partial t} + g \tag{11}$$

For the problem situation, the boundary conditions are defined as follows,

$$\frac{\partial T(0,t)}{\partial r} = 0 \tag{12}$$

$$-\frac{\partial T(1,t)}{\partial r} = m(T - T_\infty), \quad m = \frac{-2ah}{kc} \tag{13}$$

T_∞ is the dimensionless ambient temperature and m is the convection heat transfer coefficient.

The initial conditions are considered to be,

$$T(r,0) = \varphi(r) \tag{14}$$

$$\frac{\partial T(r,0)}{\partial t} = \psi(r) = \frac{g_0}{2} \tag{15}$$

These conditions can be derived from Equation (9). In Equations (14) and (15), φ and ψ are dimensionless initial condition function and dimensionless initial rate of temperature change function, respectively.

In this paper, g as an internal energy generation will be defined as [11],

$$g(r) = g_0 \exp(-\mu r) \tag{16}$$

or

$$g(r) = g_0 \exp(-\mu r) \exp(-t) \tag{17}$$

where

$$g_0 = \frac{2I_0\mu(1 - R)}{f_r} \tag{18}$$

where μ is the absorption coefficient, I_0 is the amplitude of laser density, and R is the solid surface reflectivity. This model assumes no spatial variations of g_0 in the plane perpendicular to the laser beam.

2.2. Solution Structure Theorems and Superposition Approach

One of the famous and most widely used techniques for solving some types of heat conduction equations is the superposition method. This method can be used for solving the linear heat transfer problems with non-homogenous conditions. In this method, an origin problem is split into different easier subproblems which can be integrated to take a solution to the original problem. The method of superposition relies upon the assumption that the original problem (Equation (10)) can be divided into three subproblems by setting the heat generation term (Equation (11)), the initial conditions (Equations (14) and (15)), and the boundary conditions (Equations (12) and (13)) to different values in each subproblem:

$$f(r,t) = \varphi(r) = 0 \tag{19}$$

$$f(r,t) = \psi(r) = 0 \tag{20}$$

$$\varphi(r) = \psi(r) = 0 \tag{21}$$

Solutions to these subproblems are assigned as T_1, T_2, and T_3 sequentially. Therefore, the general solution to the first equation (Equation (10)) is the sum of subproblems one through three, which is

$$T(r,t) = T_1(r,t) + T_2(r,t) + T_3(r,t)) \tag{22}$$

It can be seen that T_1, T_2, and T_3 illustrated the independent contributions of the initial rate of temperature variation, initial condition, and internal heat generation to the temperature field, respectively. Subproblems one to three can be simply solved by applying the solution structure theorems [18] once the solution to subproblem one is known. By using the solution structure theorems, the solutions of subproblems one to three can be determined as follows:

$$T_1(r,t) = F(r,t,\psi(r)) \tag{23}$$

$$T_2(r,t) = (2 + \frac{\partial}{\partial t})F(r,t,\varphi(r)) \tag{24}$$

$$T_3(r,t) = \int_0^t F(r,t-\xi,f(r,\xi))d\xi \tag{25}$$

where $T_1(r,t)$ will be obtained by using the Fourier method. $T_2(r,t)$ and $T_3(r,t)$ can be simply derived by using the solution structure theorem. It means that only the solution of subproblem one is needed to obtain the solutions to subproblems two and three. Finally, the general solution to the first heat conduction equation is the sum of subproblems one to three.

2.3. Formulation of the Problem

In the literature, there are a few non-Fourier heat conduction problems with different boundary conditions but most of them are limited to boundaries with zero temperature or the insulated boundaries. In this study, we obtain the general analytical solution to hyperbolic heat conduction in a cylinder composed of a homogenous heat conducting material with different boundary conditions at both sides: A symmetry boundary condition in the cylinder's central line and the convection boundary condition in the cylinder surface ($r = R$) with ambient.

Let us first consider the solution to subproblem one. This subproblem is solved with the condition $f(r,t) = \varphi(r) = 0$. As a result, the equation of this subproblem and the initial conditions and boundary conditions are as bellow:

$$\frac{\partial^2 T_1}{\partial t^2} + 2\frac{\partial T_1}{\partial t} = \frac{\partial^2 T_1}{\partial r^2} \tag{26}$$

$$\frac{\partial T_1(0,t)}{\partial r} = 0 \tag{27}$$

$$-\frac{\partial T(1,t)}{\partial r} = m(T - T_\infty), \quad m = \frac{-2ah}{kc} \tag{28}$$

$$T_1(r,0) = 0 \tag{29}$$

$$\frac{\partial T_1(r,0)}{\partial t} = \psi(r) \tag{30}$$

Using the Fourier series expansion theory, the general form of solution of equation is,

$$T_1(r,t) = T_\infty + \tfrac{1}{2}\left[1 - e^{-2t}\right]\int_0^1 (\psi(\zeta) - T(\infty))d\zeta +$$

$$\sum_{n=0}^{\infty} \frac{e^{-2t}}{\gamma_n}\left[\frac{\int_0^1 \zeta(\psi(\zeta) - T(\infty))J_0(\lambda_n\zeta)d\zeta}{\frac{\lambda_n^2 + m^2}{2\lambda_n^2}J_0^2(\lambda_n)}\right]\sin(\gamma_n t)J_0(\lambda_n r) \tag{31}$$

Now, by using the solution structure theorems, we have

$$T_2(r,t) = T_\infty + \int_0^1 \varphi(\zeta) - T(\infty)d\zeta + \sum_{n=1}^{\infty}\left[\frac{\int_0^1 r\varphi(\zeta) - T(\infty)J_0(\lambda_n r)d\zeta}{\frac{\lambda_n^2 - m^2}{2\lambda_n^2}J_0^2(\lambda_n)}\right]\times$$

$$\left[\frac{e^{-t}}{\gamma_n}[\sin(\gamma_n t) + \gamma_n\cos(\gamma_n t)]\right]J_0(\lambda_n r) \tag{32}$$

$$T_3(r,t) = T_\infty + \int_0^t \tfrac{1}{2}\left(1 - e^{-2(t-\xi)}\right)\left[\int_0^1 f(\zeta)d\zeta\right]d\xi +$$

$$\int_0^t\left\{\sum_{n=0}^{\infty}\frac{e^{-(t-\xi)}}{\gamma_n}\left[\frac{\int_0^1 rf(\zeta)J_0(\lambda_n r)d\zeta}{\frac{\lambda_n^2 + m^2}{2\lambda_n^2}J_0^2(\lambda_n)}\right]\sin(\gamma_n(t-\xi))J_0(\lambda_n r)\right\}d\xi \tag{33}$$

where

$$\frac{J_0(\lambda_n)}{J_1(\lambda_n)} = \frac{\lambda_n}{m} \tag{34}$$

J_0 and J_1 are the first order Bessel functions and was obtained by solving the Equation (34).
Also,

$$\gamma_n = \sqrt{\lambda_n^2 - 1} \tag{35}$$

Ultimately, the temperature field within the slab can be obtained as,

$$T(x,t) = T_1(r,t) + T_2(r,t) + T_3(r,t))$$

3. Results and Discussions

In this paper, the temperature field in a one dimensional cylinder with non-zero initial and boundary condition are examined. The values $g_0 = 100$, $\mu = 5$, $T_0 = 2.5$, and $T_\infty = 1.8$ were selected. It can be concluded that all the profiles of temperature from $T_1(r,t)$ to $T_3(r,t)$ are in the infinite series form. By using a relative error test, we can write:

$$\frac{\left|T_{n+1}(r,t) - T_n(r,t)\right|}{\left|T_{n+1}(r,t)\right|} < \varepsilon \tag{36}$$

where $T_{n+1}(r,t)$ and $T_n(r,t)$ are two consecutive partial sums for the temperature, and $\varepsilon = 10^{-6}$ is the relative error for this study. We showed that the original partial differential equation is split into three subproblems, subproblems one through three demonstrating the contributions of initial rate of change in temperature, initial condition, and internal heat generation, which are given by Equations (31) to (35), respectively.

It can be noted from these equations that all the temperature profiles from $T_1(r,t)$ to $T_3(r,t)$ are in the form of an infinite series.

Figure 2 shows the contribution of different components of temperature at various times for a one dimensional cylinder with non-zero initial and boundary condition. According to the Figure 2a, it can be seen that at small times ($t = 0.1$) up to about $r = 0.4$, the contribution of T_1 and T_3 dominate compared to T_2 which contributes little to the overall temperature. But at $r > 0.4$, all three temperature components will have the same role and less impact on the overall temperature.

Figure 2b shows that T_2 still does not have much effect on the overall temperature and acts approximately uniformly with a constant value. T_1 and T_3 still have a downward trend, but T_3, because of is related to the temperature component of internal heat generation, is dominant.

The downward trend of T_1 and T_3 will be continued to $r = 0.5$ and $r = 0.8$, respectively. After these points, significant variations were not seen. Also, in the areas near the center of the cylinder, where the source term generates the energy, the overall temperature has larger values compared to its values at $t = 0.1$.

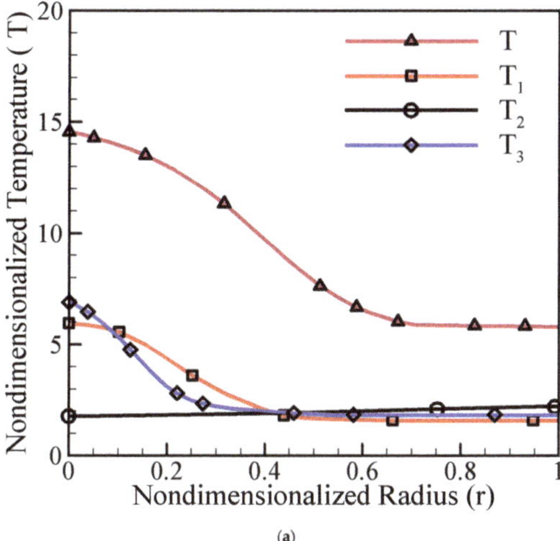

Figure 2. *Cont.*

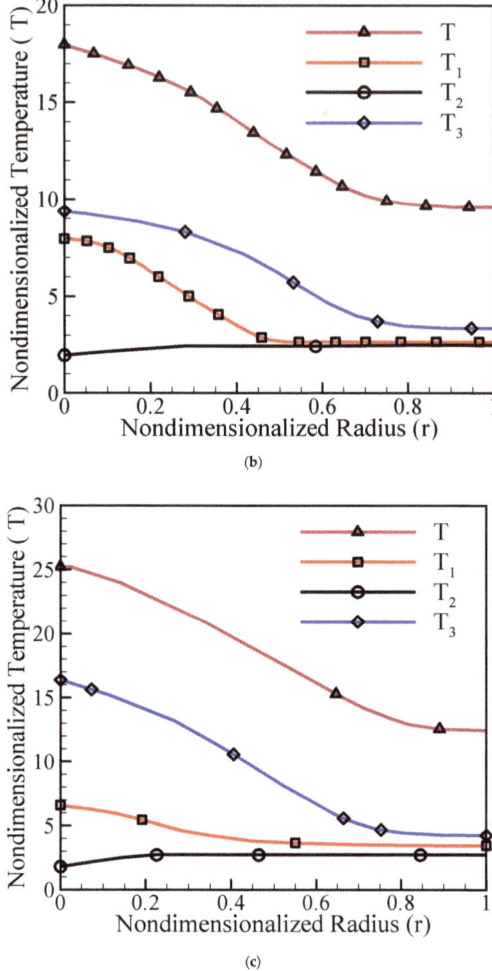

Figure 2. Contribution of temperature components at $t = 0.1$ (**a**), $t = 0.5$ (**b**), and $t = 1$ (**c**).

As time increases Figure 2c, the contribution of T_2 to the overall temperature is less. Also, the effects of T_1 on the overall temperature near to the center of the cylinder are decreased. This means that the effects of initial rate of temperature change have been remarkable, only, at the small times and with increasing the time the effects were not impressive. Due to the fact that the internal heat generation is time and space dependent, the values of T_3 in areas close to the center of the cylinder have increased dramatically and will have the greatest impact on the overall temperature (T).

Figure 3a illustrates the temperature temporal trend at low times. As time arises, temperatures of cylinder's center ($r = 0$) enhance because of absorption of more energy compared to the other places. However, the boundary of cylinder surface remains relatively unchanged and keeps its initial temperature. According to the Figure 3b, with enhancing the time, the surface boundary temperature became affected by the entering energy, hence increasing the temperatures at a slower rate occurred. The temperature throughout the cylinder will continue to increase toward equilibrium between the center and the surface of the cylinder. This trend was shown in Figure 3c.

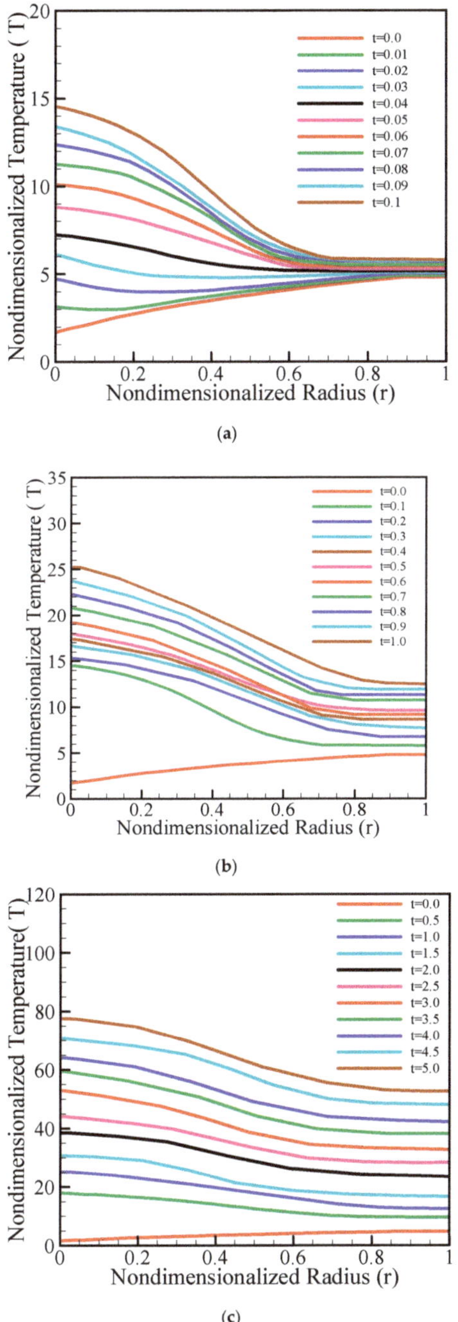

Figure 3. Temperature distributions from $t = 0$ to $t = 0.1$ (**a**), $t = 0.0$ to $t = 1$ (**b**), and $t = 1$ to $t = 5$ (**c**).

4. Conclusions

In this paper the analytical solution of non-Fourier heat conduction in a cylinder composed of a homogenous material with different boundary conditions: A symmetry boundary condition in the cylinder's central line and the convection in the cylinder surface (r = R) with ambient is investigated.

We conclude that the obtained results provide an accurate, convenient, and useful solution to the non-Fourier equation, which is usable for analyses of various engineering applications.

The key findings and conclusions from the present solution are as follows:

At small times (t = 0.1) up to about r = 0.4, the contribution of T_1 and T_3 dominate compared to T_2 contributing little to the overall temperature.

At t = 0.5, T_2 does not have much effect on the overall temperature and acts approximately uniformly with a constant value.

At t = 0.5, T_1 and T_3 have a downward trend, but T_3 is dominant.

At t = 1, the effects of T_1 on the overall temperature near to the center of the cylinder are decreased.

At low times, by enhancing the time, temperatures at the center of the cylinder (r = 0) enhance.

At big times, the temperature throughout the cylinder will continue to increase.

Author Contributions: The contributions of each author in preparing this paper has been clearly identified as bellow, In the writing the article all authors had contributions. The literature review was performed by R.K. and S.M.A. All equations were derived and checked by M.A. (Mohammad Akbari) The results and discussion was prepared by M.A. (Masoud Afrand) According to the reviewer comments, all of the authors prepared the revision format of the manuscript. Ultimately, final approval of the article was done by R.K., M.A. (Mohammad Akbari) and M.A. (Masoud Afrand).

Funding: This research received no external funding.

Conflicts of Interest: The authors of this article certify that they have NO affiliations with or involvement in any organization or entity with any financial interest or non-financial interest in the subject matter or materials discussed in the manuscript entitled "Analytical Solution of Heat Conduction in a Symmetrical Cylinder Using the Solution Structure Theorem and Superposition Technique ".

Abbreviations

Nomenclature

c	Thermal Wave Propagation Speed, m/s
c_p	Specific Heat, J/kg K
f	Total Internal Heat Generation in System
f_r	Reference Laser Power Density, W/m^2
g	Dimensionless Internal Heat Generation
g_0	Transmitted Energy Strength
$\dot{g}^m$	Internal Heat Generation, W/m^3
I_0	Laser peak Power Density, W/m^2
k	Thermal Conductivity, W/mK
h	Convection Heat Transfer Coefficient, W/m^{2}K
m	Dimensionless Convection Heat Transfer Coefficient
q^*	Heat Flux, c
q	Dimensionless Heat Flux, q^*/f_r
Q	Dimensionless Source Term
R	Surface Reflectivity of the Solid
t	Dimensionless Time, $c^2 t^*/2\alpha$
t^*	Time, s
T	Dimensionless Temperature, $kcT^*/\alpha f_r$
T_∞	Dimensionless Ambient Temperature, $kcT_\infty{}^*/\alpha f_r$
T^*	Temperature, K
$T_\infty{}^*$	Ambient Temperature, K
r^*	r-coordinate, m
r	Dimensionless Space Coordinate, $cr^*/2\alpha$

Greek symbols

α	Thermal Diffusivity $k/\rho c_p$, $\mathrm{m^2/s}$
γ_n	Eigen Value, $\gamma_\mathrm{n} = \sqrt{\lambda_n^2 - 1}$
ε	Relative Error
μ	Dimensionless Absorption Coefficient, $2c\tau\mu^*$
ρ	Density $(\mathrm{kg/m^3})$
τ_0	Relaxation Time α/c^2, s
φ	Dimensionless Initial Condition Function
ψ	Dimensionless Initial rate of Temperature Change Function
ξ	Dummy Index

References

1. Dabby, F.W.; Paek, U.-C. High-intensity laser-induced vaporization and explosion of solid material. *IEEE J. Quantum Electron.* **1972**, *8*, 106–111. [CrossRef]
2. Yeung, W.K.; Lam, T.T. Thermal analysis of anisotropic thin-film superconductors. *Adv. Electron. Packag.* **1999**, *26*, 1261–1268.
3. Lundell, J.H.; Dickey, R.R. Vaporization of graphite in the temperature range of 4000 to 4500 K. In Proceedings of the NASA Ames Research Center, Moffett Field, CA, USA, 26–28 January 1976; pp. 76–166.
4. Robin, J.E.; Nordin, P. Enhancement of CW laser melt-through of opaque solid materials by supersonic transverse gas flow. *Appl. Phys. Lett.* **1975**, *26*, 289–292. [CrossRef]
5. Chen, J.K.; Tzou, D.Y.; Beraun, J.E. Numerical investigation of ultrashort laser damage in semiconductors. *Int. J. Heat Mass Transf.* **2005**, *48*, 501–509. [CrossRef]
6. Cattaneo, C. Sur une former de l'equation de la chaleur elinant le paradoxe d'une propagation instance. *Comptes Rendus Acad. Sci.* **1958**, *247*, 431–432.
7. Vernotte, P. Les paradoxes de la theories continue de l'equation de la chaleur. *Comptes Rendus Acad. Sci.* **1958**, *246*, 3154–4155.
8. Wang, B.L.; Han, J.C.; Sun, Y.G. A finite element/finite difference scheme for the non-classical heat conduction and associated thermal stresses. *Finite Elem. Anal. Des.* **2012**, *50*, 201–206. [CrossRef]
9. Lee, H.L.; Chen, W.L.; Chang, W.J.; Wei, E.J.; Yang, Y.C. Analysis of dual-phase-lag heat conduction in short-pulse laser heating of metals with a hybrid method. *Appl. Therm. Eng.* **2013**, *52*, 275–283. [CrossRef]
10. Mishra, S.C.; Sahai, H. Analysis of non-Fourier conduction and radiation in a cylindrical medium using lattice Boltzmann method and finite volume method. *Int. J. Heat Mass Transf.* **2013**, *61*, 41–55. [CrossRef]
11. Qui, T.; Juhasz, T.; Suarez, C.; Bron, W.; Tien, C. Femto second laser heating of multi-layer II experiments. *Int. J. Heat Mass Transf.* **1994**, *37*, 2799–2808.
12. Ozisik, M.; Vick, B. Propagation and Reflection of Thermal Waves in a Finite Medium. *Int. J. Heat Mass Transf.* **1984**, *27*, 1845–1855. [CrossRef]
13. Jiang, F. Solution and analysis of hyperbolic heat propagation in hollow spherical objects. *Heat Mass Transf.* **2006**, *42*, 1083–1091. [CrossRef]
14. Moosaie, A. Non Fourier heat conduction in a finite medium with insulated boundaries and arbitrary initial condition. *Int. Commun. Heat Mass Transf.* **2008**, *35*, 103–111. [CrossRef]
15. Moosaie, A. Non Fourier heat conduction in a finite medium subjected to arbitrary non-periodic surface disturbance. *Int. Commun. Heat Mass Transf.* **2008**, *35*, 376–383. [CrossRef]
16. Ahmadikia, H.; Rismanian, M. Analytical solution of non-Fourier heat conduction problem on a fin under periodic boundary conditions. *J. Mech. Sci. Technol.* **2011**, *25*, 2919–2926. [CrossRef]
17. Bamdad, K.; Azimi, A.; Ahmadikia, H. Thermal performance analysis of arbitrary-profile fins with non-fourier heat conduction behavior. *J. Eng. Math.* **2012**, *76*, 181–193. [CrossRef]
18. lam, T.T.; Fong, E. Application of solution structure theorem to non-Fourier heat conduction problems: Analytical Approach. *Int. J. Heat Mass Transf.* **2011**, *54*, 1–11. [CrossRef]
19. Liu, F.; Chen, Q.; Kang, Z.; Pan, W.; Zhang, D.; Wang, L. Non-Fourier heat conduction in oil-in-water emulsions. *Int. J. Heat Mass Transf.* **2019**, *135*, 323–330. [CrossRef]

20. Daneshjou, K.; Bakhtiari, M.; Parsania, H.; Fakoor, M. Non-Fourier heat conduction analysis of infinite 2D orthotropic FG hollow cylinders subjected to time-dependent heat source. *Appl. Therm. Eng.* **2016**, *98*, 582–590. [CrossRef]

21. Ma, J.; Sun, Y.; Yang, J. Analytical solution of non-Fourier heat conduction in a square plate subjected to a moving laser pulse. *Int. J. Heat Mass Transf.* **2017**, *115*, 606–610. [CrossRef]

22. Wankhade, P.A.; Kundu, B.; Das, R. Establishment of non-Fourier heat conduction model for an accurate transient thermal response in wet fins. *Int. J. Heat Mass Transf.* **2018**, *126*, 911–923. [CrossRef]

23. Hana, S.; Peddieson, J. Non-Fourier heat conduction/convection in moving medium. *Int. J. Therm. Sci.* **2018**, *130*, 128–139. [CrossRef]

24. Liu, Y.; Li, L.; Lou, Q. A hyperbolic lattice Boltzmann method for simulating non-Fourier heat conduction. *Int. J. Heat Mass Transf.* **2019**, *131*, 772–780. [CrossRef]

25. Kalbasi, R.; Afrand, M.; Alsarraf, J.; Tran, M.D. Studies on optimum fins number in PCM-based heat sinks. *Energy* **2019**, *171*, 1088–1099. [CrossRef]

26. Li, Z.; Al-Rashed, A.A.; Rostamzadeh, M.; Kalbasi, R.; Shahsavar, A.; Afrand, M. Heat transfer reduction in buildings by embedding phase change material in multi-layer walls: Effects of repositioning, thermophysical properties and thickness of PCM. *Energy Convers. Manag.* **2019**, *195*, 43–56. [CrossRef]

27. Li, Z.; Shahsavar, A.; Al-Rashed, A.A.; Kalbasi, R.; Afrand, M.; Talebizadehsardari, P. Multi-objective energy and exergy optimization of different configurations of hybrid earth-air heat exchanger and building integrated photovoltaic/thermal system. *Energy Convers. Manag.* **2019**, *195*, 1098–1110. [CrossRef]

28. Nadooshan, A.A.; Kalbasi, R.; Afrand, M. Perforated fins effect on the heat transfer rate from a circular tube by using wind tunnel: An experimental view. *Heat Mass Transf.* **2018**, *54*, 3047–3057. [CrossRef]

29. Salimpour, M.R.; Kalbasi, R.; Lorenzini, G. Constructal multi-scale structure of PCM-based heat sinks. *Contin. Mech. Thermodyn.* **2017**, *29*, 477–491. [CrossRef]

30. Shanazari, E.; Kalbasi, R. Improving performance of an inverted absorber multi-effect solar still by applying exergy analysis. *Appl. Therm. Eng.* **2018**, *143*, 1–10. [CrossRef]

31. Yari, M.; Kalbasi, R.; Talebizadehsardari, P. Energetic-exergetic analysis of an air handling unit to reduce energy consumption by a novel creative idea. *Int. J. Numer. Methods Heat Fluid Flow* **2019**, *29*, 3959–3975. [CrossRef]

Article

Large-Eddy Simulation of an Asymmetric Plane Diffuser: Comparison of Different Subgrid Scale Models

Hui Tang [1,2,3], Yulong Lei [1,2,*], Xingzhong Li [1,2] and Yao Fu [1,2]

1   College of Automotive Engineering, Jilin University, Renmin Street No. 5988, Changchun 130012, China; tanghui15@mails.jlu.edu.cn (H.T.); whdxjx123@126.com (X.L.); fuy.jlu@163.com (Y.F.)
2   State Key Laboratory of Automotive Simulation and Control, Jilin University, Renmin Street No. 5988, Changchun 130012, China
3   Department of Mechanical Engineering, Osaka University, 2-1 Yamada-oka, Suita-city, Osaka 565-0871, Japan
*   Correspondence: leiyl@jlu.edu.cn

Received: 14 September 2019; Accepted: 21 October 2019; Published: 29 October 2019

Abstract: Large-eddy simulation (LES) of separated turbulent flow through an asymmetric plane diffuser is investigated. The outcome of an actual LES depends on the quality of the subgrid-scale (SGS) model, as well as the accuracy of the numerical method used to solve the equations for the resolved scales. In this paper, we focus on the influence of SGS models for LES of the diffuser flow through using a high-order finite difference method to solve the equations for the resolved scales. Six resolutions are computed to investigate the influence of mesh resolution. Four existing SGS models, a new one-equation dynamic SGS model and a direct numerical simulation (DNS) are conducted in the diffuser flow. A series of computational analyses is performed to assess the performance of different SGS models on the coarse grids. By comparison with the experiment and DNS, the results produced by the new one-equation dynamic model give better agreement with experiment and DNS than the four other existing SGS models.

Keywords: large eddy simulation; subgrid scale model; diffuser; dynamic one equation model; Vreman model; separation

---

1. Introduction

Diffusers are ubiquitous in engineering applications. Usually simple in design, they serve to increase the static pressure of a flow by reducing its velocity, albeit often with significant losses. It has a simple design, but it develops complex three-dimensional flow features. A stable separation bubble forms early in the expanding section of the diffuser and spreads across one of the two expanding walls of the diffuser. The flow eventually reattaches in a straight exhaust duct where further pressure recovery occurs. Because of its simple three-dimensional geometry and the existence of a high quality velocity dataset, this diffuser has become a popular test case for measurement of mild separation and validating numerical simulations. However, accurate measurement and prediction of the pressure-driven separation in diffusers has always been challenging in fluid mechanics. In an experimental setup of the diffuser, the first challenge is achieving spanwise homogeneity of turbulence due to the presence of sidewalls producing strong backflow. The point of separation and the extent of the flow reversal zone in diffusers are particularly sensitive to the inlet condition [1]. In a numerical simulation of the diffuser, the presence of an adverse pressure gradient and the formation of an unsteady separation bubble make this flow very sensitive and difficult to predict with numerical means.

Two canonical laboratory incompressible diffuser flows have emerged as standard test-flows in the past: the diffuser studied by Azad [2] and the diffuser studied by Obi et al. [3]. The diffuser studied by Obi et al. [3] is selected in this paper since it has several desirable features. Firstly, the fully developed channel flow is utilized as the inlet condition. Flow in a duct with smooth parallel walls was fully investigated by Kim et al. [4]. For validating the simulation of a separating flow in diffusers. it is vitally significant to know accurately the statistics of the upstream flow because the separation bubble is so sensitive to upstream conditions. Secondly, the flow in this diffuser has rich flow physics, such as pressure-driven separation from a smooth wall, subsequent reattachment, and redevelopment of the downstream boundary layer, which are challenges for large eddy simulation (LES) with subgrid scale mode (SGS) models. Thirdly, based on the mean center velocity U_c and the inlet duct of height H, the Reynolds number of the inlet channel flow is 7488. The corresponding Reynolds number based on the friction velocity is 407. A direct numerical simulation (DNS) of a diffuser flow is feasible at this Reynolds number, thus the DNS data can be used as supplementary benchmarks for assessing different SGS models [5]. Moreover, this Reynolds number is high enough that the flow is not sensitive to this parameter [6].

Actually, the diffuser studied by Obi et al. has been applied as a test flow in Reynolds-averaged Navier–Stokes (RANS) simulations, LES studies and DNS. Obi et al. [3] demonstrated that the standard $k - \epsilon$ model fails to accurately predict the separation bubble in the diffuser. Durbin [7] proposed the $k - \epsilon - v^2$ model for separated flow and validated his new model in asymmetric plane diffuser with satisfactory accuracy in comparison with measurement. Iaccarino [8] investigated the turbulent flow in an asymmetrical two-dimensional diffuser using three commercial CFD codes: CFX, Fluent, and star-CD. El Behery and Hamed [9] presented a comparative study of turbulence models performance for the separating flow in a planar asymmetric diffuser, in which the steady RANS equations for turbulent incompressible fluid flow and six turbulence closures are used. Schneider et al. [10] performed the LES and RANS calculations for two asymmetrical three-dimensional diffusers and validated LES with wall function delivering the results within the accuracy of experimental data. Abe and Ohtsuka [11] investigated high Reynolds-number complex turbulent flows in a 3D diffuser using LES and hybrid LES/RANS models. Kaltenbach et al. [6] elaborately studied the flow in a planar asymmetric diffuser using LES with the dynamic Smagorinsky model. The overall good agreement of simulation results and measurement is obtained and it is found that the SGS model plays a significant role for both mean momentum and turbulent kinetic energy balances. Schluter et al. [12] compared LESs with no subgrid model, the standard Smagorinsky model, the dynamic Smagorinsky model and the dynamic localization model in a separated plane diffuser and demonstrated the dynamic localization model performing the best agreement with measurement. Kobayashi et al. [13] tested their own coherent structures model through some complex geometries in which the asymmetric plane diffuser is included. Taghinia et al. [14] developed a one-equation subgrid scale model with a variable eddy-viscosity coefficient for LES and validated the model in complex separated and reattaching turbulent flows, i.e., the turbulent flow through an asymmetric planar diffuser. Shuai and Agarwal [15] proposed a new one-equation eddy-viscosity model from the two-equation k-kL model. The new turbulence model is used to simulate an asymmetric plane diffuser and improved the accuracy of the flow simulations compared to the one-equation Spalart–Allmaras model. DNS of turbulent flow through an asymmetric plane diffuser was performed by Ohta and Kajishima [5] using a high-order finite difference method. The DNS results from this type of field can be used as one of the benchmarks for numerical simulation schemes and SGS models.

With increasing computing power, LES is more widely used in three-dimensional, unsteady complex flows. In LES technique, turbulent motions are separated into large-scale and small-scale contributions by using a filtering operation in which the large-scale fluid motions are directly calculated, whereas the unresolved SGS motions are modeled. The outcome of an actual LES therefore depends on the quality of the SGS model, as well as the accuracy of the numerical method used to solve the equations for the resolved scales. The important interactions between the resolved large eddies and unresolved SGS eddies, in the case of fully developed isotropic homogeneous turbulent flows, can be reduced to be seen as that of the energy transfers by omitting a portion of information included in the

SGS eddies, for instance, the structural information associated to the anisotropy. It has been shown [16] that the energy transfer between SGS eddies and large eddies mainly exhibit two mechanisms: a forward energy transfer from large eddies to the SGS eddies and a backward transfer to the resolved scales, which, it seems, is much weaker in intensity. The SGS models are responsible for describing the desired dissipation or energy production effects.

In the present study, we focus on the influence of SGS models for LES of the separated turbulent flow in an asymmetric diffuser on the relative coarse mesh through using a high-order finite difference method to solve the equations for the resolved scales. Four existing SGS models, namely the Smagorinsky model [17], the dynamic Smagorinsky model [18,19], the one-equation model [20] and the one-equation dynamic model [21], and a new dynamic one-equation SGS model are investigated in diffuser flow. The Smagorinsky model, which can give a moderately accurate magnitude of energy transfer, is a simple and robust eddy viscosity model; however, it uses a priori model parameter. In the dynamic Smagorinsky model, the model coefficient is dynamically decided from the resolved scales to the unresolved SGS ranges based on an assumption of the scale invariance. However, the clipping procedure or averaging operation is still required in a homogeneous direction or in the global volume of domain. The one-equation model directly uses the information concerned with the SGS motions, i.e., the SGS kinetic energy, to decide the eddy viscosity, in which a transport equation of the SGS kinetic energy is solved and local equilibrium hypothesis of previous models can be removed. The one-equation dynamic model is a modification of the standard one-equation model, in which the production of SGS kinetic energy and the energy loss in large-scale portion are treated with different dynamic mechanisms, i.e., the production term in SGS kinetic energy transport equation is solved using a SGS model based on the resolved scales such as the dynamic Smagorinsky model, while the eddy viscosity in the filtered equation of motion is determined from information of unresolved scales—the transport equation of SGS kinetic energy. Since the test filtering operation is required in the one-equation dynamic model, all undesirable features and inconsistencies related to the test filtering operation are retained. The computation cost of solving an additional transport equation, as well as a dynamic procedure over test filter in the one-equation dynamic model is relatively huge. Thus, these properties still limit its application for the simulation of turbulent flows in complex geometries. It is necessary to continue the search for a new one-equation dynamic model that performs as well as the one-equation dynamic model, while the new model does not require any test filter and is not more expensive in terms of computational cost than the standard one-equation model.

Therefore, the goal of this study is to develop a new one-equation dynamic SGS model and examine the performance of the new SGS model and four existing SGS models on the flow prediction of pressure-driven separation in a diffuser by using a high-order finite difference method to solve the equations for the resolved scales. In addition, we examine the choice of mesh resolution of an asymmetric plane diffuser.

2. SGS Models

Applying a filter with scale Δ, and assuming the filtering operations are commuting with the operations of differentiation, the filtered Navier–Stokes (N-S) equation for LES of incompressible flows can be given as:

$$\frac{\partial \overline{u}_i}{\partial t} + \frac{\partial \overline{u}_i \overline{u}_j}{\partial x_j} = -\frac{1}{\rho}\frac{\partial \overline{p}}{\partial x_i} + \nu \frac{\partial^2 \overline{u}_i}{\partial x_j \partial x_j} - \frac{\partial \tau_{ij}}{\partial x_j}, \quad \frac{\partial \overline{u}_i}{\partial x_i} = 0. \tag{1}$$

where the SGS stress tensor is defined as $\tau_{ij} = \overline{u_i u_j} - \overline{u}_i \overline{u}_j$. A index notation is utilized to represent the components, in which the coordinates x_1, x_2 and x_3 are denoted as streamwise, wall-normal and spanwise directions, respectively, and the corresponding velocity components are presented as u, v and w. Throughout this paper, summation convention is implied for repeated index.

As Boussinesq proposed, the SGS stress can be modeled as:

$$-\frac{\partial \tau_{ij}^a}{\partial x_i} \equiv -\frac{\partial}{\partial x_i}\left(\tau_{ij} - \frac{1}{3}\tau_{kk}\delta_{ij}\right) = \frac{\partial}{\partial x_i}\left(2\nu_{sgs}\overline{S}_{ij}\right),\tag{2}$$

in which τ_{ij}^a is the anisotropic component of τ_{ij}; $\overline{S}_{ij}$ is the characteristic filtered rate-of-strain tensor and is defined as $(\partial\overline{u}_i/\partial x_j + \partial\overline{u}_j/\partial x_i)/2$; and the coefficient of proportionality ν_{sgs} is the SGS viscosity that remains to be evaluated by a SGS model.

2.1. Smagorinsky Model (SM)

Using the local equilibrium assumption, i.e., the dissipation rate of SGS energy is in balance with the production rate, the Smagorinsky model can be obtained for LES:

$$\nu_{sgs} = (C_s\Delta)^2|\overline{S}|,\tag{3}$$

in which C_s is a constant and taken to be 0.1 in this study. $|\overline{S}|$ denotes the norm of the rate-of-strain tensor and is defined as $\sqrt{2\overline{S}_{ij}\overline{S}_{ij}}$. For wall-resolved LES, to make sure that modeled SGS stresses exhibit the near-wall behavior, a damping function f_s has to be incorporated with the Smagorinsky model:

$$\nu_{sgs} = (C_s f_s \Delta)^2|\overline{S}|.\tag{4}$$

The van Driest function is employed as the damping function:

$$f_s = 1 - exp\left(-\frac{y^+}{A^+}\right),\tag{5}$$

in which the non-dimensional constant $A^+ \approx 25$ and $y^+ = yu_\tau/\nu$. For the separated flow in an asymmetric diffuser, it would be challenging to determine the friction velocity u_τ in the vicinity of a separation point due to $u_\tau = 0$. This is why few researchers try to use the Smagorinsky model for LES to simulate the diffuser flow. Even though the Smagorinsky model is applied in diffuser, its results are far from satisfactory. For example, Schluter et al. [12] found the flow separation was predicted on the upper wall instead on the inclined wall using the Smagorinsky model. In this study, an interesting method that u_τ is determined by a linear interpolation is used in Smagorinsky model.

The two-dimensional schematic of Obi et al. diffuser is shown in Figure 1, in which the expansion ratio and expansion part of the diffuser is 4.7 and $21H$, respectively. A Cartesian coordinate system with the origin on the upper wall where the inlet channel wall and the deflected wall form a corner is used to define x in the streamwise direction and y in the downward wall-normal direction. Then, y^+ in diffuser can be approximately defined as (see Appendix A):

$$y^+(x,y) = \begin{cases} Re_\tau0 \cdot \min(y, 1 - y)/H, & x \leq 0 \\ Re_\tau0 \cdot (1 - 3.7/4.7 \cdot x/H) \cdot \min[y, Y(x) - y]/H, & 0 < x \leq 21H \\ Re_\tau0/4.7 \cdot \min(y, 4.7H - y)/H, & x > 21H. \end{cases}\tag{6}$$

Here, $Re_\tau0$ denotes the Reynolds number based on the friction velocity found in the inlet of duct. $Y(x)$ means the local width of expansion in diffuser.

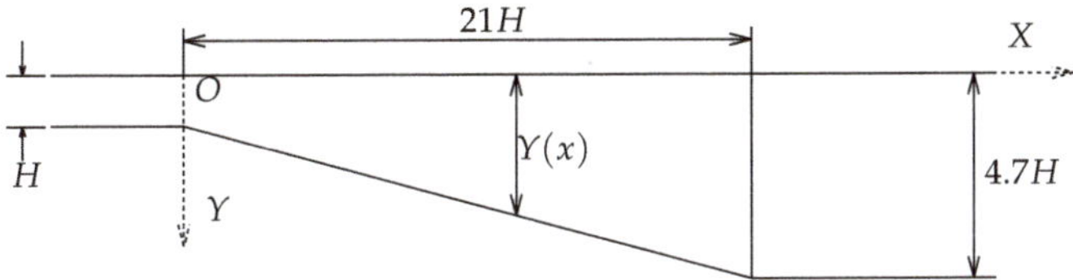

Figure 1. Schematic of an asymmetric plane diffuser.

2.2. Standard Dynamic Smagorinsky Model (DSM)

Applying the test filter on the grid-filtered N-S equations, the Germano identity can be defined as

$$L_{ij} = T_{ij} - \widetilde{\tau}_{ij} = \widetilde{\overline{u_i}\overline{u_j}} - \widetilde{\overline{u}}_i\widetilde{\overline{u}}_j, \tag{7}$$

where L_{ij} can be calculated based on the resolved scales, $T_{ij} = \widetilde{\overline{u_i u_j}} - \widetilde{\overline{u}}_i\widetilde{\overline{u}}_j$ represents the residual turbulent stress at a test-filter scale $\widetilde{\Delta}$, and can be given as

$$T_{ij} - \frac{1}{3}\delta_{ij}T_{kk} = -2C\widetilde{\Delta}^2|\widetilde{\overline{S}}|\widetilde{\overline{S}}_{ij}. \tag{8}$$

On substituting Equations (2) and (8) into Equation (7) and assuming $\overline{\Delta}$ and C are constant inside the test filter, an equation for determining C is obtained:

$$L_{ij} - \frac{1}{3}\delta_{ij}L_{kk} = -2C\overline{\Delta}^2 M_{ij}, \tag{9}$$

where

$$M_{ij} = \frac{\widetilde{\Delta}^2}{\overline{\Delta}^2}|\widetilde{\overline{S}}|\widetilde{\overline{S}}_{ij} - \widetilde{|\overline{S}|\overline{S}_{ij}}. \tag{10}$$

Minimization of the error of Equation (9) over all independent tensor components [19], as well as over some averaging region of statistical homogeneity, leads to

$$C = -\frac{1}{2\overline{\Delta}^2}\frac{\langle L_{ij}M_{ij}\rangle}{\langle M_{ij}M_{ij}\rangle}. \tag{11}$$

2.3. Standard One-Equation Model (OM)

The SGS eddy viscosity ν_{sgs} is represented as $\nu_{sgs} = C_\nu\overline{\Delta}_\nu\sqrt{k_{sgs}}$ by the dimensional analysis, where $k_{sgs} = (\overline{u_i u_i} - \overline{u}_i\overline{u}_i)/2 = \tau_{ii}/2$, i.e., the SGS kinetic energy. k_{sgs} is evaluated by solving an evolution equation

$$\frac{\partial k_{sgs}}{\partial t} + \overline{u}_j\frac{\partial k_{sgs}}{\partial x_j} = - \tau_{ij}\overline{S}_{ij} - C_\varepsilon\frac{k_{sgs}^{3/2}}{\Delta} - \varepsilon_w$$
$$+ \frac{\partial}{\partial x_j}\left[\left(C_d\Delta_\nu\sqrt{k_{sgs}} + \nu\right)\frac{\partial k_{sgs}}{\partial x_j}\right], \tag{12}$$

which was theoretically derived by the authors of [22–24]. For meeting the correct asymptotic behavior to the wall, an additional modification was proposed by Okamoto & Shima [24] for the characteristic length Δ_ν and additional dissipation term ε_w, i.e.,

$$\Delta_\nu = \frac{\overline{\Delta}}{1 + C_k\overline{\Delta}^2\overline{S}^2/k_{sgs}}, \tag{13}$$

$$\varepsilon_w = 2\nu \frac{\partial \sqrt{k_{sgs}}}{\partial x_j} \frac{\partial \sqrt{k_{sgs}}}{\partial x_j}. \tag{14}$$

Non-dimensional constants associated with one-equation model are as follow: $C_v = 0.05$, $C_\varepsilon = 0.835$, $C_d = 0.10$, and $C_k = 0.08$.

2.4. One-Equation Dynamic Model (ODM)

Kajishima and Nomachi [21] considered that the dynamic procedure (that is, Section 2.2) is suitable for the determination of energy transfer from resolved scales to SGS components since this transfer is regarded to be local and instantaneous, whereas the energy loss in large-scale portion is considered from the historic effect of SGS turbulence due to the transport. Thus, the dynamic procedure is applied to the production term in the transport equation of k_{sgs}, while the SGS eddy viscosity in the filtered N-S equation is determined by using k_{sgs} of Equation (12). The first term in the right-hand side of Equation (12) is given by

$$- \tau_{ij}\overline{S}_{ij} = C_s\overline{\Delta}^2|\overline{S}|^3, \tag{15}$$

where C_s is calculated by the dynamic procedure of Equation (11). It should be noted that C_s in the ODM model does not require any averaging over homogeneous direction or ad hoc clipping technique to avoid negative Smagorinsky constants. The negative value for C_s is acceptable and it results in the decrease in k_{sgs}. However, the computation of the test filter to evaluate the Germano Identity L_{ij} and the parameter M_{ij} is still required.

2.5. One-Equation Vreman Model (OVM)

Since the test filtering operation is required in the ODM model, all undesirable features and inconsistencies related to the test filtering operation are retained. The computation cost of solving an additional transport equation, as well as a dynamic procedure over test filter in the ODM model is relatively huge. Thus, these properties still limit its application for the simulation of turbulent flows in complex geometries. It is necessary to continue the search for a new one-equation dynamic model that performs as well as the ODM model, while not requiring any test filter and not being more expensive in terms of computational cost than the standard one-equation model.

Vreman [25] proposed a SGS model for the LES of turbulent shear flows (henceforth, referred to as Vreman model). Vreman model is essentially not more complicated than the Smagorinsky model, but is able to adequately handle turbulent as well as transitional flow. The model is expressed in first-order derivative, and does not involve test filtering, averaging, or clipping procedures. From two test cases of a transitional and turbulent mixing layer at high Reynolds number and a turbulent channel flow, Vreman model is found to be more accurate than Smagorinsky model and as good as the standard dynamic Smagorinsky model. Because of these desirable properties, it seems to be particularly pertinent for incorporating Vreman model into the ODM model to develop a new one-equation dynamic model that is suitable for LES of turbulent flows in complex geometries. In the new model, the production term in the transport equation of k_{sgs} (Equation (12)) is represented as:

$$- \tau_{ij}\overline{S}_{ij} = C_{vm}^+ \sqrt{\frac{B_\beta}{\alpha_{ij}\alpha_{ij}}}|\overline{S}|^2, \tag{16}$$

where

$$B_\beta = \beta_{11}\beta_{22} - \beta_{12}^2 + \beta_{11}\beta_{33} - \beta_{13}^2 + \beta_{22}\beta_{33} - \beta_{23}^2, \tag{17}$$

$$\beta_{ij} = \Delta_m^2\alpha_{mi}\alpha_{mj}, \quad \alpha_{ij} = \frac{\partial \overline{u}_j}{\partial x_i}, \tag{18}$$

$$C_{vm}^+ = \begin{cases} C_{vm}|\overline{\Omega}|/|\overline{S}|, & |\overline{\Omega}|/|\overline{S}| < 1, \\ C_{vm}, & |\overline{\Omega}|/|\overline{S}| \geq 1. \end{cases} \tag{19}$$

The eddy viscosity is obtained from solving Equations (12) and (16) simultaneously. We call the new SGS eddy viscosity model as one-equation Vreman model. Note that a fixed coefficient $C_{vm}(=0.025)$ is used in Vreman model, while in our new one-equation dynamic model a modified coefficient C_{vm}^+ is introduced for taking excess SGS production rate where both $|\overline{\Omega}|$ and $|\overline{S}|$ are large. Actually, to further improve the performance of Vreman model, Park et al. [26] and You and Moin [27,28] proposed different dynamic procedures to dynamically determine the parameter C_{vm}, in which single-level or double-level test filters are employed.

3. Computational Method

In this study, to make much more persuasive comparison of SGS models performance for diffuser flow and expel the other factors, which can interface the observation of comparison, the aforementioned SGS models in LES and a DNS are numerically implemented through the same set of numerical methods in an asymmetric planar diffuser. The results of DNS are used as benchmarks for SGS models as well. The configuration of the diffuser, as shown in Figure 2, and Reynolds number $Re_c = 7488$ based on the mean center velocity U_c found in the inlet duct of height H match previous experiments performed by Obi et al. [3,29].

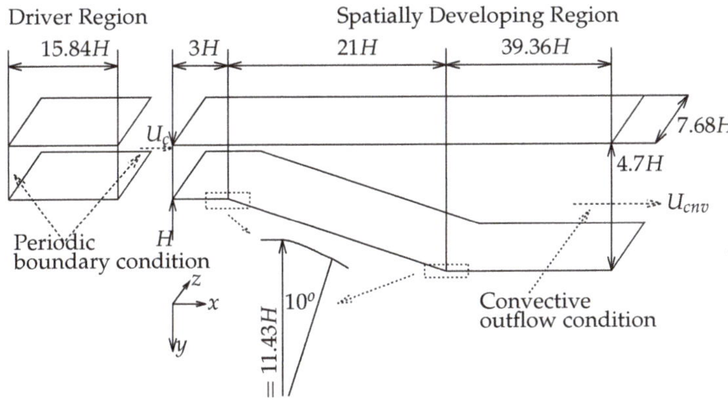

Figure 2. Three-dimensional schematic of Obi et al. diffuser.

3.1. Domain Size and Boundary Condition

There are two computational regions sandwiched by two walls, called the driver and the spatially developing region, as given in Figure 2. The expansion ratio of the diffuser is 4.7, the inclination of the wall (the lower wall) is 10 degrees and the spanwide extend is 7.68 H. The driver region length is 15.84 H and the upstream channel, expansion part and downstream extension of diffuser are 3 H, 21 H and 39.36 H, respectively. Both the joins between the expansion part and upstream/downstream parts are rounded with radius of 11.43 H, which are the same as in experiment. A Cartesian coordinate system with the origin on the upper wall where the inlet channel wall and the deflected wall form a corner is used to define x in the streamwise direction, y in the downward wall-normal direction, and z in the spanwise direction.

The flow in two regions are numerically computed simultaneously with periodic and no-slip boundary conditions for the velocity in spanwise direction and wall-normal direction, respectively. There, it should be noted that the boundary condition in the spanwise direction between simulations and the experiment performed by Obi et al. [3] is different, i.e., periodic (homogeneous) boundary condition in simulations but closed (rectangular duct) boundary condition in the experiment. To some extent, this kind of difference will affect the agreement between the simulations and the experiment, which is specifically discussed in the following. The fully-developed turbulent channel flow that is treated as inflow for the spatially-developing region are generated by the driver region with the periodic boundary condition in streamwise direction. The pressure gradients for driver region in streamwise direction are controlled to keep the flow rate a constant. Velocity profiles at outflow boundary condition for spatially-developing region are determined by a convective outflow condition

$$\frac{du}{dt} = -u_{cnv}\frac{du}{dx},$$ (20)

in which convective velocity u_{cvn} is initialized with a constant bulk velocity to keep the flow rate in spatially-developing region. Finally, the pressure gradients at spatially-developing region are obtained automatically.

3.2. Computational Meshes

To assess the influence of the mesh resolution and find relatively coarser mesh for the asymmetric planar diffuser, simulations were performed on seven different collocated meshes, as given in Table 1. Here, u_τ is the wall friction velocity which is found at the inlet dust, and the corresponding Reynolds number based on u_τ is $Re_\tau = 407$. In the DNS and from M1 to M5, uniform meshes are applied in the streamwise and spanwise direction for the driver and spatially-developing region, in which the mesh gradually becomes coarser from M1 to M5 in the streamwise direction while the grids remain the same in the other two directions. The nonuniform mesh are used in the streamwise direction for M6 in spatially-developing region and designed such that the spacing gradually increases from the diffuser throat toward the downstream join and the spacing gradually decreases from the inlet dust to the diffuser throat. This kind of design is necessary to resolve the sharp mean gradients in the diffuser throat for LES when the whole grids are relatively coarse. To resolve the boundary layers, the meshes in the wall normal direction distribute non-uniformly more densely near walls for all cases in both the driver and spatially-developing region. What is more, not only the boundary layers upstream but also downstream of the diffuser should be resolved in wall normal direction.

Table 1. Mesh resolution of simulations : N_x, N_y, and N_z denote the number of cells in streamwise, wall-normal and spanwise direction, respectively. Δx^+, Δy^+, and Δz^+ are grid spacing used in simulations normalized by v/u_τ. Super/subscripts of 1 and 2 represent driver and spatially-developing region, respectively.

Case	N_x^1	N_x^2	N_y	N_z	Δx_1^+	Δx_2^+	Δy_{min}^+	Δz^+
DNS	320	1200	160	320	20.15	20.15	0.50	9.77
M1	160	600	80	160	40.30	40.30	1.04	19.54
M2	128	480	80	160	50.38	50.38	1.04	19.54
M3	128	360	80	160	50.38	67.17	1.04	19.54
M4	128	300	80	160	50.38	80.6	1.04	19.54
M5	128	480	160	160	50.38	50.38	0.50	19.54
M6	128	372	80	160	50.38	50.38–100.76	1.04	19.54

3.3. Solution Strategy

A fourth-order central finite-difference discretization scheme is used for the incompressible (filtered) continuity equation and (filtered) Navier–Stokes equation in the (LES) DNS, in which the fractional method is selected for coupling the continuity equation and the pressure field, the second-order Adams–Bashforth method is used to the convective term and viscous term, and the backward Euler method to pressure term. A fourth-order central difference schemes is applied in the Smagorinsky model. For the dynamic procedure, the test filter is used in the streamwise direction and spanwise direction with second-order accuracy and the test-to-grid filter ratio $\tilde{\Delta}/\Delta = 2$. For the transport equation of the SGS kinetic energy, Crank–Nicolson method is utilized to dissipation term, the second-order Adams–Bashforth method to convective and diffusive terms. The initialization data of k_{sgs} is solved from $k_{sgs} = \left(\nu_{sgs}/C_\nu \overline{\Delta}\right)^2$ using the results of ν_{sgs} from the dynamic Smagorinsky model. The present numerical method and computer program have been tested extensively in several turbulent flows [5,30–32].

4. Results and Discussion

All simulations were computed on an NEC SX-8R supercomputer of Cybermedia Center, Osaka University with the time step $dt = 0.0495H/U_c$. The great mass of the total effort of calculation was spent on solving the Poisson equation through the residual cutting method [33]. All simulations were run until the flow fields were fully developed and the first-order and second-order statistics exhibited adequate convergence. All results were collected by time averaging and spatial averaging in the spanwise direction. To allow a good comparison of simulation results and experimental measurements, the data associated with vertical cross-sections $x/H = 9.2, 15.2, 19.2$ and 25.2 in the spatially-developing region, which match the location used in previous experiments, are combined into one plot. Note that the results of our DNS for the diffuser agree quite well with the DNS results of Ohta & Kajishima [5]. For the validation discussed in this section, we restrict ourselves to use our own DNS as a comparison with three sets of LESs.

4.1. Comparison of Mesh Resolution

Figures 3–6 show the axial mean velocity and axial Reynolds stress profiles for the mesh sensitivity study. All simulations used the LES with the standard Smagorinsky model due to its low computational cost. In Figures 3 and 4, profiles of U and $\langle u'u' \rangle$ agree well with the measurement for the mesh resolution of M1 and M2, while the situation for the mesh resolution of M3 and M4 is reversed, i.e., the deviation between simulations and experiment is large. Thus, the mesh resolution of M1 and M2 rather than M3 and M4 is acceptable for the LES of the diffuser. Furthermore, differences between M1 and M2 are not apparent, although mesh of M2 is relatively coarser than M1. In Figure 4, M1 and M2 both underpredict the peak value of Reynolds stress $\langle u'u' \rangle$ and represent a large deviation in the region close to the inclined wall compared with the experimental data. Since this deviation does not decrease much with increasing mesh resolution, we suspect the standard Smagorinsky model itself and the blocking effect of the sidewall boundary layers in the experiment to contribute to this disagreement. Through the previous analysis mesh resolution of M2 and the grid spacing $\Delta x^+ = 50.38$ is therefore used as a benchmark mesh and a benchmark axial grid spacing in the inlet of the diffuser, respectively. Based on the benchmark mesh and axial grid spacing, a nonuniform mesh (that is, M6) is designed and tested. In Figures 5 and 6, we compare the profiles of U and $\langle u'u' \rangle$ from simulation and experiment for the different wall-normal mesh resolution and the nonuniform mesh, which shows a similar trend with M1 and M2 in Figures 3 and 4. With increasing wall-normal mesh refinement from M2 and M5 an increasing agreement of the LES with the experimental data can be stated, especially in the region close to the walls, while this improvement is quite small. Even the coarsest mesh simulation, i.e., M6, demonstrates a good overall agreement with the experimental data. Thus, for the study of different SGS model performance for separating flow in the diffuser, the coarsest mesh M6 is used since the influence of the SGS model on the results is largest on this mesh.

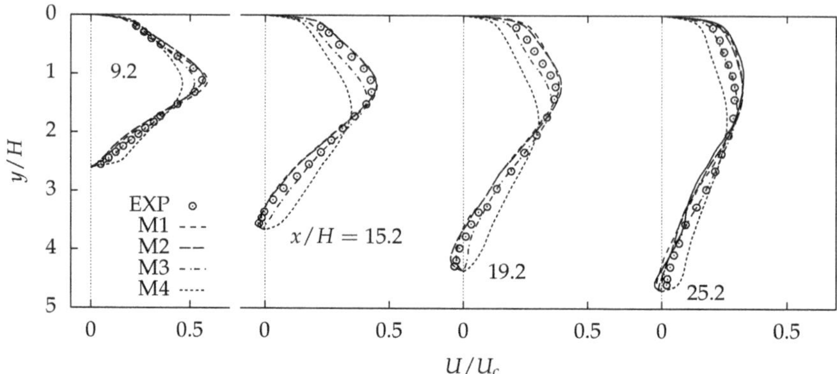

Figure 3. Comparison of LES with different mesh resolution in the streamwise direction and experimental results of Obi in terms of mean streamwise velocity profiles.

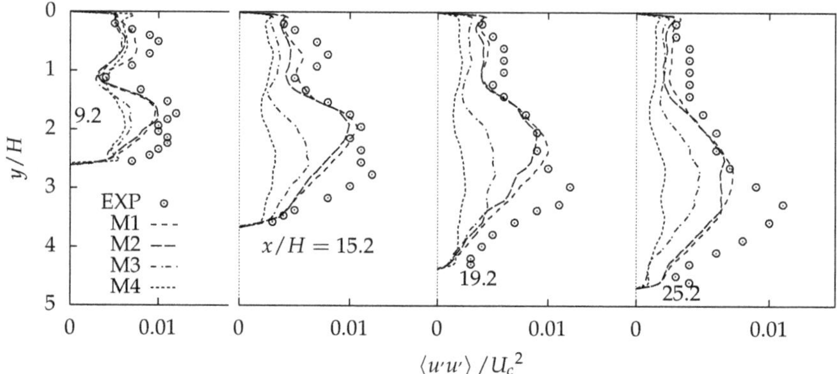

Figure 4. Comparison of LES with different mesh resolution in streamwise direction and experimental results of Obi in terms of axial Reynolds stress profiles.

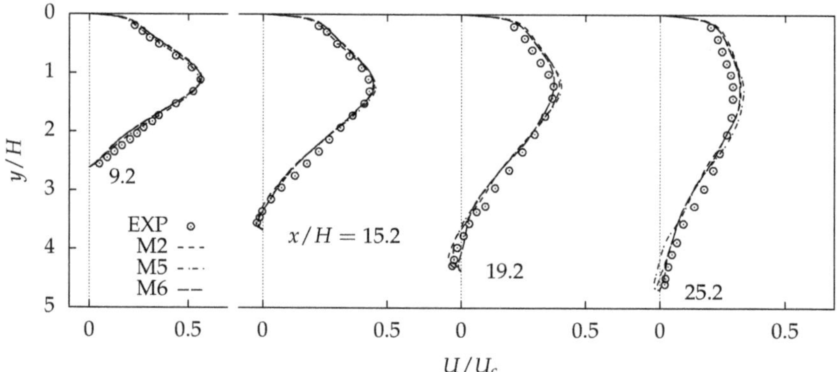

Figure 5. Comparison of LES with different mesh resolution and experimental results of Obi in terms of axial mean velocity profiles.

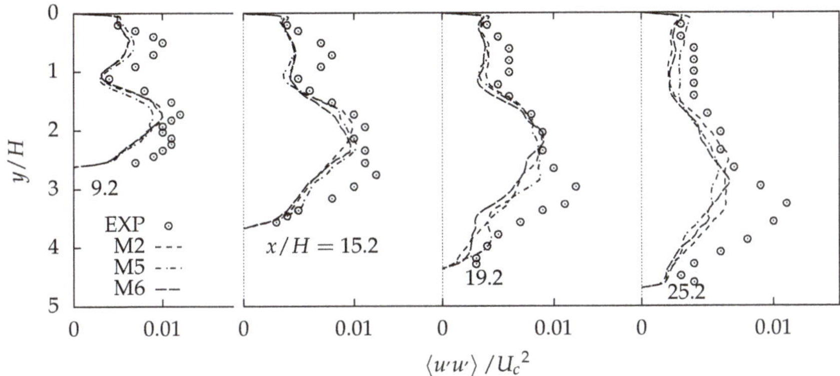

Figure 6. Comparison of LES with different mesh resolution and experimental results of Obi in terms of axial Reynolds stress profiles.

4.2. Comparison of Subgrid Modeling

The DNS as well as five sets of LESs, i.e., SM, DSM, OM, ODM, and VOM, were computed on an NEC SX-8R supercomputer of Cybermedia Center (CMC), Osaka University. The great mass of the total effort of calculation was spent on solving the Poisson equation through the residual cutting method [33]. The time step used for these computation is about $dt = 0.0495H/U_c$. On a node of CMC, 13.188 s CPU time are needed to advance the computation one time step for the DNS, 1.071 s CPU time for the LES with SM model, 1.254 s CPU time for the LES with DSM model, 1.278 s CPU time for the LES with OM model, 1.391 s CPU time for the LES with ODM model, and 1.288 s CPU time for the LES with VOM model.

4.2.1. Comparison of Mean Properties

The profiles of the mean velocity for the streamwise direction and the wall-normal direction non-dimensionalized by the mean center velocity at the inlet duct are shown in Figures 7 and 8, respectively, where five sets of LESs corresponding to the SM model, DSM model, OM model, ODM model and our OVM model are compared with DNS and experimental data of Obi. Overall, in Figure 7, the agreement of profiles of streamwise mean velocity between five LESs and experiment is quite good at all locations. The slight advance of LES velocity profile corresponding to OVM model in comparison with experimental values exists in the vicinity of the flat wall, whereas the situation adjacent to the inclined wall is reversed, i.e., the LES velocity profiles exhibit the slight lag in comparison with the measurement. This situation of velocity profiles for OVM model compared with the experiment is similar to that of DNS, but the latter exhibits more obvious difference. It seems the SM, DSM, OM and ODM models show a better overall agreement with the measurement than OVM model and DNS. Meanwhile, LESs exhibit better overall match with experiment than DNS. However, as discussed previously, agreement with the measurement is not a first priority. The results of DNS are should be used as benchmarks for SGS eddy viscosity models as well. From this perspective, the performance of OVM model in the prediction of mean streamwise velocity is slightly better than the four other SGS models since the OVM model agrees well with DNS database. Actually, the blocking effect of the sidewall boundary layers in experiment is much more pronounced for mean velocity in the wall-normal direction and turbulent stresses, which are discussed below. According to the mean velocity in the wall-normal direction from DNS and LESs, see Figure 8, the mean flow is directed from the straight (upper) wall towards the inclined wall in the flow filed except for some small regions adjacent to the inclined wall. Furthermore, the peak velocity V_{max} of DNS and LESs is located in the center region of the flow filed. However, the strong flows toward the inclined wall and the straight wall are both observed in experimental data. It has been shown [34] that the influence of the side wall of the channel on the flow generated a secondary flow in the laboratory experiment, being markedly so especially in the case of a low Reynolds number. However, the agreement between simulations and measurement is good in the vicinity of the inclined wall. Compared with DNS, five SGS eddy viscosity models overpredict the mean velocity V near the diffuser throat $x/H = 9.2$ and underpredict V in the other regions $x/H = 15.2$, 19.2 and 25.2, except for the location at 25.2 corresponding to the DSM, ODM and OVM models. Clearly, overall, the best agreement is observed between the OVM model and DNS among five SGS eddy viscosity models.

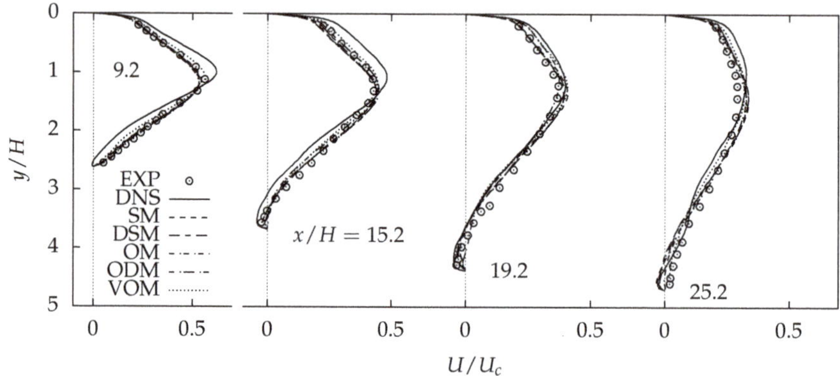

Figure 7. Comparison of LES with different SGS models, DNS and experimental results of Obi in terms of axial mean velocity profiles.

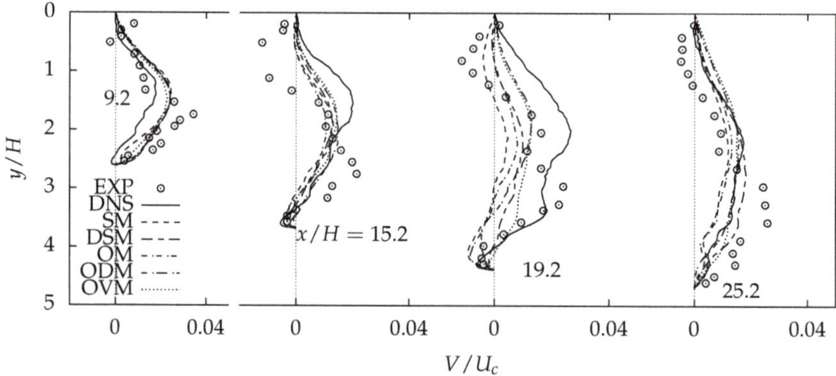

Figure 8. Comparison of LES with different SGS models, DNS and experimental results of Obi in terms of wall-normal mean velocity profiles.

4.2.2. Comparison of Turbulent Stresses and Resolved Turbulent Kinetic Energy

In Figures 9–12, we compare the corresponding profiles of the turbulent shear stress $\langle u'v' \rangle$, and turbulent normal stresses $\langle u'u' \rangle$, $\langle v'v' \rangle$, $\langle w'w' \rangle$ from simulations and experiment, where the turbulent stresses normalized by the square of mean center velocity U_c, $<>$ refers to time averaging and spatial averaging in the homogenous direction. Note that the simulation results of three SGS eddy viscosity models are the summation of the resolvable portion and SGS portion, which coincides with the DNS and experiment. It has been shown [6] that experimental errors are higher for turbulent stresses than the mean flow velocities, in particular at the regions in which measurement volumes are large in comparison with the local gradients of turbulent stresses. However, overall, good agreement between the computations and experiment is observed at locations in the vicinity of either walls. As shown in Figure 9, turbulent shear stresses from three SGS eddy viscosity models, the DNS and experiment all exhibit a characteristic shape with a double peak. The locations of the peak value of all three SGS eddy viscosity models agree well with the DNS, whereas they are higher compared with measurement. All five SGS eddy viscosity models underpredict the peak value of turbulent shear stress compared with DNS, in which the performance of the DSM and ODM models are better than that of the SM and OM models, the OVM model demonstrates the best agreement. This can be seen especially at the location of $x/H = 15.2$ and 19.2. Compared with measurement, all five SGS eddy viscosity models show a better agreement in the region close to the straight wall than the inclined wall. In the center region of the diffuser, the SM, OM, DSM and ODM models underpredict the turbulent shear stress in comparison with experiment, whereas the performance of the OVM model is a little bit of opposite. As shown in Figure 10, profiles of $\langle u'u' \rangle$ for the five SGS eddy viscosity models exhibit more prominent difference in the center region of the diffuser than adjacent to either walls, where the deviation between the SM model and the DNS or measurement is largest. Compared with the DNS as well as measurement, the SM, OM and DSM models underpredict the value of $\langle u'u' \rangle$ at all locations, except for some of small regions near the diffuser throat and further downstream. The slight advance of $\langle u'u' \rangle$ from the OVM model compared with experiment is observed in the region of flow interior. Obviously, not only the overall but also the local agreement between the OVM and ODM models and DNS / experiment is better than other three SGS eddy viscosity models. With respect to profiles of $\langle v'v' \rangle$ and $\langle w'w' \rangle$, an overall better agreement between the LESs and the DNS is observed compared with that of turbulent shear stress and normal stress $\langle u'u' \rangle$. A characteristic double-peak shape can still be seen for profiles in Figures 11 and 12. The location of the peak value moves away from the wall into the flow interior with increasing distance from the diffuser throat. While the peak value of the turbulent stress $\langle v'v' \rangle$ from DNS as well as LESs gravely deviates from the measurement, locations of the peak value are close to each other and coincide with the measurement, except for the location at $x/H = 9.2$. In comparison with the DNS, SM model, DSM model, OM model and ODM model underpredict $\langle v'v' \rangle$ and $\langle w'w' \rangle$ at all locations, except for some of small regions adjacent to straight wall, whereas the OVM model slightly overpredicts both of them in the vicinity of the inclined wall. Overall, the OVM model agree better with the DNS than other four SGS eddy viscosity models. Figure 13 shows the profiles of the resolved turbulent kinetic energy K non-dimensionalized by the square of the mean center velocity. All five SGS eddy viscosity models capture the locations as well as the magnitude of the double-peak in K. Compared with the DNS, the five SGS models accurately predict the near-wall peak of the straight wall except for the location at $x/H = 25.2$, but underpredict the larger peak value at all location. In particular, the SM model and the OM model do not correctly capture the location of the larger peak for the locations at $x/H = 15.2$. The OVM model agrees well with the DNS with respect to the location as well as the value of the double-peak in the resolved turbulent kinetic energy.

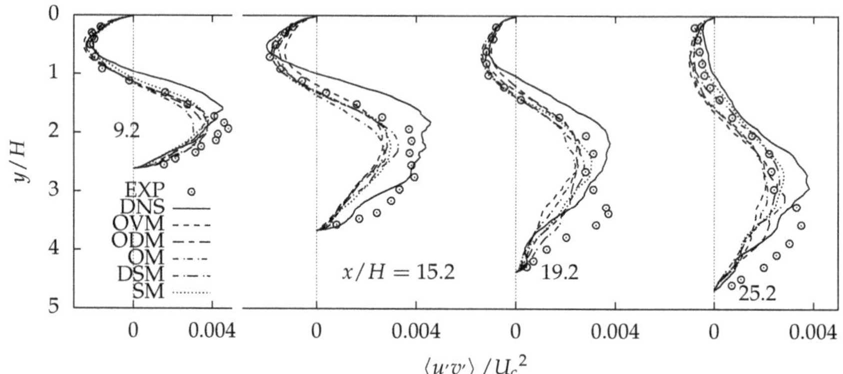

Figure 9. Comparison of LES with different SGS models, DNS and experimental results of Obi in terms of Reynolds shear stress profiles.

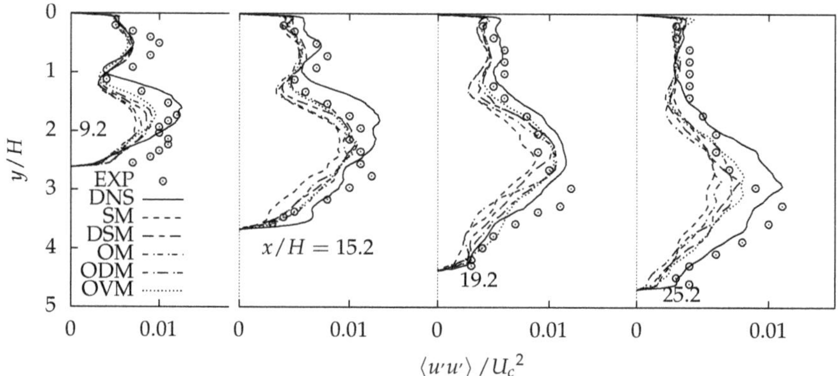

Figure 10. Comparison of LES with different SGS models, DNS and experimental results of Obi in terms of Reynolds stress $\langle u'u'\rangle / U_c^2$ profiles.

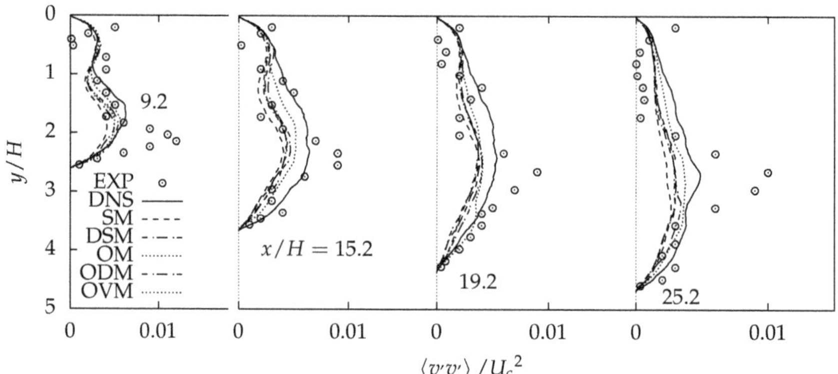

Figure 11. Comparison of LES with different SGS models, DNS and experimental results of Obi in terms of Reynolds stress $\langle v'v'\rangle / U_c^2$ profiles.

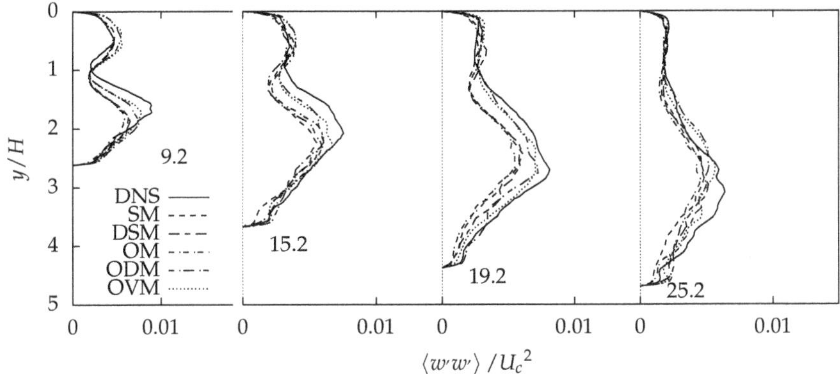

Figure 12. Comparison of LES with different SGS models, DNS and experimental results of Obi in terms of Reynolds stress $\langle w'w' \rangle / U_c^2$ profiles.

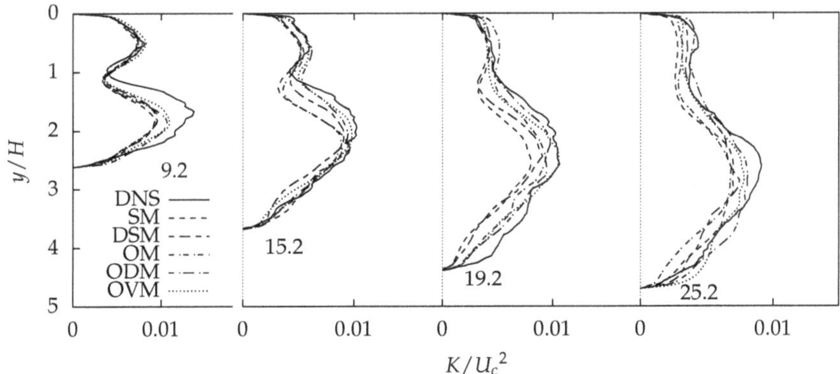

Figure 13. Comparison of LES with different SGS models, DNS and experimental results of Obi in terms of resolved turbulent kinetic energy profiles.

5. Conclusions

The turbulent flow through a planar asymmetric diffuser was investigated numerically using LES with a high-order finite difference method to solve the equations for the resolved scales. Six resolutions were conducted for investigating the influence of mesh resolution in the turbulent diffuser flow. Four existing SGS models, a new one-equation dynamic model and a DNS were performed for LES of turbulent diffuser flow. The performance of five different SGS models was compared with published experiments and our DNS results. Firstly, LES on a coarse grid using the standard Smagorinsky model and well-designed wall-functions is able to predict the three-dimensional separated diffuser flow with fair accuracy at reasonable cost. Secondly, the ODM model and OVM model gave better agreement with experiments and DNS than the SM model, OM model and DSM model. Thirdly, unlike the ODM model, our OVM model, which does not require any test filtering and is not more expensive in terms of computational cost than the standard one-equation model, gave considerable improvement of the prediction accuracy even with moderate grid resolution in comparison with the four other SGS models.

Author Contributions: H.T. and Y.L. conceived the original ideal; H.T. wrote and edited the manuscript; and Y.L., X.L. and Y.F. supervised the study.

Funding: This research was funded by the project of Chinese National Natural Science Foundation (grant number: 51575220), the project of the Key Scientific and Technological Project of Jilin Province (grant numbers:

20160519008JH and 20170204073GX), and the project of National Key R&D Program of China (grant number: 2016YFB0101402).

Acknowledgments: The authors thank Takeo Kajishima and Kei Okabayashi for their help with the numerical method proposed and simulations conducted.

Conflicts of Interest: The authors declare no conflict of interest.

Abbreviations

The following abbreviations are used in this manuscript:

CFD	Computational fluid dynamics
DNS	Direct numerical simulation
DSM	Standard dynamic Smagorinsky model
LES	Large eddy simulation
N-S	Navier–Stokes
ODM	One-equation dynamic model
OM	Standard one-equation model
OVM	One-equation Vreman model
RANS	Reynolds-averaged Navier–Stokes simulation
SGS	subgrid scale
SM	Smagorinsky model

Appendix A. Determination of the Distance from the Wall Measured in Wall Units in the Diffuser Flow for Smagorinsky Model

Using a linear interpolation, let us assume the friction velocity $u_\tau(x)$ in the diffuser is as follows:

$$u_\tau(x) = \begin{cases} u_\tau 0, & x \leq 0 \\ u_\tau 0 + x(u_\tau e - u_\tau 0)/H, & 0 < x \leq 21H \\ u_\tau e, & x > 21H, \end{cases} \tag{A1}$$

where $u_\tau 0$ denotes the inlet friction velocity, 407; and $u_\tau e$ denotes the friction velocity in down stream extension.

Reynolds number Re_m based on the bulk velocity U_m is constant in the diffuser because of the continuity:

$$\frac{U_m H}{\nu} = \frac{\frac{U_m}{4.7} 4.7H}{\nu}. \tag{A2}$$

Thus, Re_τ is also constant due to the Dean's suggested correction [35] $Re_\tau = 0.191 Re_m^{0.875}$, which was validated by Kim et al. [4] in fully developed channel flow. Then, Re_τ in diffuser can be approximately expressed as:

$$Re_\tau 0 = \frac{u_\tau 0 \cdot H}{\nu} = \frac{u_\tau e \cdot 4.7H}{\nu} = 407, \tag{A3}$$

thus we get $u_\tau e = u_\tau 0/4.7$. Using Equation (A1), y^+ can be approximately defined as:

$$y^+(x,y) = \begin{cases} Re_\tau 0 \cdot \min(y, 1 - y)/H, & x \leq 0 \\ Re_\tau 0 \cdot (1 - 3.7/4.7 \cdot x/H) \cdot \min[y, Y(x) - y]/H, & 0 < x \leq 21H \\ Re_\tau 0/4.7 \cdot \min(y, 4.7H - y)/H, & x > 21H. \end{cases} \tag{A4}$$

References

1. Sayles, E.L.; Eaton, J.K. Sensitivity of an asymmetric, three-dimensional diffuser to inlet condition perturbations. *Int. J. Heat Fluid Flow* **2014**, *49*, 100–107. [CrossRef]
2. Azad, R.S. Turbulent flow in a conical diffuser: A review. *Exp. Fluid Sci.* **1996**, *13*, 318–337. [CrossRef]

3. Obi, S.; Aoki, K.; Masuda, S. Experimental and computational study of turbulent separating flow in an asymmetric plane diffuser. In Proceedings of the Ninth Symposium on Turbulent Shear Flows, Hyoto, Japan, 16–18 August 1993; p. 305-1.

4. Kim, J.; Moin, P.; Moser, R. Turbulence statistics in fully developed channel flow at low Reynolds number. *J. Fluid Mech.* **1987**, *177*, 133–166. [CrossRef]

5. Ohta, T.; Kajishima, T. Analysis of non-steady separated turbulent flow in an asymmetric plane diffuser by direct numerical simulations. *J. Fluid Sci. Technol.* **2010**, *5*, 515–527. [CrossRef]

6. Kaltenbach, H.J.; Fatica, M.; Mittal, R.; Lund, T.S.; Moin, P. Study of flow in a planar asymmetric diffuser using large-eddy simulation. *J. Fluid Mech.* **1999**, *390*, 151–185. [CrossRef]

7. Durbin, P.A. Separated flow computations with the k-epsilon-v-squared model. *AIAA J.* **1995**, *33*, 659–664. [CrossRef]

8. Iaccarino, G. Predictions of a turbulent separated flow using commercial CFD codes. *J. Fluids Eng.* **2001**, *123*, 819–828. [CrossRef]

9. El-Behery, S.M.; Hamed, M.H. A comparative study of turbulence models performance for separating flow in a planar asymmetric diffuser. *Comput. Fluids* **2011**, *44*, 248–257. [CrossRef]

10. Schneider, H.; von Terzi, D.; Bauer, H.J.; Rodi, W. Reliable and accurate prediction of three-dimensional separation in asymmetric diffusers using large-eddy simulation. *J. Fluids Eng.* **2010**, *132*, 031101. [CrossRef]

11. Abe, K.I.; Ohtsuka, T. An investigation of LES and Hybrid LES/RANS models for predicting 3-D diffuser flow. *Int. J. Heat Fluid Flow* **2010**, *31*, 833–844. [CrossRef]

12. Schluter, J.; Wu, X.; Pitsch, H. Large-Eddy Ssimulations of a Separated Plane Diffuser. In Proceedings of the 43rd AIAA Aerospace Sciences Meeting and Exhibit, Reno, NV, USA, 10–13 January 2005; p. 672.

13. Kobayashi, H.; Ham, F.; Wu, X. Application of a local SGS model based on coherent structures to complex geometries. *Int. J. Heat Fluid Flow* **2008**, *29*, 640–653. [CrossRef]

14. Taghinia, J.; Rahman, M.M.; Siikonen, T.; Agarwal, R.K. One-equation sub-grid scale model with variable eddy-viscosity coefficient. *Comput. Fluids* **2015**, *107*, 155–164. [CrossRef]

15. Shuai, S.; Agarwal, R.K. A One-Equation Turbulence Model Based on k-kL Closure. In Proceedings of the AIAA Scitech 2019 Forum, San Diego, CA, USA, 7–11 January 2019; p. 1879.

16. Sagaut, P. *Large Eddy Simulation for Incompressible Flows: An Introduction*; Springer Science & Business Media: Berlin/Heidelberg, Geramny, 2006.

17. Smagorinsky, J. General circulation experiments with the primitive equations: I. The basic experiment. *Mon. Weather Rev.* **1963**, *91*, 99–164. [CrossRef]

18. Germano, M.; Piomelli, U.; Moin, P.; Cabot, W.H. A dynamic subgrid-scale eddy viscosity model. *Phys. Fluids A* **1991**, *3*, 1760–1765. [CrossRef]

19. Lilly, D.K. A proposed modification of the Germano subgrid-scale closure method. *Phys. Fluids A* **1992**, *4*, 633–635. [CrossRef]

20. Yoshizawa, A. A statistically-derived subgrid model for the large-eddy simulation of turbulence. *Phys. Fluids* **1982**, *25*, 1532–1538. [CrossRef]

21. Kajishima, T.; Nomachi, T. One-equation subgrid scale model using dynamic procedure for the energy production. *J. Appl. Mech.* **2006**, *73*, 368–373. [CrossRef]

22. Yoshizawa, A.; Horiuti, K. A statistically-derived subgrid-scale kinetic energy model for the large-eddy simulation of turbulent flows. *J. Phys. Soc. Jpn.* **1985**, *54*, 2834–2839. [CrossRef]

23. Horiuti, K. Large eddy simulation of turbulent channel flow by one-equation modeling. *J. Phys. Soc. Jpn.* **1985**, *54*, 2855–2865. [CrossRef]

24. Okamoto, M.; Shima, N. Investigation for the One-Eaquation-Type Subgrid Model with Eddy-Viscosity Expression Including the Shear-Damping Effect. *JSME Int. J. Ser. B* **1999**, *42*, 154–161. [CrossRef]

25. Vreman, A.W. An eddy-viscosity subgrid-scale model for turbulent shear flow: Algebraic theory and applications. *Phys. Fluids* **2004**, *16*, 3670–3681. [CrossRef]

26. Park, N.; Lee, S.; Lee, J.; Choi, H. A dynamic subgrid-scale eddy viscosity model with a global model coefficient. *Phys. Fluids* **2006**, *18*, 125109. [CrossRef]

27. You, D.; Moin, P. A dynamic global-coefficient subgrid-scale eddy-viscosity model for large-eddy simulation in complex geometries. *Phys. Fluids* **2007**, *19*, 065110. [CrossRef]

28. You, D.; Moin, P. A dynamic global-coefficient subgrid-scale model for large-eddy simulation of turbulent scalar transport in complex geometries. *Phys. Fluids* **2009**, *21*, 045109. [CrossRef]

29. Obi, S.; Ohizumi, H.; Aoki, K.; Masuda, S. Turbulent separation control in a plane asymmetric diffuser by periodic perturbation. *Eng. Turbul. Model. Exp.* **1993**, 633–642. [CrossRef]
30. Kajishima, T.; Ohta, T.; Okazaki, K.; Miyake, Y. High-order finite-difference method for incompressible flows using collocated grid system. *JSME Int. J. Ser. B* **1998**, *41*, 830–839. [CrossRef]
31. Ohta, T. DNS of turbulent channel flow subjected to the acceleration. *Trans. Jpn. Soc. Mech. Eng. Ser. B* **2004**, *70*, 1–8. [CrossRef]
32. Tang, H.; Lei, Y.; Fu, Y. Noise Reduction Mechanisms of an Airfoil with Trailing Edge Serrations at Low Mach Number. *Appl. Sci.* **2019**, *9*, 3784. [CrossRef]
33. Tamura, A.; Kikuchi, K.; Takahashi, T. Residual cutting method for elliptic boundary value problems. *J. Comput. Phys.* **1997**, *137*, 247–264. [CrossRef]
34. Buice, C.U.; Eaton, J.K. Experimental Investigation of Flow Through an Asymmetric Plane Diffuser. *Center for Turbulence Research Annual Research Briefs* **1995**, *117–120*. [CrossRef]
35. Dean, R.B. Reynolds number dependence of skin friction and other bulk flow variables in two-dimensional rectangular duct flow. *J. Fluids Eng.* **1978**, *100*, 215–223. [CrossRef]

 *symmetry*

Article

MHD Flow and Heat Transfer in Sodium Alginate Fluid with Thermal Radiation and Porosity Effects: Fractional Model of Atangana–Baleanu Derivative of Non-Local and Non-Singular Kernel

Arshad Khan [1], Dolat Khan [2], Ilyas Khan [3,*], Muhammad Taj [4], Imran Ullah [5], Abdullah Mohammed Aldawsari [6], Phatiphat Thounthong [7] and Kottakkaran Sooppy Nisar [8]

1   Institute of Computer Science and Information Technology, The University of Agriculture, Peshawar 25130, Pakistan; arsh_math@yahoo.com
2   Department of Mathematics, City University of Science and Information Technology, Peshawar 25130, Khyber Pakhtunkhwa, Pakistan; dolat.ddk@gmail.com
3   Faculty of Mathematics and Statistics, Ton Duc Thang University, Ho Chi Minh City 72915, Vietnam
4   Department of Mathematics, University of Azad Jammu and Kashmir, Muzaffarabad 13100, Pakistan; Muhammad_taj75@yahoo.com
5   College of Civil Engineering, National University of Sciences and Technology Islamabad, Islamabad 44000, Pakistan; ullahimran14@gmail.com
6   Department of Chemistry, College of Arts and Science-Wadi Al-Dawaser, Prince Sattam bin Abdulaziz University, 11991 Al-kharj, Saudi Arabia; abdullah.aldawsari@psau.edu.sa
7   Renewable Energy Research Centre, Department of Teacher Training in Electrical Engineering, Faculty of Technical Education, King Mongkut's University of Technology North Bangkok, Bangkok 10800, Thailand; phatiphat.t@fte.kmutnb.ac.th
8   Department of Mathematics, College of Arts and Sciences, Wadi Aldawaser, Prince Sattam bin Abdulaziz University, 11991 Al-Kharj, Saudi Arabia; n.sooppy@psau.edu.sa
*   Correspondence: ilyaskhan@tdtu.edu.vn

Received: 4 July 2019; Accepted: 9 October 2019; Published: 15 October 2019

Abstract: Heat transfer analysis in an unsteady magnetohydrodynamic (MHD) flow of generalized Casson fluid over a vertical plate is analyzed. The medium is porous, accepting Darcy's resistance. The plate is oscillating in its plane with a cosine type of oscillation. Sodium alginate (SA–NaAlg) is taken as a specific example of Casson fluid. The fractional model of SA–NaAlg fluid using the Atangana–Baleanu fractional derivative (ABFD) of the non-local and non-singular kernel has been examined. The ABFD definition was based on the Mittag–Leffler function, and promises an improved description of the dynamics of the system with the memory effects. Exact solutions in the case of ABFD are obtained via the Laplace transform and compared graphically. The influence of embedded parameters on the velocity field is sketched and discussed. A comparison of the Atangana–Baleanu fractional model with an ordinary model is made. It is observed that the velocity and temperature profile for the Atangana–Baleanu fractional model are less than that of the ordinary model. The Atangana–Baleanu fractional model reduced the velocity profile up to 45.76% and temperature profile up to 13.74% compared to an ordinary model.

Keywords: SA–NaAlg fluid; MHD; porosity; fractional model; Atangana–Baleanu derivative

1. Introduction

Due to the relevance of non-Newtonian fluids in some of the optimization processing of food items, the heat transfer phenomenon is an important research area. Non-Newtonian fluids are investigated much more, but Casson fluids among them have been investigated by a small number of researchers.

This model was introduced by Casson in 1959 [1]. This fluid has great importance in many fields of industries and medicines; see for examples, Qasim and Ahmad [2], Shehzad et al. [3], and Qasim and Noreen [4]. The Casson fluid model plays a vital role in many materials such as chocolate and blood, according to Mukhopadhyay and Mandal [5]. The metachronal coordination between the beating cilia in a cylindrical tube is one of the factors under which inspiration, and some mathematical formulations, is made to analyze Casson fluid flow, according to Siddiqui et al. [6]. A system of equations for non-linear flow problems are modeled with the help of axisymmetric cylindrical coordinates. Simplification is taken place by inducting long wavelength and low Reynolds number assumptions.

The mathematical model for Casson fluid for mixed convection taking a stretching sheet was studied by Hayat et al. [7]. The velocity and temperature fields are calculated by solving partial differential equations. The homotopy analysis method is incorporated. The Atangana–Baleanu fractional derivative (ABFD) definition was based on the Mittag–Leffler function and promises an improved description of the dynamics of the system with the memory effects by Asjad et al. [8]. The mathematical model for some geological formation is often called an aquifer, which is commonly used to solve subsurface motion by a fractional-order derivative, as in Atangana and Baleanu [9]. It was an alternative to Caputo-Fabrizio [10]. A relation between the solution obtained by the Atangana–Baleanu fractional derivative and through experimental methods is highlighted. Exact solutions were obtained by integral transforms.

Aliyu et al. [11] studied the model of transmission of vector-borne diseases and cure using the Atangana–Baleanu fractional operator in a Caputo sense. They also discussed the existence and uniqueness via numerical simulations. This model was taken into consideration by the Atangana–Baleanu fractional operator in the Caputo sense (ABC) with non-singular and non-local kernels. The Picard–Lindel method is adopted. The same method is used by Koca [12].

The effectiveness of this model is seen by a huge reduction in the disease growth rate. A comparison of heat transfer enhancement in fractional nanofluids with ordinary nanofluids is made by Azhar et al. [13]. Water is taken as a base fluid mixed with fractional nanofluids. A closed form of solution for motion and temperature is calculated and graphically underlined with a higher heat enhancement rate than ordinary fluids. The influence of memory on nanofluid behavior is difficult to elaborate by classical nanofluids. This can be easy to describe by fractional nanofluids with Caputo time derivatives. Such investigations are made by Fetecau et al. [14]. Closed-form results are obtained and presented in a Wright function. Heat enhancement was found to be lower for fractional nanofluids.

The Caputo–Fabrizio and Atangana–Baleanu derivatives are those fractional derivatives that transformed the classical model to a generalized model. Comparison investigations were made by Sheikh et al. [15]. Karaagac [16] applied two step Adams Bashforth method for time fractional tricomi equation with non-local and non-singular Kernel. Furthermore, Ali et al. [17] investigated MHD flow for free convection generalized Walter's-B fluid. The exact solution is gained by the Laplace technique. The flow is over a static vertical plate. After that, Saqib et al. [18] studied the application of Atangana–Baleanu fractional derivative to MHD channel flow. The fluid is taken through a porous medium. Abro et al. [19] investigated the MHD convection flow for Maxwell fluid as the application of the Atangana–Baleanu fractional derivative. The fluid is taken over a vertical plate in a porous medium. The same author in [20] investigated the exact solution for MHD flow of nanofluids via Atangana–Baleanu and Caputo–Fabrizio fractional derivatives. The general analytical solutions are expressed in the layout of Mittage Leffler function and generalized Mittage Leffler function. The comparison of new fractional derivative operators involving exponential and Mittag–Leffler kernels is investigated by Yavuz and Ozdemir [21]. The comparison was investigated between Caputo–Fabrizio and Atangana–Baleanu fractional derivatives. Furthermore, Imran et al. [22] investigated the comprehensive report on the MHD convective flow of Atangana–Baleanu and Caputo–Fabrizio fractional derivatives. The viscous fluid is subject to generalized boundary conditions. Abro and Khan [23] examined the study of Atangana–Baleanu fractional derivatives for the MHD flow of fractional Newtonian fluid embedded in a porous medium. Electrically conducting viscous fluid is considered in their study.

The flow of generalized second-grade fluid between parallel plates with the Riemann–Liouville fractional derivative model was investigated by Wenchang and Mingyu [24]. They acquired the exact analytical solution using the Laplace transform and Fourier transform. The flow of second-order fluid induced by a plate moving impulsively with fractional anomalous diffusion was investigated by Mingyu and Wenchang [25]. The Rayleigh–Stokes problem for a fractional second-grade fluid was studied by Shen et al. [26]. Fractional Laplace transforms and Fourier sine transform was employed to obtain the exact solution. Exact analytical solution for the unsteady flow of a generalized Maxwell fluid between two circular cylinders was determined via Laplace and Hankel transforms by Mahmood et al. [27]. Recently, Shen et al. [28] studied fractional Maxwell viscoelastic nanofluid for various particle shapes. A Caputo time-fractional derivative was implemented by Zhang et al. [29] to acquire the numerical and analytical solutions for the problem of the 2D flow of Maxwell fluid under a variable pressure gradian gradient. They used the separation of the variables method to acquire the analytical solution, while for numerical solution, the finite difference method was used. Aman et al. [30] studied fractional Maxwell fluid with a second-order slip effect for the exact analytical solution. Jan et al. [31] determined the solution for Brinkman-type nanofluid using an Atangana–Baleanu fractional model. Owolabi and Atangana [32] analyzed the numerical simulation of the Adams–Bash forth scheme using Atangana–Baleanu Caputo fractional derivatives. Saad et al. [33] established the numerical solutions for the fractional Fisher's type equation with Atangana–Baleanu fractional derivatives. They employed the spectral collocation method based on Chebyshev approximations. In this research work, the spectral collocation method was implemented for the first time to solve the non-linear equation with Atangana–Baleanu derivatives. Some plenteous literature regarding Atangana–Baleanu derivatives and analytical solution can be found in Saqib et al. [34], Abro et al. [35], and Hristov [36] and the references therein.

Due to the importance of MHD in fluids, many researchers have taken MHD in their studies. In recent years, Khan and Alqahtan [37] investigated the MHD effects for nanofluids in a permeable channel with porosity. The solution is obtained by using Laplace transformation. Further, in the same year, Asif et al. [38] studied the unsteady flow of fractional fluid between two parallel walls with arbitrary wall shear stress. The influence of MHD slip flow of Casson fluid along a non-linear permeable stretching cylinder saturated in a porous medium with chemical reaction, viscous dissipation, and heat generation or absorption is investigated by Ullah et al. [39]. Khan et al. [40] investigated the MHD with heat transfer. A generalized modeling is carried out for the proposed problem by using the new idea of a fractional derivative, i.e., Atangana–Baleanu and Caputo Fabrizio. After that, this idea is used by Gul et al. [41] for the study of forced convection carbon nanotube nanofluid flow passing over a thin needle. Atangana and Alqahtani [42] studied the Caputo fractional derivative for analysis of the spread of river blindness disease. Furthermore, the idea of fractional derivatives was examined by Gomez and Atangana [43] with the power law and the Mittag–Leffler kernel applied to the non-linear Baggs–Freedman model, while Muhammad and Atangana [44] examined for dynamics of Ebola disease, and Khan et al. [45] studied the analytical solution of the hyperbolic telegraph equation, using the natural transform decomposition method. Some other related references dealing with fluid motion, heat transfer, or fractional derivatives are given in [46–53].

On the basis of the above literature, this work aims to use the Atangana–Baleanu fractional derivative (ABFD) of the non-singular and non-local kernel for SA (Sodium alginate) fluid. The Laplace transform method is used to get the exact solution, which is graphically plotted via Mathcad-15 software and discussed in detail.

2. Mathematical Framing of the Problem

Let us consider the unsteady flow of Casson fluid over a vertical plate, i.e., the plate is taken perpendicular to the $y - axis$, with perpendicular employed magnetic field to the flow of the fluid. The plate is oscillating, and the medium is porous. The heat flux is also taken into consideration. Thermal radiation effect is also considered. Initially, both the plate and fluid are at rest. After $t > 0$, the plate started oscillation in its plane. The fluid is electrically conducting there by the Maxwell equation:

$$\text{div}\mathbf{B} = 0, \ \text{Curl}\mathbf{E} = -\frac{\partial \mathbf{B}}{\partial t}, \ \text{Curl}\mathbf{B} = \mu_e \mathbf{J}. \tag{1}$$

By using Ohm's law:

$$\mathbf{J} = \sigma(\mathbf{E} + \mathbf{V} \times \mathbf{B}), \tag{2}$$

The magnetic field $\mathbf{B}$ is normal to $\mathbf{V}$. The Reynolds number is so small that the flow is laminar. Hence:

$$\frac{1}{\rho}\mathbf{J} \times \mathbf{B} = \frac{\sigma}{\rho}[(\mathbf{V} \times \mathbf{B}_0) \times \mathbf{B}_0] = -\frac{\sigma B_0^2 \mathbf{V}}{\rho}. \tag{3}$$

Keeping in mind the above assumptions, the governing equation of momentum and energy are given as (Khalid et al. [46]):

$$\rho\frac{\partial u(y,t)}{\partial t} = \mu\left(1 + \frac{1}{\gamma}\right)\frac{\partial^2 u(y,t)}{\partial y^2} - \left(\sigma B_0^2 + \left(1 + \frac{1}{\gamma}\right)\frac{\mu}{k_1}\right)u(y,t) + \rho\beta g(T - T_\infty) \tag{4}$$

$$\rho c_p\frac{\partial T(y,t)}{\partial t} = k\frac{\partial^2 T(y,t)}{\partial y^2} - \frac{\partial q_r}{\partial y}, \tag{5}$$

where T and u represent temperature and velocity. ρ, μ, β, σ, B_0, k_1, g, c_p, k, q_r and γ are the density, dynamic viscosity, thermal expansion coefficient, heat source parameter, magnetic parameter, porosity parameter, gravitation acceleration, heat capacity, thermal conductivity, heat flux, and Casson fluid parameter. Following Makinde and Mhone [47] and Cogley et al. [48], the fluid used is thin with a low density and radiative heat flux given by $\frac{\partial q_r}{\partial y} = 4\alpha_0^2(T - T_0)$.

The physical boundary conditions are:

$$\begin{aligned} T(y, 0) &= T_\infty, \ T(0, t) = T_w, \ T(\infty, t) = T_\infty \\ u(y, 0) &= 0, \ \ u(0, t) = H(t)\cos(\omega t), \ u(\infty, t) = 0. \end{aligned} \tag{6}$$

For non-dementalization, the following dimensionless variables are introduced.

$$u^* = \frac{u}{U_0}, \quad y^* = \frac{U_0}{v}y, \quad t^* = \frac{U_0^2}{v}t, \quad \theta = \frac{T - T_\infty}{T_w - T_\infty}. \tag{7}$$

By using Equation (4), the dimensionless form of Equations (1)–(3) are: (asterisk * is omitted for convenience):

$$\frac{\partial u(y,t)}{\partial t} = \left(1 + \frac{1}{\gamma}\right)\frac{\partial^2 u(y,t)}{\partial y^2} - Hu(y,t) + Gr\theta(y,t), \tag{8}$$

$$Pr\frac{\partial \theta(y,t)}{\partial t} = \frac{\partial^2 \theta(y,t)}{\partial y^2} + N^2\theta(y,t), \tag{9}$$

$$\begin{aligned} \theta(y, 0) &= 0, \ \theta(0, t) = 1, \ \theta(\infty, t) = 0 \\ u(y, 0) &= 0, \ \ u(0, t) = H(t)\cos(\omega t), \ u(\infty, t) = 0, \end{aligned} \tag{10}$$

where:

$$Gr = \frac{g\beta v(T_w - T_\infty)}{U_0^3}, \ M = \frac{\sigma B_0^2 v}{\rho U_0^2}, \ K = \frac{v^2}{k_1 U_0^2}, \ \frac{1}{a_1} = 1 + \frac{1}{\gamma},$$

$$H = \frac{K}{a_1} + M, \ Pr = \frac{\mu c_p}{k}, \ N^2 = \frac{4\alpha_1^2 v^2}{kU_0^2},$$

where M, K, Pr, Gr, α_1 and N represent the magnetic, porosity, Prandtl number, Grashof number, mean radiation absorption, and radiation parameter, respectively.

With the intention of converting the ordinal time derivative to the Atangana–Baleanu fractional time derivative, Equations (5) and (6) reduced to:

$$D_t^\alpha u(y,t) = \left(1 + \frac{1}{\gamma}\right)\frac{\partial^2 u(y,t)}{\partial y^2} - Hu(y,t) + Gr\theta(y,t), \tag{11}$$

$$PrD_t^\alpha \theta(y,t) = \frac{\partial^2 \theta(y,t)}{\partial y^2} + N^2\theta(y,t). \tag{12}$$

The Atangana–Baleanu fractional time derivative is defined as:

$$D_t^\alpha f(y,t) = \frac{N(\alpha)}{1-\alpha}\int_0^t E_\alpha\left(-\alpha\frac{(t-\tau)^\alpha}{1-\alpha}\right)f'(y,t)d\tau, \tag{13}$$

where $N(\alpha)$ is the normalization function, $N(1) = N(0) = 1$ and $\alpha \in (0,\,1)$. where $E_\alpha = \sum_{n=0}^\infty \frac{z^n}{\Gamma(\alpha n+b)}$ is the Mittag–Leffler function. After Laplace transform, Equation (13) becomes:

$$L\{D_t^\alpha f(t)\} = \frac{q^\alpha L\{f(t)\} - q^{\alpha-1}f(0)}{q^\alpha(1-\alpha) + \alpha} \tag{14}$$

where q is represent the Laplace transform operator.

3. Problem Solution, Skin Friction, and Nusselt Number

Taking the Laplace transform of Equations (8) and (9) and by using Equation (7), we get:

$$\overline{\theta}(y,q) = \left(\frac{1}{q^{1-\alpha}}\right)\frac{1}{q^\alpha}e^{-y\sqrt{a_5}\sqrt{\frac{q^\alpha+a_6}{q^\alpha+a_4}}} \tag{15}$$

$$\overline{u}(y,q) = \left(\frac{q}{q^2+\omega^2} - \frac{Gr}{q}\frac{q^\alpha+a_4}{a_7q^\alpha+a_8}\right)e^{-y\sqrt{a_1a_2}\sqrt{\frac{q^\alpha+a_3}{q^\alpha+a_4}}} + \frac{Gr}{q}\frac{q^\alpha+a_4}{a_7q^\alpha+a_8}e^{-y\sqrt{a_5}\sqrt{\frac{q^\alpha+a_6}{q^\alpha+a_4}}} \tag{16}$$

where:

$$a_2 = \frac{1+H(1-\alpha)}{1-\alpha},\ a_3 = \frac{H\alpha}{1+H(1-\alpha)},\ a_4 = \frac{\alpha}{1-\alpha},\ a_5 = \frac{Pr+N^2(1-\alpha)}{1-\alpha},$$
$$a_6 = \frac{N^2\alpha}{Pr+N^2(1-\alpha)},\ a_7 = a_2 - \frac{a_5}{a_1},\ a_8 = a_2a_3 - \frac{a_5a_6}{a_1}.$$

Note that in Equations (15) and (16), the bar on θ and u shows the Laplace transformed function. After taking the Laplace inverse, we get:

$$\theta(y,t) = \int_0^t h(t-q,\alpha)\psi(y\sqrt{a_5},t,0,a_6,a_4)dq, \tag{17}$$

where:

$$\overline{\psi}(y,q,a,b,c) = \frac{1}{q^\alpha+a}e^{-y\sqrt{\frac{q^\alpha+b}{q^\alpha+c}}},\ \psi(y,t,a,b,c) = \int_0^\infty \psi(y,t,a,b,c)g(u,t)du$$

where:

$$\psi(y,t,a,b,c) = L^{-1}\{\overline{\psi}(y,q,a,b,c)\}$$

$$\psi(y,t,a,b,c) = e^{-at-y} - \frac{y\sqrt{b-c}}{2\sqrt{\pi}} \int_0^\infty \int_0^t \frac{e^{-at}}{\sqrt{t}} \times \exp\left(at - ct - \frac{y^2}{4u} - u\right) I_1\left(2\sqrt{(b-c)ut}\right) dt\,du,$$

$$L^{-1}\left(\frac{1}{q^{1-\alpha}}\right) = h(t,\alpha) = \frac{1}{t^\alpha \Gamma(1-\alpha)}, \quad g(u,t) = L^{-1}\{\exp[-uq^{-\alpha}]\} = t^{-1}\varphi(0,-\alpha,-ut^{-\alpha}),$$

Here, φ is the Wright function, and is defined as:

$$\varphi(k_2,-\alpha,\tau) = \sum_{n=0}^\infty \frac{\tau^n}{n!\,\Gamma(k_2 - n\alpha)}$$

Then, the velocity profile becomes:

$$\overline{\psi_1}(y,a,b,c,d,q) = \exp\left(-y\sqrt{\frac{aq^\alpha + b}{cq^\alpha + d}}\right), \quad \overline{\psi_2}(a,b,c,q) = \frac{aq^\alpha + b}{q(q^\alpha + c)}$$

$$u(y,t) = \cos\omega t * \psi_1\left(y\sqrt{a_1 a_2}, 1, 1, a_3, a_4, t\right) - \frac{Gr}{a_7}\psi_2\left(1, a_4, \frac{a_8}{a_7}, t\right) * \left[\begin{array}{c} \psi_1\left(y\sqrt{a_5}, 1, 1, a_6, a_4, t\right) \\ -\psi_1\left(y\sqrt{a_1 a_2}, 1, 1, a_3, a_4, t\right) \end{array}\right] \tag{18}$$

where:

$$\psi_1(y,a,b,c,d,t) = t^{-1}\varphi(0,-\alpha,ut^{-\alpha}) * \psi_3(y,a,b,c,t),$$

$$\psi_3(y,a,b,c,t) = \frac{a}{c}e^{-\frac{t}{c}} \int_0^\infty erfc\left(\frac{t}{2z}\right) - e^{-\frac{lz}{c}} I_0\left(2\sqrt{(a-cb)tz}\right) dz$$

$$+ \frac{b}{c}\int_0^t \int_0^\infty erfc\left(\frac{t}{2z}\right) - e^{-\frac{lz+s}{c}} I_0\left(2\sqrt{(a-cb)tz}\right) ds\,dz,$$

$$\psi_2(a,b,c,t) = aE_\alpha(-ct) - b[a - E_\alpha(-ct)]$$

$E_\alpha = \sum_{n=0}^\infty \frac{z^n}{\Gamma(\alpha n + b)}$ is the Mittag–Leffler function.

Note that for deep understanding of ABFD, on may refer to the excellent articles [35,53]. However, for the detailed solution procedure of the problem, one can refer to research works [15,18,20,23,31].

4. Skin Friction and Nusselt Number

Expressions for Nusselt number and skin friction are calculated from Equations (14) and (15) by using the relation from Khan et al. [49]:

$$Nu = -\left.\frac{\partial T(y,t)}{\partial y}\right|_{y=0} \tag{19}$$

$$C_f = \mu\left(1 + \frac{1}{\gamma}\right)\left.\frac{\partial u(y,t)}{\partial y}\right|_{y=0} \tag{20}$$

5. Discussion

To study the influence of many embedded parameters on temperature and velocity, the graphs are plotted by using the Mathcad-15 software. Figure 1 shows the physical sketch of the problem. Figure 2, Figure 3, Figure 4, Figure 5, Figure 6, Figure 7, Figure 8, Figure 9 are prepared to highlight the influence of time-fractional parameter $0 < \alpha < 1$ on fluid flow and temperature. The time-fractional parameter α controls the velocity as well as the temperature profile. Sodium alginate (SA–NaAlg) is taken as a specific example of Casson fluid. The physical sketch of the problem is provided in Figure 1.

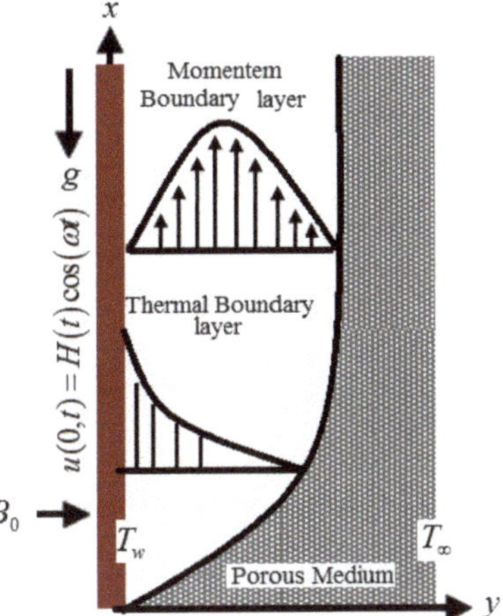

Figure 1. Physical sketch of the problem with the coordinates system.

Figure 2 is plotted for the influence of Casson parameter γ on velocity, which shows that if the value of the Casson parameter is increased, the fluid velocity increases. This is because of the circumstance that with a large value of γ, the yield stress falls through, and the boundary layer thickness reduces. Figure 3 investigates the impact of Grashof number Gr on velocity. The greater value Gr leads to an increase in the velocity of the fluid. Physically, the increase of Gr leads to an increase in the bouncy force, and as a result, the velocity of the fluid increases.

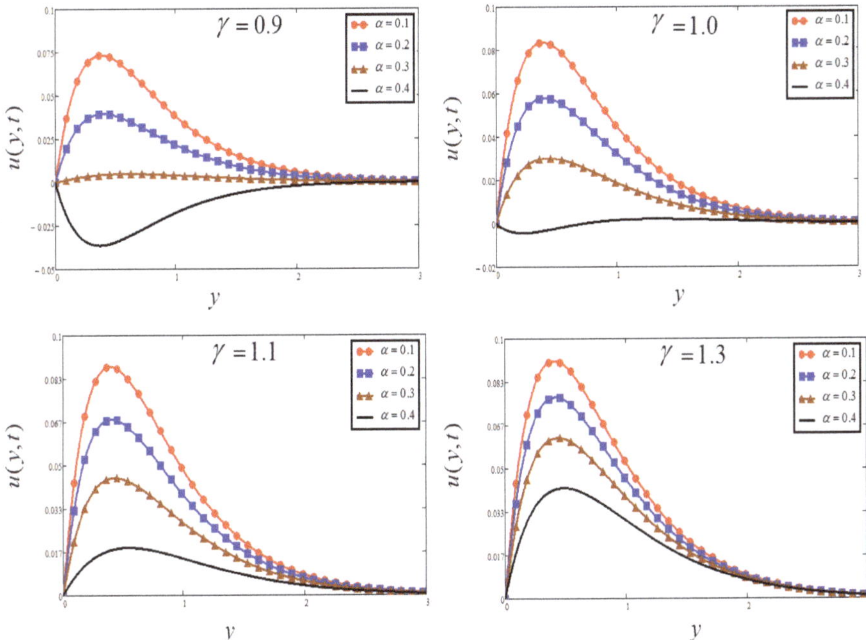

Figure 2. Plots of velocity for four different γ when $K = 0.5$, $\mathrm{Pr} = 7.2$, $M = 1$, $Gr = 10$, $N = 0.5$, and $t = 10$.

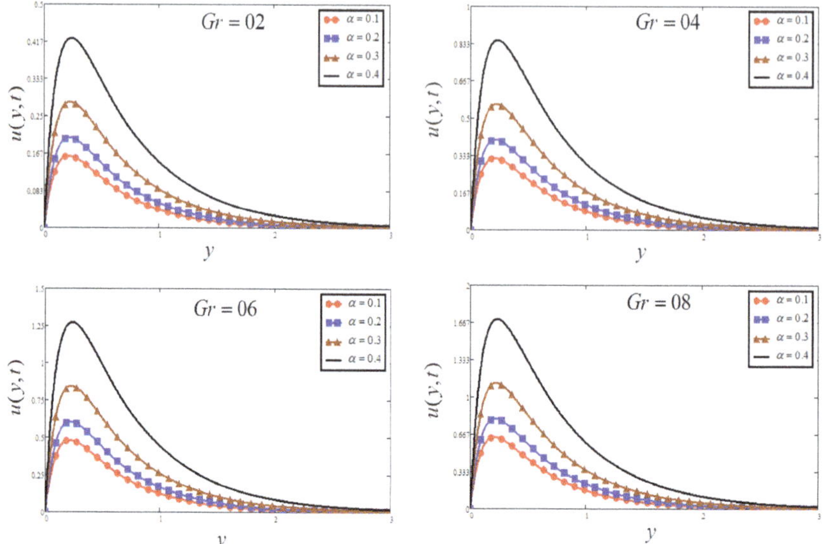

Figure 3. Plots of velocity for four different Gr when $K = 0.5$, $Pr = 7.2$, $\gamma = 0.3$, $M = 1, N = 0.5$, and $t = 10$.

The effect of porosity parameter K against the velocity profile is investigated in Figure 4. To increase the value of the porosity parameter K, first we need to decrease the flow of fluid. Physically, the resistance of the porous medium is depressed, which raises the momentum development of the flow regime, and finally accelerates the velocity of the fluid. Figure 5 shows the influence of M on flow of the fluid; the rise in M decreases the flow of fluid. It is physically true because the increase of M means to increase the frictional force (Lorentz force), which leads to a decrease in the velocity of the fluid.

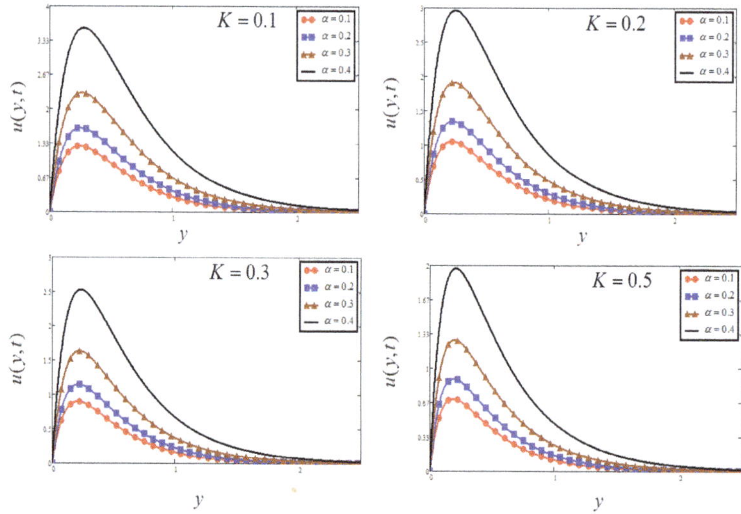

Figure 4. Plots of velocity for four different K when $Pr = 7.2$, $\gamma = 0.3$, $M = 1$, $Gr = 10.N = 0.5$, and $t = 10$.

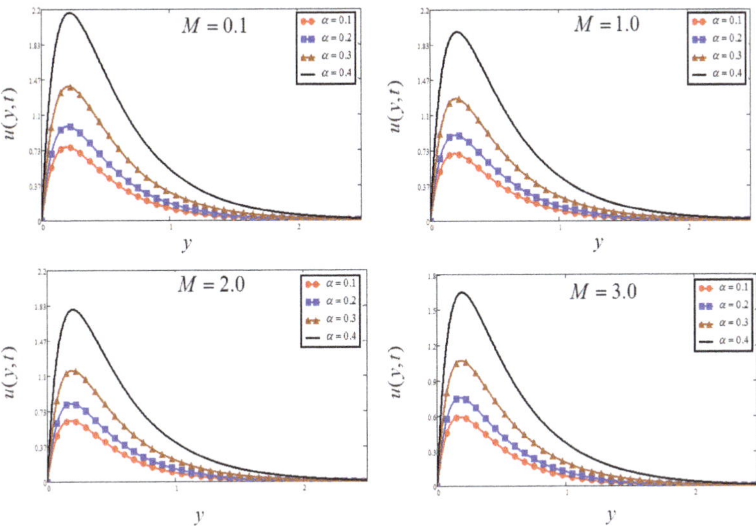

Figure 5. Plots of velocity for four different M values when $K = 0.5$, $\text{Pr} = 7.2$, $\gamma = 0.3$, $Gr = 10$, $N = 0.5$, and $t = 10$.

The influence of radiation parameter N is highlighted in Figure 6. For the higher value of N, the fluid velocity decreases. Physically, the increase in radiation parameter means the release of heat energy from the flow region, and so the fluid temperature decreases as the thermal boundary layer thickness become thinner. Figures 7 and 8 illustrate the influence of Prandtl number Pr on velocity and temperature respectively, the increase of Pr causing a decrease in the temperature and as a result a decrease in the fluid velocity. The small degree of thermal diffusion causes expanding in velocity boundary layer width. Pr controls the comparative thickness of the momentum and thermal boundary layers in the heat transfer problems. Subsequently, Pr can be applied to develop the percentage of cooling.

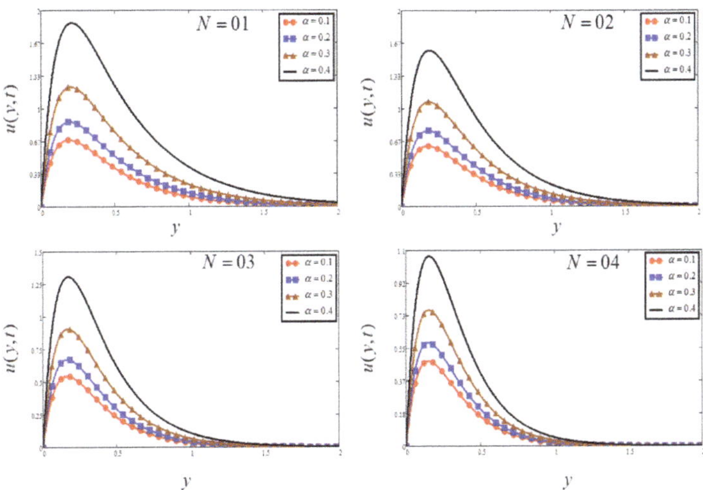

Figure 6. Plots of velocity for four different N values when $K = 0.5$, $\text{Pr} = 7.2$, $\gamma = 0.3$, $M = 1$, $Gr = 10$, and $t = 10$.

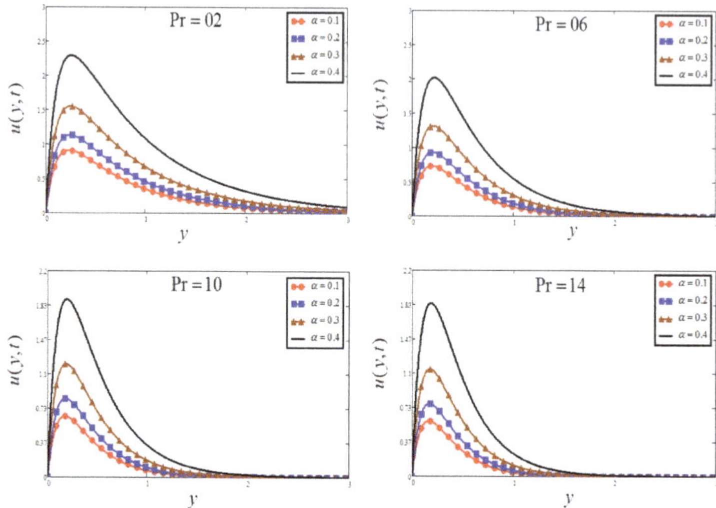

Figure 7. Plots of velocity for four different Pr values when $K = 0.5$, $\gamma = 0.3$, $M = 1$, $Gr = 10$, $N = 0.5$, and $t = 10$.

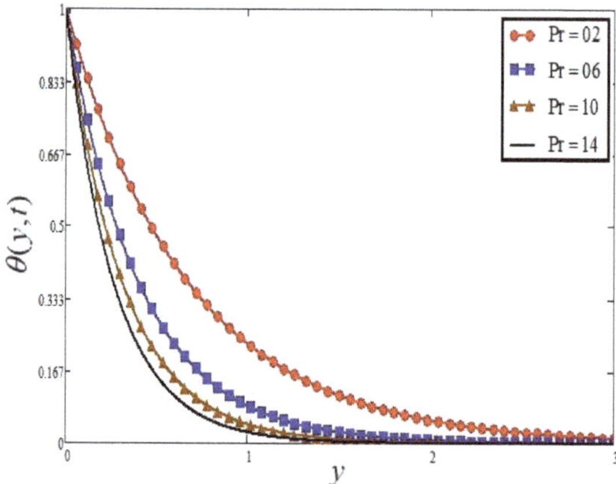

Figure 8. Plots of temperature for four different Pr values when $N = 0.5$, $\alpha = 0.1$ and $t = 10$.

Figures 9 and 10 illustrate the outcome of time t on temperature and velocity profiles. Figure 11 investigates the effect of time fraction derivative parameter α on temperature. It is investigated that the time-fractional derivative parameter controls the temperature profile.

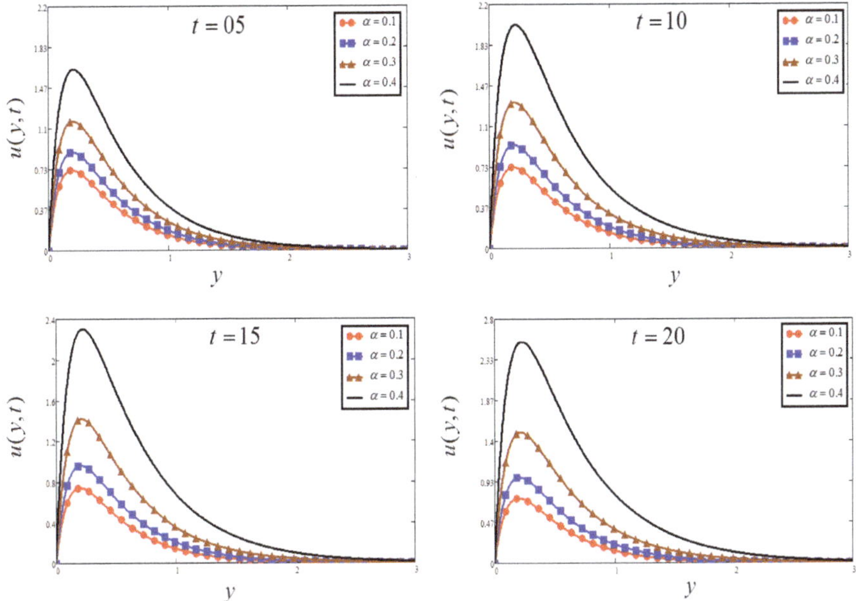

Figure 9. Plots of velocity for four different t values when $K = 0.5$, $\mathrm{Pr} = 7.2$, $\gamma = 0.3$, $M = 1$, $Gr = 10$, and $N = 0.5$.

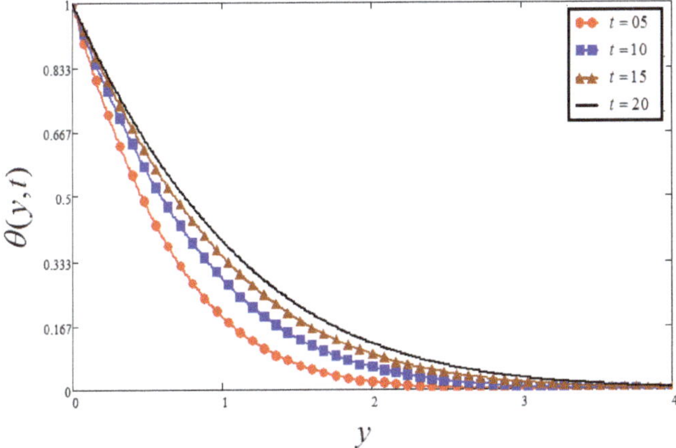

Figure 10. Plots of temperature for four different t values when $\mathrm{Pr} = 7.2$, $\alpha = 0.1$, and $N = 0.5$.

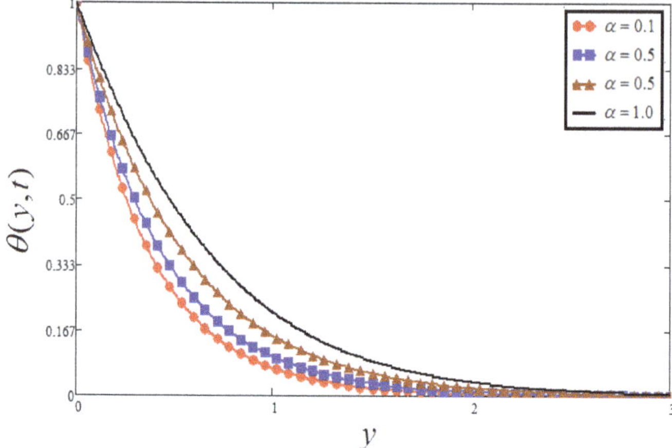

Figure 11. Plots of temperature for four different α values when $\text{Pr} = 7.2$, $N = 0.5$, and $t = 10$.

A comparison of the Atangana–Baleanu fractional model with an ordinary model is investigated in Figures 12 and 13 for velocity and temperature, respectively. For both cases, it is detected that the temperature and velocity profile for the Atangana–Baleanu fractional model is less than that of the ordinary model.

Note that all these graphs of velocity are plotted for the phase angle ωt equal to 90 degrees, and this value cosine is zero; therefore, all the graphs of velocity (Figures 2–7, Figure 9, Figure 12) have a unique pattern of velocity. That is, at the plate surface $y = 0$, the fluid is at rest or there is no motion in the fluid, and for large values of the independent variable y, the fluid velocity decays, and as y approaches infinity (further bigger values of y), the fluid motion disappears and velocity tends to zero. This physical pattern of the graphs of velocity agrees with the imposed condition on velocity given in Equation (6). Similarly, the unique style of all the graphs of temperature, that is temperature at $y = 0$, is 1, and far away from the plate surface; that is, for larger values of y, the temperature decays and tends to zero as y tends to infinity.

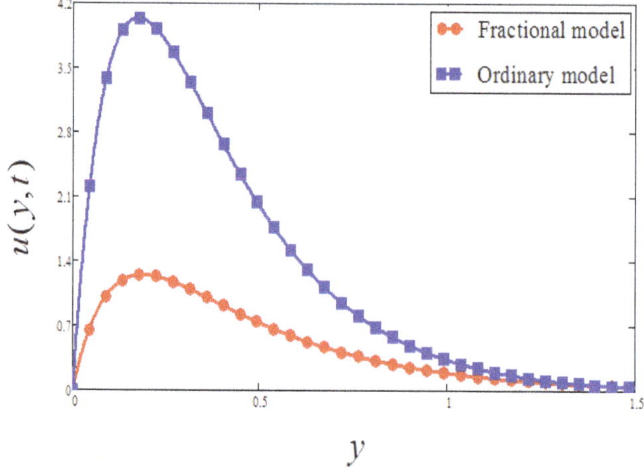

Figure 12. Comparison of fractional SA fluid and ordinary SA fluid (velocity) when $K = 0.5$, $\text{Pr} = 7.2$, $\gamma = 0.3$, $M = 1$, $Gr = 10\,N = 0.5$, and $t = 10$.

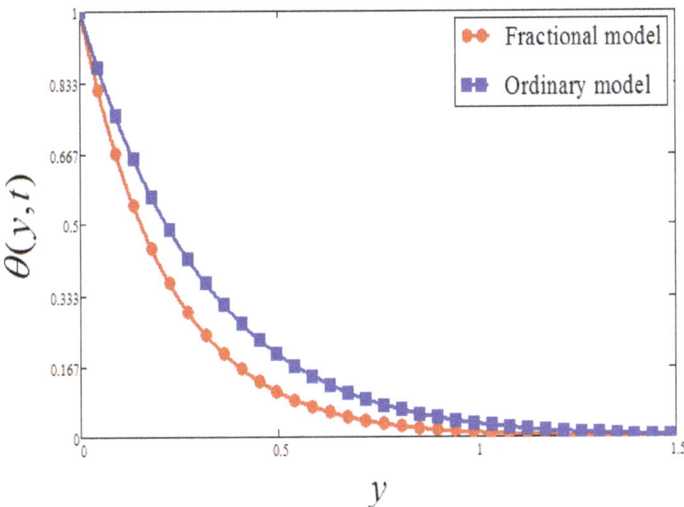

Figure 13. Comparison of fractional SA fluid and ordinary SA fluid (temperature) when $Pr = 7.2$, $\alpha = 0.1$, $N = 0.5$, and $t = 10$.

Tables 1 and 2 are plotted to show the variation of different parameters on Nusselt number and skin fraction. Table 1 clarified that the Nusselt number is increased when Pr, and N and are increased, while an increase in fractional derivative parameter α and t decreased the Nusselt number. The behavior of the present results is identical with the published results of Khan et al. [49] and Ali et al. [50]. Table 2 shows the impact of the deferent parameter on skin fraction. It is observed that t, α, Gr, and Pr have a positive (increasing) impact on the skin fraction, while M, N, γ, and K and show a negative (decreasing) impact on the skin fraction. This behavior of skin fraction against different parameters is identical with the published results of Mackolil and Mahanthesh [51].

Table 1. Variation of Nusselt number.

t	Pr	N	α	Nu
1	0.7	0.5	0.7	0.858
2				0.739
3				0.684
	7			2.297
	9			2.592
		1.5		1.641
		2.5		2.585
			0.8	0.795
			0.9	0.713

Table 2. Variation of skin-friction.

t	α	M	Pr	N	γ	Gr	K	C_f
0.3	0.5	0.4	0.7	0.5	0.5	0.1	1.5	0.366
1.3								0.510
1								0.476
	0.7							0.411
	0.9							0.488
		8.4						0.275
		19.4						0.218
			7.2					0.451
			9.2					0.916
				2.5				0.245
				3.5				0.168
					1.5			0.159
					2.5			0.123
						0.3		1.097
						0.5		1.829
							4.5	0.268
							10.5	0.194

6. Conclusions

In this attempt, the exact solution for the heat transfer analysis in unsteady MHD flow of fractional SA fluid past a vertical flat plate is obtained. Generalized Casson fluid model is obtained using the Atangana–Baleanu fractional derivative (ABFD) of the non-local and non-singular kernel. The exact solution for velocity and temperature are obtained by applying the Laplace transform method, and then, the influence of various embedded parameters are presented graphically. Some important outcomes are:

- ➢ Velocity rises for a large value of Gr, γ, and t.
- ➢ Velocity reduces for a large value of M, Pr, N, and K.
- ➢ Temperature is increased by increasing t and α, while decreasing with the increase of Pr.
- ➢ The temperature and velocity of the fractional fluid model converge faster compared to an ordinary fluid model.
- ➢ The Atangana–Baleanu fractional model reduced the velocity profile up to 45.76% and temperature profile up to 13.74% compared to an ordinary model.

Suggestions for future research work.

- ➢ The researchers extend this work for different kind of nanofluids.
- ➢ The authors also can take this model in different geometries.

Author Contributions: A.K., D.K., and I.K. modeled the problem; M.T., I.U., and A.M.A. solved the problem. P.T. and K.S.N. computed the results and wrote the main manuscript text. All authors reviewed the manuscript.

Funding: This research received no external funding

Conflicts of Interest: The authors declare no conflict of interest.

References

1. Casson, N. A flow equation for the pigment oil suspensions of the printing ink type. In *Rheology of Disperse Systems*; Pergamon Press: New York, NY, USA, 1959; Volume 84, p. e102.
2. Qasim, M.; Ahmad, B. Numerical solution for the Blasius flow in Casson fluid with viscous dissipation and convective boundary conditions. *Heat Transf. Res.* **2015**, *46*, 689–697. [CrossRef]
3. Shehzad, S.A.; Hayat, T.; Qasim, M.; Asghar, S. Effects of mass transfer on MHD flow of Casson fluid with chemical reaction and suction. *Braz. J. Chem. Eng.* **2013**, *30*, 187–195. [CrossRef]
4. Qasim, M.; Noreen, S. Heat transfer in the boundary layer flow of a Casson fluid over a permeable shrinking sheet with viscous dissipation. *Eur. Phys. J. Plus* **2014**, *129*, 7. [CrossRef]
5. Mukhopadhyay, S.; Mandal, I.C. Boundary layer flow and heat transfer of a Casson fluid past a symmetric porous wedge with surface heat flux. *Chin. Phys. B* **2014**, *23*, 044702. [CrossRef]
6. Siddiqui, A.M.; Farooq, A.A.; Rana, M.A. A mathematical model for the flow of a Casson fluid due to metachronal beating of cilia in a tube. *Sci. World J.* **2015**, *2015*, 487819. [CrossRef] [PubMed]
7. Hayat, T.; Shehzad, S.A.; Alsaedi, A.; Alhothuali, M.S. Mixed convection stagnation point flow of Casson fluid with convective boundary conditions. *Chin. Phys. Lett.* **2012**, *29*, 114704. [CrossRef]
8. Asjad, M.I.; Miraj, F.; Khan, I. Soret effects on simultaneous heat and mass transfer in MHD viscous fluid through a porous medium with uniform heat flux and Atangana-Baleanu fractional derivative approach. *Eur. Phys. J. Plus* **2018**, *133*, 224. [CrossRef]
9. Atangana, A.; Baleanu, D. Caputo-Fabrizio derivative applied to groundwater flow within confined aquifer. *J. Eng. Mech.* **2017**, *143*, D4016005. [CrossRef]
10. Atangana, A.; Goufo, E.F.D. The Caputo-Fabrizio fractional derivative applied to a singular perturbation problem. *Int. J. Math. Model. Numer. Optim.* **2019**, *9*, 241–253.
11. Aliyu, A.I.; Inc, M.; Yusuf, A.; Baleanu, D. A fractional model of vertical transmission and cure of vector-borne diseases pertaining to the Atangana–Baleanu fractional derivatives. *Chaos Solitons Fractals* **2018**, *116*, 268–277. [CrossRef]
12. Koca, I. Modelling the spread of Ebola virus with Atangana-Baleanu fractional operators. *Eur. Phys. J. Plus* **2018**, *133*, 100. [CrossRef]
13. Azhar, W.A.; Vieru, D.; Fetecau, C. Free convection flow of some fractional nanofluids over a moving vertical plate with uniform heat flux and heat source. *Phys. Fluids* **2017**, *29*, 082001. [CrossRef]
14. Fetecau, C.; Vieru, D.; Azhar, W. Natural convection flow of fractional nanofluids over an isothermal vertical plate with thermal radiation. *Appl. Sci.* **2017**, *7*, 247. [CrossRef]
15. Sheikh, N.A.; Ali, F.; Khan, I.; Gohar, M.; Saqib, M. On the applications of nanofluids to enhance the performance of solar collectors: A comparative analysis of Atangana–Baleanu and Caputo–Fabrizio fractional models. *Eur. Phy. J. Plus* **2017**, *132*, 540. [CrossRef]
16. Karaagac, B. Two step Adams Bashforth method for time fractional Tricomi equation with non-local and non-singular Kernel. *Chaos Solitons Fractals* **2019**, *128*, 234–241. [CrossRef]
17. Ali, F.; Saqib, M.; Khan, I.; Sheikh, N.A. Application of Caputo-Fabrizio derivatives to MHD free convection flow of generalized Walters'-B fluid model. *Eur. Phys. J. Plus* **2016**, *131*, 377. [CrossRef]
18. Saqib, M.; Khan, I.; Shafie, S. Application of Atangana–Baleanu fractional derivative to MHD channel flow of CMC-based-CNT's nanofluid through a porous medium. *Chaos Solitons Fractals* **2018**, *116*, 79–85. [CrossRef]
19. Abro, K.A.; Khan, I.; Tassaddiq, A. Application of Atangana-Baleanu fractional derivative to convection flow of MHD Maxwell fluid in a porous medium over a vertical plate. *Math. Model. Nat. Phenom.* **2018**, *13*, 1. [CrossRef]
20. Abro, K.A.; Khan, I.; Nisar, K.S.; Alsagri, A.S. Efects of carbon nanotubes on Magnetohydrodynamic flow of methanol based nanofluids via Atangana-Baleanu and Caputo-fabrizio fractional derivatives. *Therm. Sci.* **2019**, *23*, 883–898.
21. Yavuz, M.; Özdemir, N. Comparing the new fractional derivative operators involving exponential and Mittag-Leffler kernel. *Discret. Contin. Dyn. Syst. S* **2019**, *13*, 995–1006. [CrossRef]
22. Imran, M.A.; Aleem, M.; Riaz, M.B.; Ali, R.; Khan, I. A comprehensive report on convective flow of fractional (ABC) and (CF) MHD viscous fluid subject to generalized boundary conditions. *Chaos Solitons Fractals* **2019**, *118*, 274–289. [CrossRef]

23. Abro, K.A.; Khan, I. MHD flow of fractional Newtonian fluid embedded in a porous medium via Atangana-Baleanu fractional derivatives. *Discret. Contin. Dyn. Syst. S* **2019**. [CrossRef]
24. Wenchang, T.; Mingyu, X. Unsteady flows of a generalized second grade fluid with the fractional derivative model between two parallel plates. *Acta Mech.* **2004**, *20*, 471–476. [CrossRef]
25. Xu, M.; Tan, W. Theoretical analysis of the velocity field, stress field and vortex sheet of generalized second order fluid with fractional anomalous diffusion. *Sci. China Ser. A Math.* **2001**, *44*, 1387–1399. [CrossRef]
26. Shen, F.; Tan, W.; Zhao, Y.; Masuoka, T. The Rayleigh–Stokes problem for a heated generalized second grade fluid with fractional derivative model. *Nonlinear Anal. Real World Appl.* **2006**, *7*, 1072–1080. [CrossRef]
27. Mahmood, A.; Parveen, S.; Ara, A.; Khan, N.A. Exact analytic solutions for the unsteady flow of a non-Newtonian fluid between two cylinders with fractional derivative model. *Commun. Nonlinear Sci. Numer. Simul.* **2009**, *14*, 3309–3319. [CrossRef]
28. Shen, M.; Chen, S.; Liu, F. Unsteady MHD flow and heat transfer of fractional Maxwell viscoelastic nanofluid with Cattaneo heat flux and different particle shapes. *Chin. J. Phys.* **2018**, *56*, 1199–1211. [CrossRef]
29. Zhang, Y.; Zhao, H.; Liu, F.; Bai, Y. Analytical and numerical solutions of the unsteady 2D flow of MHD fractional Maxwell fluid induced by variable pressure gradient. *Comput. Math. Appl.* **2018**, *75*, 965–980. [CrossRef]
30. Aman, S.; Al-Mdallal, Q.; Khan, I. Heat transfer and second order slip effect on MHD flow of fractional Maxwell fluid in a porous medium. *J. King Saud Univ. Sci.* **2018**. [CrossRef]
31. Jan, S.A.A.; Ali, F.; Sheikh, N.A.; Khan, I.; Saqib, M.; Gohar, M. Engine oil based generalized brinkman-type nano-liquid with molybdenum disulphide nanoparticles of spherical shape: Atangana-Baleanu fractional model. *Numer. Methods Part. Differ. Equ.* **2018**, *34*, 1472–1488. [CrossRef]
32. Owolabi, K.M.; Atangana, A. On the formulation of Adams-Bashforth scheme with Atangana-Baleanu-Caputo fractional derivative to model chaotic problems. *Chaos Interdiscip. J. Nonlinear Sci.* **2019**, *29*, 023111. [CrossRef] [PubMed]
33. Saad, K.M.; Khader, M.M.; Gómez-Aguilar, J.F.; Baleanu, D. Numerical solutions of the fractional Fisher's type equations with Atangana-Baleanu fractional derivative by using spectral collocation methods. *Chaos Interdiscip. J. Nonlinear Sci.* **2019**, *29*, 023116. [CrossRef] [PubMed]
34. Saqib, M.; Khan, I.; Shafie, S. New Direction of Atangana–Baleanu Fractional Derivative with Mittag-Leffler Kernel for Non-Newtonian Channel Flow. In *Fractional Derivatives with Mittag-Leffler Kernel*; Springer: Cham, Switzerland, 2019; pp. 253–268.
35. Abro, K.A.; Chandio, A.D.; Abro, I.A.; Khan, I. Dual thermal analysis of magnetohydrodynamic flow of nanofluids via modern approaches of Caputo–Fabrizio and Atangana–Baleanu fractional derivatives embedded in porous medium. *J. Therm. Anal. Calorim.* **2019**, *135*, 2197–2207. [CrossRef]
36. Hristov, J. On the Atangana–Baleanu Derivative and Its Relation to the Fading Memory Concept: The Diffusion Equation Formulation. In *Fractional Derivatives with Mittag-Leffler Kernel*; Springer: Cham, Switzerland, 2019; pp. 175–193.
37. Khan, I.; Alqahtani, A.M. MHD Nanofluids in a Permeable Channel with Porosity. *Symmetry* **2019**, *11*, 378. [CrossRef]
38. Asif, M.; Ul Haq, S.; Islam, S.; Abdullah Alkanhal, T.; Khan, Z.A.; Khan, I.; Nisar, K.S. Unsteady Flow of Fractional Fluid between Two Parallel Walls with Arbitrary Wall Shear Stress Using Caputo–Fabrizio Derivative. *Symmetry* **2019**, *11*, 449. [CrossRef]
39. Ullah, I.; Abdullah Alkanhal, T.; Shafie, S.; Nisar, K.S.; Khan, I.; Makinde, O.D. MHD Slip Flow of Casson Fluid along a Nonlinear Permeable Stretching Cylinder Saturated in a Porous Medium with Chemical Reaction, Viscous Dissipation, and Heat Generation/Absorption. *Symmetry* **2019**, *11*, 531. [CrossRef]
40. Khan, A.; Ali Abro, K.; Tassaddiq, A.; Khan, I. Atangana–Baleanu and Caputo Fabrizio analysis of fractional derivatives for heat and mass transfer of second grade fluids over a vertical plate: A comparative study. *Entropy* **2017**, *19*, 279. [CrossRef]
41. Gul, T.; Khan, M.A.; Noman, W.; Khan, I.; Abdullah Alkanhal, T.; Tlili, I. Fractional order forced convection carbon nanotube nanofluid flow passing over a thin needle. *Symmetry* **2019**, *11*, 312. [CrossRef]
42. Atangana, A.; Alqahtani, R. Modelling the spread of river blindness disease via the Caputo fractional derivative and the beta-derivative. *Entropy* **2016**, *18*, 40. [CrossRef]
43. Gómez-Aguilar, J.; Atangana, A. Fractional Derivatives with the Power-Law and the Mittag–Leffler Kernel Applied to the Nonlinear Baggs–Freedman Model. *Fractal Fract.* **2019**, *2*, 10. [CrossRef]

44. Muhammad Altaf, K.; Atangana, A. Dynamics of Ebola Disease in the Framework of Different Fractional Derivatives. *Entropy* **2019**, *21*, 303. [CrossRef]
45. Khan, H.; Shah, R.; Baleanu, D.; Kumam, P.; Arif, M. Analytical Solution of Fractional-Order Hyperbolic Telegraph Equation, Using Natural Transform Decomposition Method. *Electronics* **2019**, *8*, 1015. [CrossRef]
46. Khalid, A.; Khan, I.; Khan, A.; Shafie, S. Unsteady MHD free convection flow of Casson fluid past over an oscillating vertical plate embedded in a porous medium. *Eng. Sci. Technol. Int. J.* **2015**, *18*, 309–317. [CrossRef]
47. Makinde, O.D.; Mhone, P.Y. Heat transfer to MHD oscillatory flow in a channel filled with porous medium. *Rom. J. Phys.* **2015**, *931*, 9–10.
48. Cogley, A.C.; Gilles, S.E.; Vincenti, W.G. Differential approximation for radiative transfer in a nongrey gas near equilibrium. *AIAA J.* **1968**, *6*, 551–553. [CrossRef]
49. Khan, A.; Khan, D.; Khan, I.; Ali, F.; ul Karim, F.; Imran, M. MHD flow of Sodium Alginate-based Casson type nanofluid passing through a porous medium with Newtonian heating. *Sci. Rep.* **2018**, *8*, 8645. [CrossRef]
50. Ali, F.; Sheikh, N.A.; Khan, I.; Saqib, M. Solutions with Wright function for time fractional free convection flow of Casson fluid. *Arab. J. Sci. Eng.* **2017**, *42*, 2565–2572. [CrossRef]
51. Mackolil, J.; Mahanthesh, B. Exact and Statistical computations of radiated flow of Nano and Casson fluids under heat and mass flux conditions. *J. Comput. Des. Eng.* **2019**. [CrossRef]
52. Caputo, M.; Fabrizio, M. A new definition of fractional derivative without singular kernel. *Progr. Fract. Differ. Appl.* **2015**, *1*, 1–13.
53. Atangana, A.; Dumitru, B. New fractional derivatives with nonlocal and non-sin- gular kernel: Theory and application to heat transfer model. *Therm. Sci.* **2016**, *20*, 763–769. [CrossRef]

Article

Experimental Study on Plasma Flow Control of Symmetric Flying Wing Based on Two Kinds of Scaling Models

Like Xie [1], Hua Liang [1,*], Menghu Han [2], Zhongguo Niu [3], Biao Wei [1], Zhi Su [1] and Bingliang Tang [1]

1 Science and technology on plasma dynamics laboratory, Air Force Engineering University, Xi'an 710038, China; xielike1011@139.com (L.X.); rourou39153118@126.com (B.W.); zhisuemail@163.com (Z.S.); tangbingliang123@163.com (B.T.)
2 Air Force Harbin Flight Academy, Harbin 150000, China; gratetigerhan@163.com
3 AVIC Aerodynaiviics Research Institute, Harbin 150001, China; nzg9527@163.com
* Correspondence: lianghua82702@126.com

Received: 8 September 2019; Accepted: 6 October 2019; Published: 9 October 2019

Abstract: The symmetric flying wing has a simple structure and a high lift-to-drag ratio. Due to its complicated surface design, the flow field flowing through its surface is also complex and variable, and the three-dimensional effect is obvious. In order to verify the effect of microsecond pulse plasma flow control on the symmetric flying wing, two different sizes of scaling models were selected. The discharge energy was analyzed, and the force and moment characteristics of the two flying wings and the particle image velocimetry (PIV) results on their surface flow field were compared to obtain the following conclusions. The microsecond pulse surface dielectric barrier discharge energy density is independent of the actuator length but increases with the actuation voltage. After actuation, the stall angle of attack of the small flying wing is delayed by 4°, the maximum lift coefficient is increased by 30.9%, and the drag coefficient can be reduced by 17.3%. After the large flying wing is actuated, the stall angle of attack is delayed by 4°, the maximum lift coefficient is increased by 15.1%, but the drag coefficient is increased. The test results of PIV in the flow field of different sections indicate that the stall separation on the surface of the symmetric flying wing starts first from the outer side, and then the separation area begins to appear on the inner side as the angle of attack increases.

Keywords: symmetric flying wing; plasma flow control; energy; stall; dimensionless frequency; particle image velocimetry

---

1. Introduction

Compared with the conventional layout aircraft, the flying wing has no flat tail and vertical tail, and the wing and the fuselage are highly integrated, so it is subjected to greater lift and less drag [1,2]. High aerodynamic efficiency, light weight structure, large loading space, and good stealth performance are the characteristics of the flying wing [3,4]. As the use of the air field is getting deeper, in future development, the aircraft of the flying wing layout seems to be the seed candidate [5]. However, just like the aircraft of the conventional layout, after the angle of attack exceeds the critical value, the surface boundary layer of the flying wing will be separated, the lift will drop sharply, and its maneuverability will deteriorate [6]. The flying wing is similar to a delta wing aircraft. At moderate angles of attack, both sides of the wing are subjected to a pair of asymmetric flow axial vortex systems. At high angles of attack, the left and right asymmetrical vortices are formed in the leeward area of the fore body, resulting in yawing and rolling moments, affecting its maneuverability [7,8].

Through the technical means of flow control, improving the stall characteristics of the wing surface and then improving the aerodynamic performance of the aircraft at a high angle of attack have become an important research direction in the field of aviation [9,10]. At present, there are many research methods to improve the flow field by active flow control, such as acoustic excitation [11], oscillating jet [12], synthetic jet [13], blow and suction air [14], microelectromechanical system (MEMS) [15], and plasma flow control [16]. Plasma flow control has many types of research in the field of flow control because of its simple structure, no moving parts, rapid response and flexible function [17–20]. The plasma flow control technology is used to delay the flow separation under the low speed condition to improve the stall angle of attack and is also used to reduce the shock angle under the high speed condition to weaken the shock wave [21,22]. The common methods of using plasma to control the flow field are alternating current dielectric barrier discharge (ac-DBD) and nanosecond dielectric barrier discharge (ns-DBD). When the incoming flow velocity is low (less than 0.4 Ma), the ac-DBD control effect is good, mainly by generating body force in the flow field, inducing the near-wall airflow to accelerate, thereby suppressing flow separation [23]. When the incoming flow velocity is higher, it is more suitable for ns-DBD, which produces instantaneous heating in the flow field, and can effectively suppress airfoil flow separation at a large flow velocity (0.8 Ma) [24,25].

Servant et al. proposed a method for optimizing aerodynamic design, which was successfully applied to the design of 3D flying wing by parameterizing complex surfaces and removing invalid noise data by partial differential equation (PED) and response surface methodology (RSM) methods [26]. Huber et al. studied and analyzed the interaction between the vortex structure and the vortex on the upper surface of the DLR-F17 wing model by numerical simulation to evaluate the aerodynamic behavior in the flight envelope in order to study the requirements for stability and maneuverability [27]. Xu et al. used the synthetic jet control method to improve the lateral aerodynamic characteristics of the asymmetric vortex of the flying wing and control the lateral aerodynamic force through the interaction of the asymmetric vortex [28]. Han et al. studied the actuation frequency of ns-DBD to improve the aerodynamic performance of the flying wing under different Reynolds numbers. The results show that there is an optimal actuation frequency and it is more effective to delay the breakdown of the leading edge vortex at low frequencies [29]. Yao et al. studied the optimal actuation position of the ns-DBD to improve the aerodynamic performance of the flying wing. When plasma actuation is applied to the leading edge of the inner and middle wing, the control effect is obvious. The actuation effect is best when the corresponding dimensionless frequency is 1 under low frequency conditions, and the variation of actuation voltage has little effect on the effect of the lift increasing [30].

In this paper, the microsecond pulse surface dielectric barrier discharge (μs-DBD) actuation method is used to carry out flow control experiment on two flying wings with different sizes. The actuation effect of μs-DBD plasma flow control on improving the aerodynamic performance of the flying wing is compared. The energy of μs-DBD is analyzed in this paper. The optimal actuation frequency is studied from the dimensionless frequency by force measurement, and the variation of the flow field on the surface of the symmetric flying wing is analyzed by particle image velocimetry (PIV).

2. Experimental Setup

The experiment was carried out in Harbin AVIC Aerodynaiviics Research Institute, and two wind tunnels were selected for experiments. The wind tunnel in Figure 1 is FL-5, which was an open low-speed wind tunnel with an experimental section size of 1.5 m × 1.95 m (diameter × length), a maximum wind speed of 53 m/s, and a turbulence intensity of 1%. Figure 2 shows the FL-51 wind tunnel, which is a single-loop continuous wind tunnel with replaceable open/closed ports. The test section was 11 m × 4.5 m × 3.5 m (length × width × height), and the maximum airflow velocity of the closed experimental section was 100 m/s; the turbulence intensity was 0.10%.

Figure 1. FL-5 wind tunnel open test section.

Figure 2. FL-51 wind tunnel closed test section.

The two kinds of scaling flying wing models are shown in Figures 3 and 4. The small flying wing was double "W" shape, the leading edge sweep angle was 35°, the spanwise length was 0.953 m, the fuselage length was 0.386 m, and the average aerodynamic chord length was 0.214 m. The geometric parameters of the large flying wing were 2.5 times those of the small flying wing.

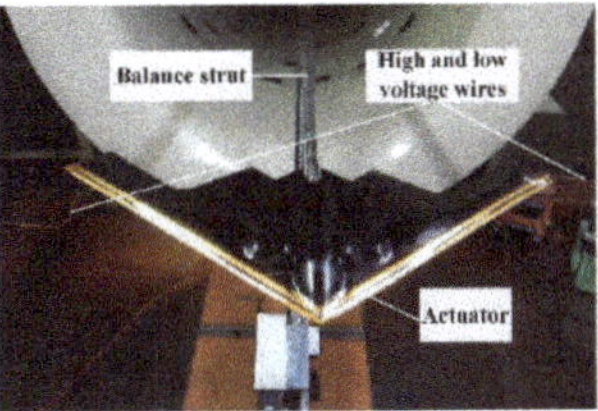

Figure 3. Small flying wing model.

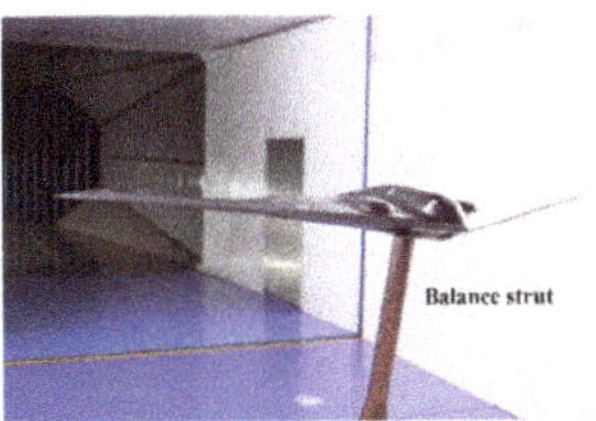

Figure 4. Large flying wing model.

An adjustable parameter microsecond pulse power supply was used for the power supply of the plasma actuator. The schematic circuit diagram is shown in Figure 5. The input AC voltage was between 0–220 V, and a relatively stable DC voltage was obtained through the full bridge rectifier (BR) and the voltage regulator capacitor C_1. Then, it charged the primary energy storage capacitor C_2 through the charging inductor L and the diode D_1. The voltage of the primary storage capacitor C_2 was about 1.4 times of the input voltage by C_1. When the semiconductor switching insulated gate bipolar transistor (IGBT) was working, C_2 performs pulse discharge. A positive high voltage pulse was generated by potential transformer (PT) boost and diode D_3 unidirectional conduction. The output actuation voltage (peak voltage) was adjustable from 0 to 10 kV, and the pulse frequency was adjustable from 0 to 2 kHz. The waveform of the no-load maximum output voltage is shown in Figure 6. Each single pulse duration time was microsecond magnitude. When the circuit exported a high-voltage pulse, the actuator performed a single pulse discharge. The number of high-voltage pulses produced in a second was the pulse repetition frequency.

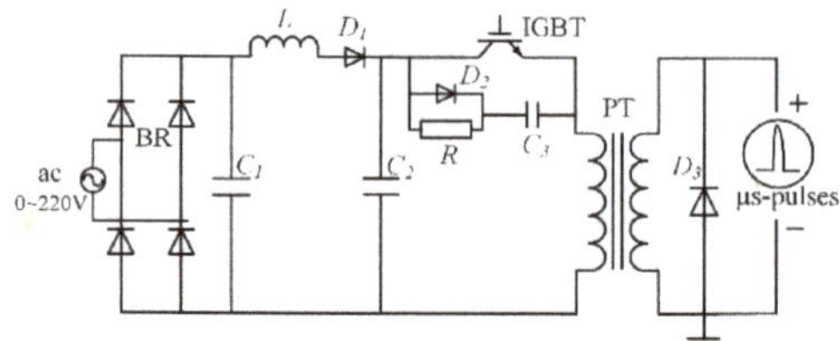

Figure 5. Microsecond pulse power supply schematic circuit diagram.

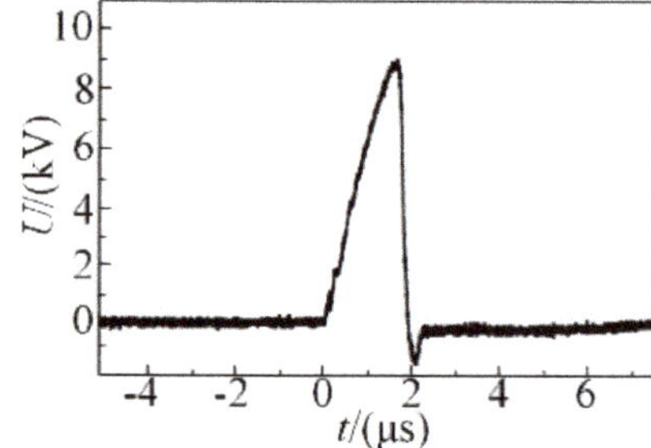

Figure 6. Waveform of no-load maximum output voltage (U).

The actuator in the experiment was composed of exposed electrode, covered electrode, and insulation dielectric, as shown in Figure 7. The covered electrode was 3 mm wide and 0.06 mm thick and its lower edge was aligned with the leading edge of the flying wing. The exposed electrode was 2 mm wide and 0.06 mm thick and the upper edge was aligned with the leading edge of the symmetric flying wing. The insulation dielectric was made of polyimide, with a thickness of 0.18 mm, a dielectric constant of 3.5, and could withstand a high voltage of 15 kV. It was placed between the two electrodes to separate them.

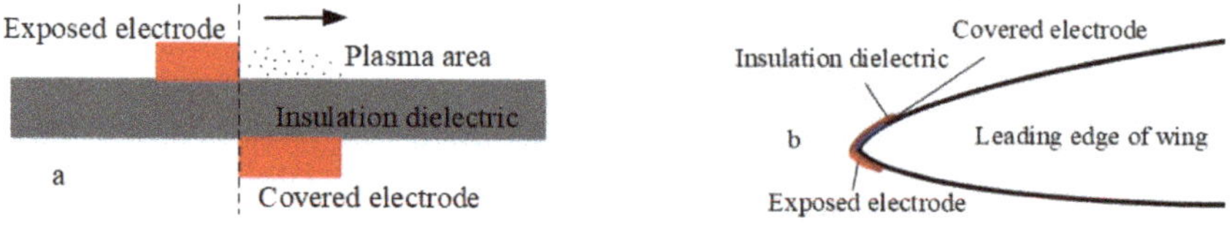

Figure 7. Surface dielectric barrier discharge actuator. (**a**): structure; (**b**): position layout.

The experiment used the PIV velocity measurement system to measure the flow field parameters by the non-contact method and study the characteristics of the plasma actuation inducing flow field. Figure 8 shows the placement of the PIV test system. The laser was an integrated double Nd:YAG laser with a single pulse energy of up to 500 mJ and a wavelength of 532 nm. The CCD camera has a pixel resolution of 16 MP (4904 × 3280 pixels), a grayscale resolution of 12 bits, and an image acquisition frequency of 3.2 fps. The system was synchronized using a programmable time controller (PTU) with a control signal time resolution of 0.3 ns. PIV data acquisition and processing were carried out by Davis 8.3 software. The tracer particles were produced by pressure atomization. The particle medium was olive oil, and the tracer particles were about 1 μm in diameter. In the time scale of microseconds, two images were taken in succession. The images captured the oil mist particles applied to the flow field. The position of the particles in the second image was obtained by image correlation theory, so as to calculate the velocity of particles in the flow field by the position of particles in two images.

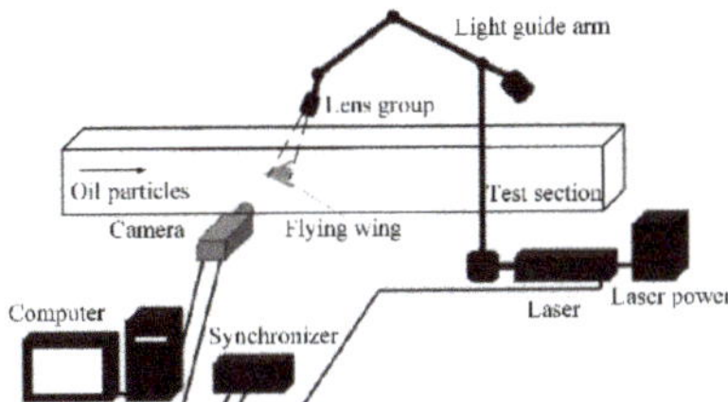

Figure 8. Particle image velocimetry (PIV) test system layout diagram.

The wind tunnel force measurement was measured by a rod-type strain balance, and the model was supported by a single-strut. The test photographs are shown as Figures 3 and 4. The rod-type six-component strain balance was installed inside the model, and the model is connected with the abdominal support rod through a balance. VXI (VMEbus Extension for Instrumentation) data acquisition system was used for balance force data acquisition.

The small and the large flying wings were tested in two different size wind tunnels, FL-5 and FL-51, respectively. The wind speed of FL-5 was a constant 30 m/s. The wind speed of FL-51 was a constant 75 m/s. The test conditions are shown in Table 1.

Table 1. The test conditions.

Model	Aspect Ratio	Wind Tunnel	Velocity	Force Measurement	PIV Measurement
Small flying wing	2.47	FL-5	30 m/s	Lift, drag, and moment	Sections 1–3
Large flying wing	2.47	FL-51	75 m/s	Lift, drag, and moment	Section 1

3. Experimental Results

3.1. Analysis of Discharge and Energy

3.1.1. Effect of Actuation Length on Discharge Energy

Figure 9a shows the relationship between the length of the actuator and the instantaneous power of the discharge. The corresponding actuator lengths of P1, P2, P3, and P4 were 27 cm, 40 cm, 56 cm, and 112 cm, respectively. It can be seen that with the increase of the length of the actuator, the peak power of the discharge increased. However, the peak power was not proportional to the length of the actuator. As shown in Figure 9b, when the length of the actuator increased from P1 to P2, the peak power changed greatly. The length was increased by 48%, while the peak power was increased by 33%. From P2 to P4, the length was increased by 180% but the peak power was increased by only 47%. Figure 9b also shows the relationship between discharge single pulse energy and actuator length. It can be seen that the total change trend of discharge energy was approximately proportional to the increase of length. The calculation result shows that the discharge energy densities per unit length of P1, P2, P3, and P4 were 17 mJ/m, 19 mJ/m, 18 mJ/m, and 16 mJ/m, respectively, so the pulse energy per unit length was independent of the length of the actuator.

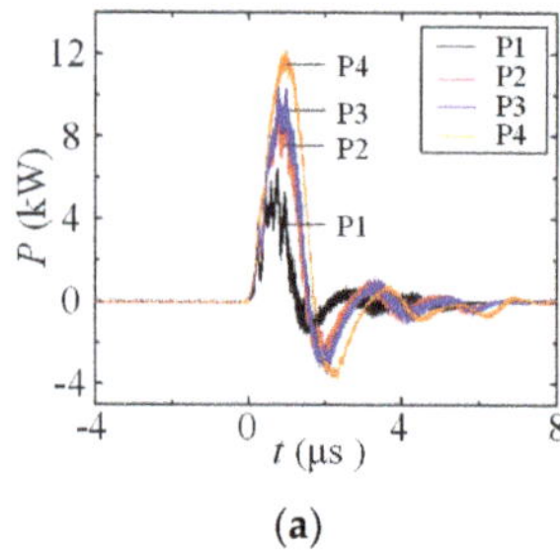 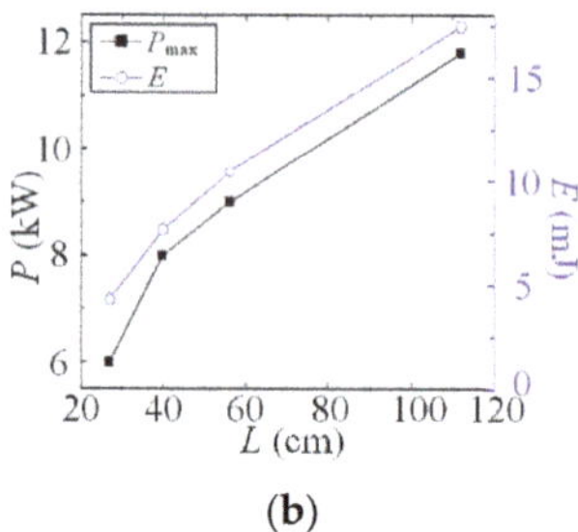

Figure 9. Variation of instantaneous power and energy at different actuator lengths. (**a**): comparison of instantaneous power; (**b**) comparison of energy.

3.1.2. Effect of Actuation Voltage on Discharge Energy

Figure 10a shows the relationship between the actuation voltage and the instantaneous power of the discharge. As the discharge voltage increased, the discharge power increased, the instantaneous power peak also increased, and the time scale of the discharge power remained substantially unchanged. The actuation voltage was below 8 kV, the power was small, and the discharge intensity was relatively low. When the actuation voltage was above 8 kV, the discharge intensity was high. This is because the electric field strength increased and more ionized positive and negative particles were directed to the two poles of the actuation under the action of the electric field force. The current increased more, and the power output power increased significantly. Figure 10b also shows the relationship between the actuation voltage and the peak power and discharge energy. It can be seen that the variation of discharge energy was similar to that of peak power. As the actuation voltage increased, both the peak power and the discharge energy increased.

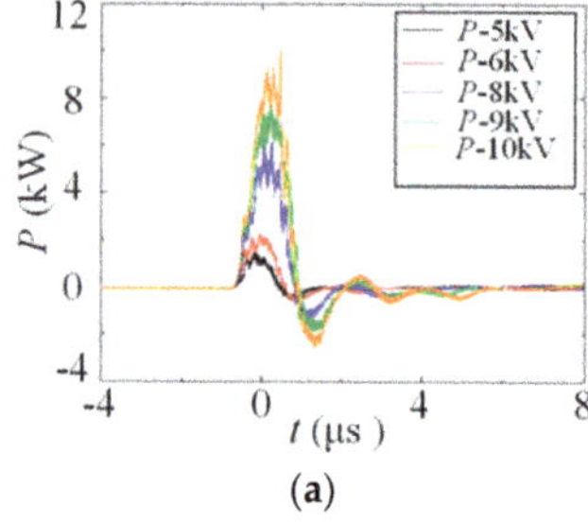 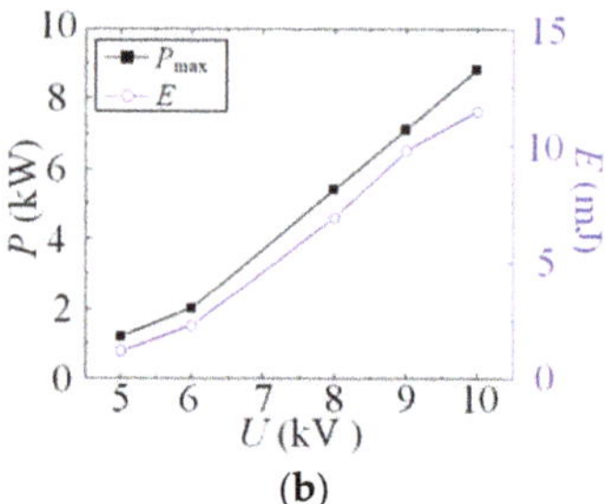

Figure 10. Variation of instantaneous power and energy at different actuator voltages. (**a**): comparison of instantaneous power; (**b**) comparison of energy.

3.1.3. Effect of Actuation Frequency on Discharge Energy

Figure 11 shows the relationship between the actuation frequency and the discharge instantaneous power. It can be seen that when the actuation voltage was the same, the discharge instantaneous power curves at different actuation frequencies coincided well, indicating that the discharge power was basically unaffected by the actuation frequency, and the single pulse energy of the discharge was independent of the actuation frequency. It also shows that the effect of frequency change on the flow field was not affected by the discharge, but the result of the unsteady disturbance was caused by different actuation frequencies in the flow field.

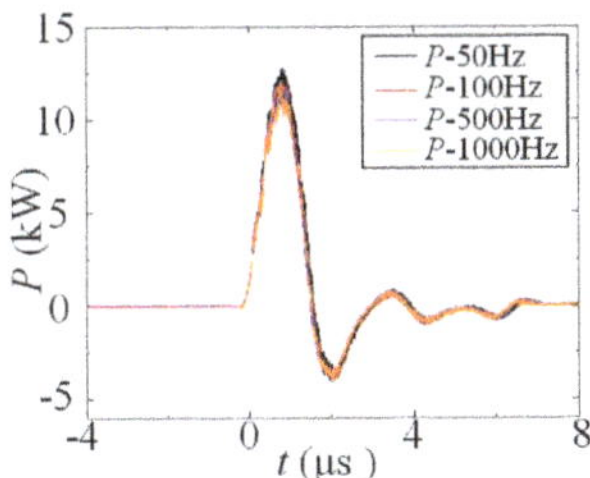

Figure 11. Variation of instantaneous power at different actuator frequencies.

3.2. Force Characteristics

When the airflow flows through the surface of the symmetric flying wing at a certain speed, the velocity in its vertical direction is different, which is the most obvious near the wall, and there is a thin boundary layer. When the angle of attack of the symmetric flying wing exceeds the critical value, the boundary layer is separated, and the separation vortex seriously disturbs the flow field, thus the aerodynamic performance of the symmetric flying wing is poor. The plasma flow control effect of two flying wing models under different actuation frequencies is studied in this paper. According to the similarity criterion of Strouhal, the dimensionless frequency formula is $F^+ = f \times l/U_\infty$, in which f is the output actuation frequency of the power supply, l is the average aerodynamic chord length of the symmetric flying wing model, and U_∞ is the incoming flow velocity. In order to ensure that the dimensionless frequency was consistent, the incoming flow velocity of the large flying wing was 2.5 times that of the small flying wing. When the incoming flow velocity of the small flying wing was 30 m/s, the incoming flow velocity the large flying wing was 75 m/s. The dimensional frequencies F^+ were 0.36, 0.71, 1.07, 2.14, 3.57 and 7.13 with corresponding actuation frequencies f of 50 Hz, 100 Hz, 150 Hz, 300 Hz, 500 Hz, and 1000 Hz, respectively, to study the effect of the unsteady actuation on the lift and drag of the two-scale flying wing model.

Figure 12 shows the lift, drag, and pitch moment coefficient curves of the small flying wing at different dimensionless frequencies when the incoming flow velocity was 30 m/s. Figure 12a shows that when the dimensionless frequency was between 0.71 and 2.14, the effect of actuation increase is obvious. The optimal dimensionless frequency was 1.07 and the maximum lift coefficient increased from 0.81 to 1.06, which was increased by 30.9%. The stall angle of attack was delayed from 15° to 19°. The maximum lift coefficient increased by 28.5% and 25.3% under the dimensional frequency actuations of 0.71 and 2.14, respectively, and the stall angle of attack was delayed by 4°. When the dimensionless frequency was greater than 1, with the increase of the corresponding actuation frequency, the increasing effect was reduced and the ability to suppress the flow separation was reduced. Figure 12b shows the variation trend of the drag coefficient with the dimensionless frequency. When the angle of attack was between 12° and 17°, the drag coefficient was gradually reduced with the increase of the dimensionless frequency, indicating that the drag reduction effect was good at high frequency actuation when the angle of attack was greater than 12°. For low frequency actuation, it can be seen that when the angle of attack was greater than 17°, the drag coefficient was greater than that without actuation, which indicates that the low-frequency actuation had a negative effect on the drag when the attack angle was greater than 17°, and the drag increased. At the angle of attack of 15°, the dimensionless frequency was 1.07, the lift coefficient could be increased by 15.1%, the drag coefficient could be reduced by 17.3%, and the effect of lift increasing and drag decreasing was good.

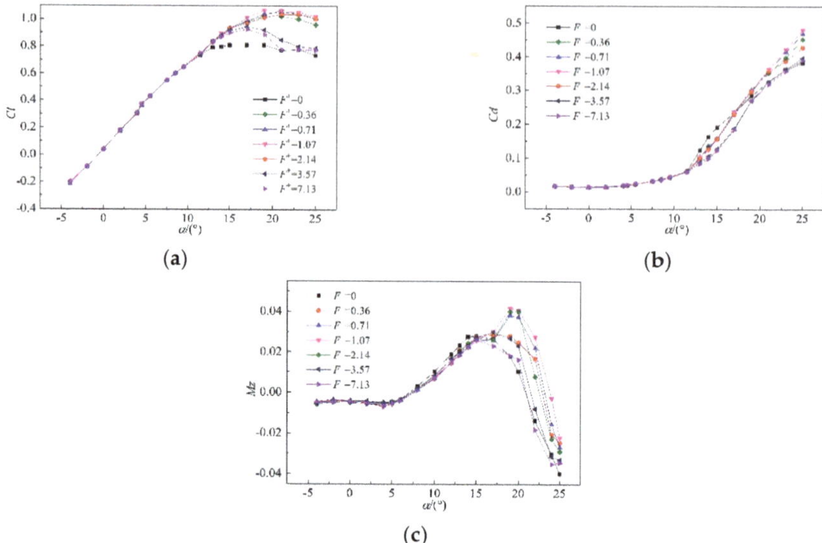

Figure 12. Lift, drag, and pitch moment coefficient curves of the small flying wing at different dimensionless frequencies (U_∞ = 30 m/s, FL-5). (**a**): Lift coefficient curves; (**b**) Drag coefficient curves; (**c**) Pitch moment coefficient curves.

Figure 13 shows the lift, drag, and pitch moment coefficient curves of the large flying wing at different dimensionless frequencies when the flow velocity was 75 m/s. With the increase of model size and incoming flow velocity, the changing trend of lift coefficient curve without actuation corresponding $F^+ = 0$ was the same as that at 30 m/s, and the stall angle of attack was 15°. The microsecond pulse plasma actuation could also improve the aerodynamic performance of the large flying wing surface and delay the stall separation. It can be seen from Figure 13a that the actuation effect was more obvious when the dimensionless frequency was between 0.71 and 2.14, and the optimal dimensionless frequency was 1.07. The maximum lift coefficients corresponding to the dimensionless frequencies of 0.71, 1.07, and 2.14 were increased by 10.4%, 15.1%, and 10%, respectively, and the stall angles of attack were delayed by 3°, 4° and 3°, respectively. As can be seen in conjunction with Figures 12a and 13a, the effect of high frequency actuation to improve the aerodynamic performance of the symmetric flying wing surface was lower than that of low frequency. It can be seen from Figure 13b that the air resistance of the large flying wing was affected by the actuation frequency. When the actuation was performed under low frequency condition, the drag coefficient was greater than that without plasma actuation, but the drag coefficient under the high frequency actuation condition was lower than that of no plasma actuation.

For different sizes of flying wings, the trends of the lift and drag coefficient were basically consistent at different actuation frequencies, with the same unit intensity actuation carried out at different incoming flow velocities. When the dimensionless frequency F^+ was equal to 1, the actuation effect was most obvious and the similarity criterion was matched. The increase of the maximum lift coefficient of the large flying wing was less than that of the small flying wing. Because the incoming flow velocity increased, the separation vorticity in the flow field became stronger. It was necessary to inject more unit intensity energy into the flow field to suppress the separation of the surface boundary layer.

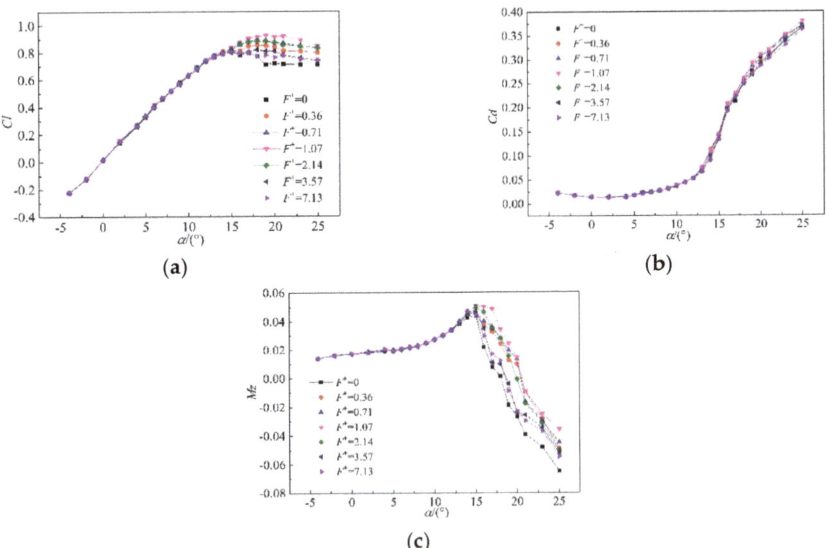

Figure 13. Lift, drag, and pitch moment coefficient curves of the large flying wing at different dimensionless frequencies (U_∞ = 75 m/s, FL-51). (**a**): Lift coefficient curves; (**b**) Drag coefficient curves; (**c**) Pitch moment coefficient curves.

Figure 12c shows the pitch moment coefficient curve at different dimensionless frequencies with an incoming flow velocity of 30 m/s. When there was no plasma actuation, the pitch moment coefficient increased as the angle of attack increased under the stall angle of attack. When the stall angle of attack was increased to 15°, the symmetric flying wing was completely stalled and the pitch moment reached the maximum value, with the two corresponding to each other. After the actuation was applied, the pitch moment coefficient could still be increased until the angle of attack was 19°. Since the actuation still had the lift increment after the 15° stall angle of attack and the pitch moment of the symmetric flying wing continued to increase, the flow separation was effectively suppressed. The dimensionless frequencies were 0.71, 1.07 and 2.14; the pitch moment coefficient increased significantly, indicating that the lift increment was significant. Figure 13c shows the pitch moment coefficient curve at different dimensionless frequencies with an incoming flow velocity of 75 m/s. When the dimensionless frequency F^+ was equal to 1, the inflection point of the pitch moment coefficient curve was delayed by 2°, while in other dimensionless frequencies, the inflection point was the same as that of no plasma actuation. On the one hand, there was an optimal actuation frequency when the unsteady actuation dimensionless frequency was equal to 1, which was related to the incoming flow velocity, the feature size of the symmetric flying wing model. On the other hand, the effect of flow control was different from the same actuation intensity at different inflow speeds.

The Reynolds number of the small flying wing corresponding to the incoming flow velocity of 30 m/s was 4.6×10^5, and the Reynolds number of the large flying wing corresponding to the incoming flow velocity of 75 m/s was 2.9×10^6. As shown in Figure 14, with the increase of the dimensionless frequency, the variation trend of the lift coefficient increment under different Reynolds numbers was consistent. The lift coefficient increment increased first and then decreased. The increment of the plasma actuation at low Reynolds number was more obvious. It can be seen that the optimal dimensionless frequency of flow control at low velocity was independent of the Reynolds number. However, the higher the Reynolds number, the greater the inertial force in the flow field, resulting in the flow field not being easy to control after stalling. Therefore, the effect of the same intensity disturbance on controlling the flow field stall separation at different incoming flow velocities was different. Increasing the intensity of unsteady actuation to improve the lift coefficient of the symmetric flying wing model at a higher Reynolds number and increasing the pitching moment coefficient are worthy of further study.

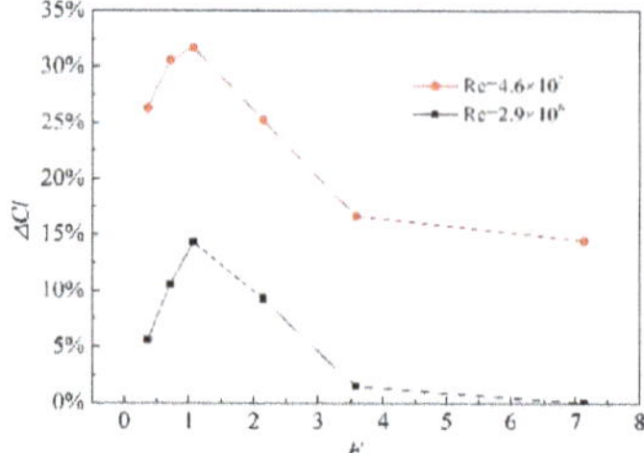

Figure 14. Lift coefficient increment at different dimensionless actuation frequencies in incoming flow velocities of 30 m/s and 75 m/s.

Han et al. studied ns-DBD plasma actuation used for aerodynamic control on the small flying wing. Their results show that the ns-DBD plasma actuator offers tremendous potential as an active flow control device to enhance the aerodynamic performance of the present model. There exists an optimal actuation frequency (f = 0.2 kHz) to reach maximum lift coefficient value. Given the high pulsed frequency of f = 1 kHz, an obvious decrease in the drag coefficient is observed. The results indicate that a 44.5% increase in the lift coefficient, a 34.2% decrease in the drag coefficient and a 22.4% increase in the maximum lift-to-drag ratio could be achieved as compared with the baseline case. In this article, μs-DBD plasma actuation was used for controlling the aerodynamics performance of two different scaling flying wings. The optimal frequency of the plasma actuation was 150 Hz and the corresponding dimensionless frequency F^+ was 1.07. The effect of low frequency actuation was better than that of higher frequency. After actuation, the stall angle of attack of the small flying wing was delayed by 4°, the maximum lift coefficient was increased by 30.9%, and the drag coefficient could be reduced by 17.3%. After the large flying wing was actuated, the stall angle of attack was delayed by 4°, the maximum lift coefficient was increased by 15.1%, but the drag coefficient was increased.

3.3. Flow Field Characteristics

The structure of the symmetric flying wing is complex, and the flow field flowing through its surface is also very complicated. Through the PIV test method, the state of the surface flow fields of the two flying wing models is clearly and intuitively observed. Three cross-sections were taken in the chord direction and the span direction of the symmetric flying wing. As shown in Figure 15, the cross-sections 1 and 2 were located at 15% l and 32% l from the left wing tip, respectively, and cross-section 3 was located at 40% c from the forefront of the left wing.

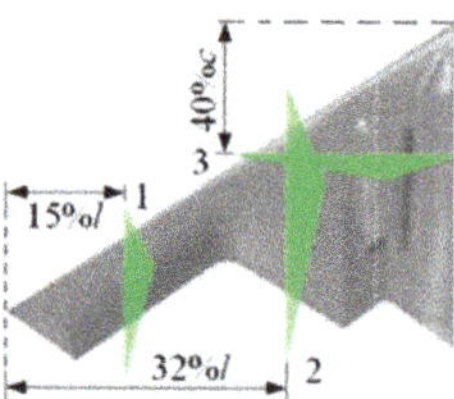

Figure 15. Schematic diagram of PIV test cross-section.

The flow field of measured cross-section 1 is shown in Figure 16. When the angle of attack (AOA) was 15°, the flow field measurement results in Figure 16a show that the flow separation had developed to the leading edge of the wing. After the plasma actuation was applied, the separation was completely suppressed and the leading edge airflow velocity increased slightly as shown in Figure 16b. When the angle of attack was 18°, the flow separation on the upper surface of the wing was very serious, and a large area of reflux appeared, as shown in Figure 16c. After the actuation, the flow separation could no longer be completely suppressed, as shown in Figure 16d; a large separation bubble was formed on the upper surface of the wing, and the flow velocity of the leading edge of the wing was slightly higher than that of no plasma actuation.

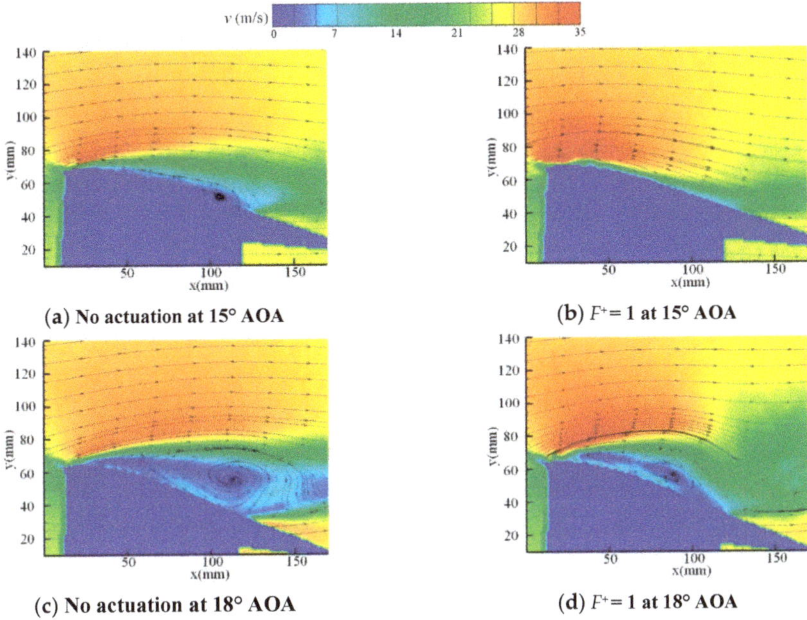

Figure 16. Time-averaged flow field of measured cross-section 1 of the small flying wing (U_∞ = 30 m/s, FL-5).

Cross-section 2 was closer to the wing root and was in the middle wing. Figure 17 shows the flow field velocity cloud diagram and streamline of cross-section 2. As shown in Figure 17a, when the angle of attack was 15°, the separation just occured at the trailing edge, which was different from that of cross-section 1. Because of the sweepback effect, the separation of cross-section 1 had developed to the leading edge of the wing at the angle of attack of 15°. As shown in Figure 17c, when the angle of attack increased to 18°, the flow separation was very serious and developed to the leading edge of the wing; the actuated flow field is shown in Figure 17d. The boundary layer of the separated flow field was reattached. Compared with cross-section 1, the separation at 18° angle of attack on the middle wing surface of the symmetric flying wing was effectively controlled. When the angle of attack was 20°, the flow separation shown in Figure 17e was more serious. As shown in Figure 17f, after plasma actuation, the large separation of wing surface was improved, but the separation of symmetric flying wing surface could not be completely suppressed. At the angle of attack of 22°, plasma actuation could no longer completely inhibit the flow separation.Although the separation area decreased, the separation still occured at the front edge. The velocity of the flow field at the leading edge increased because of actuation.

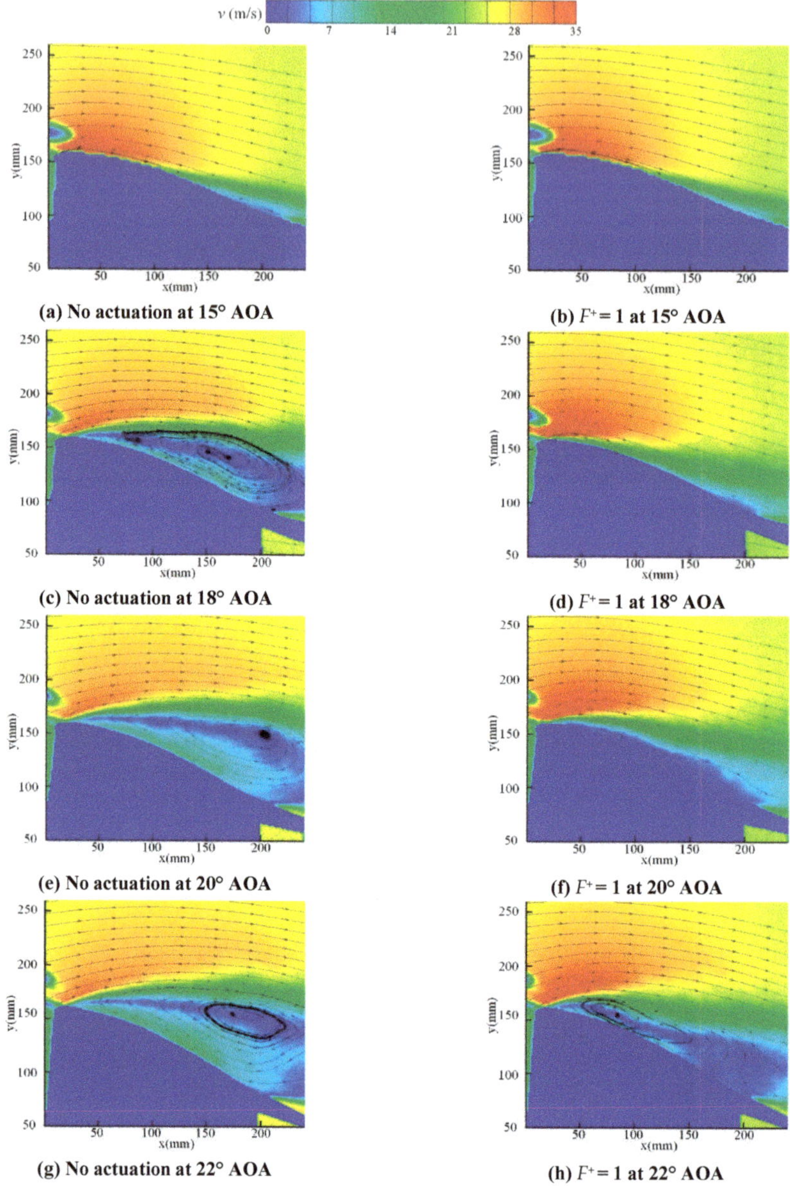

Figure 17. Time-averaged flow field of measured cross-section 2 of the small flying wing ($U_\infty = 30$ m/s, FL-5).

The force measurement experiment shows that when the stall angle of attack of the symmetric flying wing model was 15°, the flow field of cross-section 1 had completely stalled, while the flow field of cross-section 2 had only a slight stall at the trailing edge. When the angle of attack was 18°, both cross-sections had stalled. It can be seen that as the angle of attack increased, the stall on the surface of the symmetric flying wing first occurred on both sides and then moved to the inside.

The effect of plasma actuation was to inject energy into the flow field so that the separation area at the stall angle of attack was reattached.

It can be seen from the time-average flow field diagram of measured cross-section 3 in Figure 18 that there was lateral flow on the upper wing surface of the symmetric flying wing model, and its velocity was small. As the angle of attack increased, the range of lateral flow gradually increased. Stall separation was likely to occur when the lateral flow area is too large on the symmetric flying wing surface. It can be seen from Figure 18a that in no-plasma-actuation state, the outer edge airflow flowed through the edge of the wing to the inner edge of the wing, which occurred at x > 260 mm. However, after actuation it was advanced to x > 240 mm, the velocity at the outer edge of the wing had a slight increase, the flow trend toward the airfoil increased, and the point source flowing in the direction of the wing root was advanced from x = 150 mm to x = 140 mm. When the angle of attack was 18°, it can be seen from Figure 18c that there was nearly no tendency for the airflow to flow to the wing near it. After the plasma actuation was applied, the tendency of the airflow to the wing at 18° angle of attack was advanced to 270 mm. and the lateral flow velocity at x = 300–370 mm was increased. When the angle of attack was increased to 22°, the control ability of the plasma actuation to the lateral flow was weakened, as shown in Figure 18f.

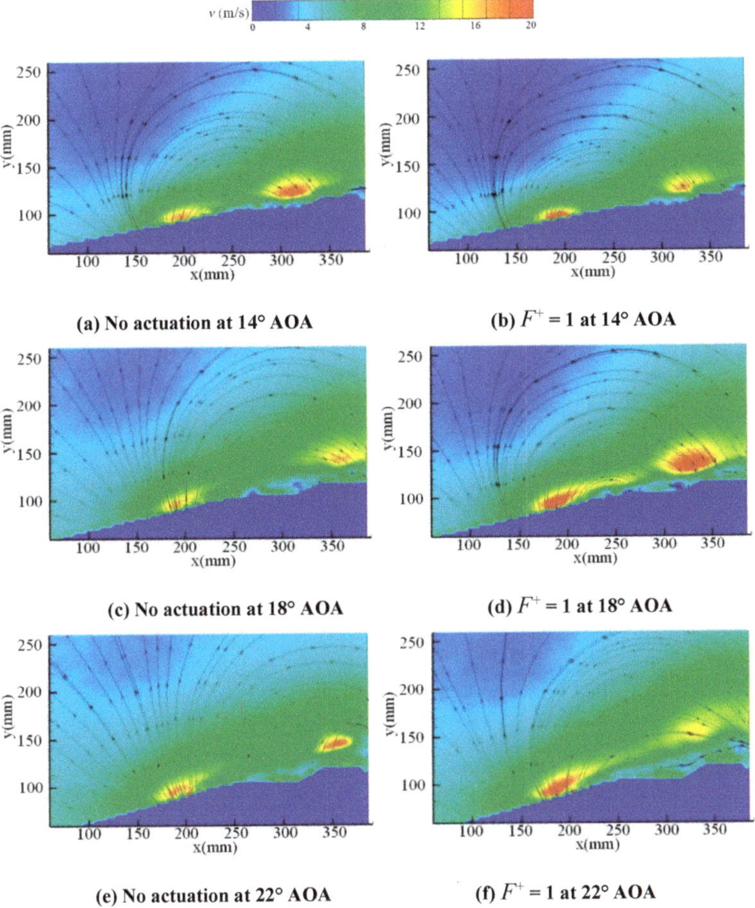

Figure 18. Time-averaged flow field of measured cross-section 3 of the small flying wing ($U_\infty = 30$ m/s, FL-5).

Figure 19 shows the flow field of the large flying wing at the incoming flow velocity of 75 m/s. When the angle of attack was small, the flow field was attached to the surface of the symmetric flying wing without flow separation just as the phenomenon shown in Figure 16. Due to the large incoming flow velocity, the flow velocity near the stagnation point of the leading edge of the symmetric flying wing was relatively low. As shown in Figure 19a, the flow field was affected by the lateral vortex at 12° angle of attack, so it was unstable at the leading edge of the symmetric flying wing and the lift coefficient was smaller than that of the small flying wing at the same angle of attack. At the 14° stall angle of attack, as shown in Figure 19c, the flow separation of the large flying wing in cross-section 1 was severe. After the plasma actuation, as shown in Figure 19d, the flow separation was suppressed and the control effect was good.

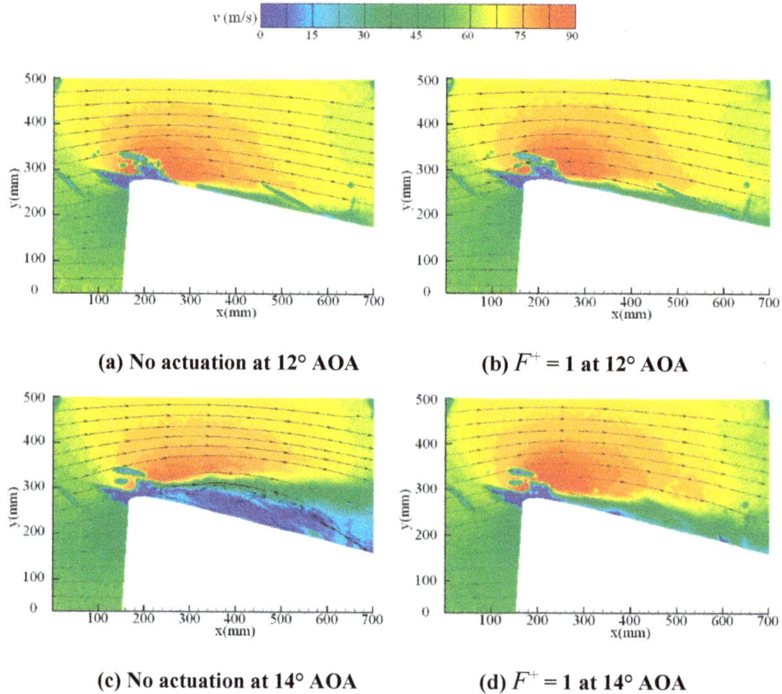

(a) No actuation at 12° AOA

(b) $F^+ = 1$ at 12° AOA

(c) No actuation at 14° AOA

(d) $F^+ = 1$ at 14° AOA

Figure 19. Time-averaged flow field of measured cross-section 1 of the large flying wing ($U_\infty = 75$m/s, FL-51).

4. Conclusions

The discharge energy of μs-DBD is to demonstrate that different length actuators mounted on the small flying wing and the large flying wing have similar energy density. The single energy intensity is positively correlated with input voltage but has nothing to do with the pulse repetition frequency. So, other interference factors were eliminated when the dimensionless frequency is contrasted and analyzed for two symmetric flying wings based on two kinds of scale models. Force and flow field characteristics of the two symmetric flying wings are the topic of the article.

The large flying wing is 2.5 times the size of the small flying wing. The purpose of this paper is to study whether the optimal dimensionless frequency of two flying wing scale models at different incoming flow velocities is consistent by measuring the lift, drag, moment, and surface flow field variation of the two scaling flying wing model before and after μs-DBD plasma flow control actuation. Through the corresponding experiments, the following conclusions can be drawn. The discharge energy results show that the longer the length of the μs-DBD actuator is, the greater the discharge output energy is, but the energy density is basically unchanged and is independent of the length of the actuator. With the increase of actuation voltage, the output energy and the energy density increase. The single pulse energy of the microsecond pulse at different actuation frequencies is basically unchanged and has nothing to do with actuation frequency.

The force measurement and flow field PIV of the small and large flying wing models are compared by experiments. The results show that the stall angles of attack at the corresponding Reynolds numbers of 4.6×10^5 and 2.9×10^6 are both 15°. When the actuation frequency of plasma flow control is 150 Hz, the unsteady actuation effect is the best, and the corresponding dimensionless frequency F^+ is 1.07, which is appropriate for the Strouhal similarity criterion. After actuation of the small flying wing, the stall angle of attack is delayed by 4°, the maximum lift coefficient is increased by 30.9% and the drag coefficient can be reduced by 17.3%. The effect of lift increase and drag reduction is good. For a large flying wing, the stall angle of attack is delayed by 4°, the maximum lift coefficient is increased by 15.1%, and the drag coefficient is increased. Similarly, the inflection point of the pitch moment is delayed at the optimal dimensionless frequency. In addition, the effect of low frequency actuation is better than that at a high frequency. The inertia force of the incoming flow at a low Reynolds number is small, and the plasma actuation effect is obvious, but at a high Reynolds number, the ability to promote the complete reattachment of the separation area is not enough due to the limitation of actuation intensity. The PIV test results of the flow field at different cross-sections show that the stall separation on the surface of the symmetric flying wing begins from the outer side, and then with the increase of the angle of attack the separation area begins to appear on the inside. The plasma flow control can not only delay the separation of the longitudinal boundary layer but also slow down the movement of the lateral vortex and inject momentum and energy into the flow field, thus effectively increasing the lift and reducing the drag.

Author Contributions: Conceptualization, H.L. and L.X.; Methodology, M.H.; Formal Analysis, B.W.; Investigation, Z.S.; Data Curation, Z.N.; Writing-Review & Editing, B.T.

Funding: This research was founded by the *National Natural Science Foundation of China* (Grant Nos. 11802341 and 11472306), *Open Project of State Key Laboratory of Aerodynamics* in 2018 (Grant No. SKLA20180207).

Acknowledgments: Thanks for the experimental platform and technical support provided by *AVIC Aerodynaiviics Research Institute*.

References

1. Wang, F. The Comparison of Aerodynamic and Stability Characteristics between Conventional and Blended Wing Body Aircraft. Master's Thesis, Cranfield University, Cranfield, UK, 2012.
2. Patil, M.J.; Hodges, D.H. Flight Dynamics of Highly Flexible Flying Wings. *J. Aircr.* **2006**, *43*, 1790–1799. [CrossRef]
3. Zhang, L.; Zhou, Z.; Xu, X.P.; Wang, H.B. Comparison on aerodynamic and stealthy performance of flying wing unmanned aerial vehicle with three conformal intake inlets. *J. Aerosp. Power* **2015**, *30*, 1651–1660.
4. Chen, S.; Lyu, Z.; Kenway, G.K.; Martins, J.R. Aerodynamic Shape Optimization of the Common Research Model Wing-Body-Tail Configuration. In Proceedings of the AIAA Aerospace Sciences Meeting, Kissimmee, FL, USA, 5–9 January 2015.
5. Nangia, R.; Palmer, M. Flying-Wings (Blended Wing Bodies) with Aft & Forward Sweep, Relating Design Camber & Twist to Longitudinal Control. In Proceedings of the AIAA Atmospheric Flight Mechanics Conference & Exhibit, Monterey, CA, USA, 5–8 August 2002.
6. Mardanpour, P.; Richards, P.W.; Nabipour, O.; Hodges, D.H. Effect of multiple engine placement on aeroelastic trim and stability of flying wing aircraft. *J. Fluids Struct.* **2014**, *44*, 67–86. [CrossRef]
7. Huang, A.; Folk, C.; Silva, C.; Christensen, B.; Chen, Y.F.; Ho, C.M.; Jiang, F.; Grosjean, C.; Tai, Y.C. Applications of MEMS devices to delta wing aircraft: From concept development to transonic flight test. In Proceedings of the 39th Aerospace Sciences Meeting and Exhibit, Reno, NV, USA, 8–11 January 2001.

8. Roos, F.W. Microblowing for High-Angle-of-Attack Vortex Flow Control on a Fighter Aircraft. *J. Aircr.* **2001**, *38*, 454–457. [CrossRef]

9. Patel, M.P.; Ng, T.T.; Vasudevan, S.; Corke, T.C.; He, C. Plasma Actuators for Hingeless Aerodynamic Control of an Unmanned Air Vehicle. *J. Aircr.* **2007**, *44*, 1264–1274. [CrossRef]

10. Su, Z.; Li, J.; Liang, H.; Zheng, B.R.; Wei, B.; Chen, J.; Xie, L.K. UAV flight test of plasma slats and ailerons with microsecond dielectric barrier discharge. *Chin. Phys. B* **2018**, *27*, 459–468. [CrossRef]

11. Hsiao, F.B.; Liu, C.F.; Shyu, J.Y. Control of wall-separated flow by internal acoustic excitation. *AIAA J.* **2012**, *28*, 1440–1446. [CrossRef]

12. Krishnappa, S.; Jogi, N.; Nguyen, L.D.; Gudmundsson, S.; MacKunis, W.T.; Golubev, V.V. Towards Experimental Validation of Robust Control of Gust-induced Airfoil Limit Cycle Oscillations Using Synthetic Jet Actuators. In Proceedings of the AIAA Fluid Dynamics Conference, Washington, DC, USA, 13–17 June 2016.

13. Zong, H.; Kotsonis, M. Effect of slotted exit orifice on performance of plasma synthetic jet actuator. *Exp. Fluids* **2017**, *58*, 17. [CrossRef]

14. Campbell, J.F. Augmentation of Vortex Lift by Spanwise Blowing. *J. Aircr.* **1976**, *13*, 727–732. [CrossRef]

15. Joseph, P.; Amandolese, X.; Edouard, C.; Aider, J.L. Flow control using MEMS pulsed micro-jets on the Ahmed body. *Exp. Fluids* **2013**, *54*, 1–12. [CrossRef]

16. Little, J. High-Lift Airfoil Separation Control with Dielectric Barrier Discharge Plasma Actuators. Ph.D. Thesis, The Ohio State University, Columbus, OH, USA, 2010.

17. Yun, W.U.; Yinghong, L.I. Progress in Research of Plasma-assisted Flow Control, Ignition and Combustion. *High Volt. Eng.* **2014**, *40*, 2024–2038.

18. Little, J.; Takashima, K.; Nishihara, M.; Adamovich, I.; Samimy, M. Separation Control with Nanosecond-Pulse-Driven Dielectric Barrier Discharge Plasma Actuators. *AIAA J.* **2012**, *50*, 350–365. [CrossRef]

19. Tang, M.; Wu, Y.; Wang, H.; Guo, S.; Sun, Z.; Sheng, J. Characterization of transverse plasma jet and its effects on ramp induced separation. *Exp. Therm. Fluid Sci.* **2018**, *99*, 584–594.

20. Zheng, B.; Xue, M.; Ke, X.; Ge, C.; Wang, Y.; Liu, F.; Luo, S. Unsteady Vortex Structure Induced by a Trielectrode Sliding Discharge Plasma Actuator. *AIAA J.* **2018**, *57*, 467–471. [CrossRef]

21. Sidorenko, A.; Budovsky, A.; Pushkarev, A.; Maslov, A. Flight testing of DBD plasma separation control system. In Proceedings of the AIAA Aerospace Sciences Meeting & Exhibit, Reno, NV, USA, 7–10 January 2008.

22. Wang, J.; Li, Y.; Xing, F. Investigation on oblique shock wave control by arc discharge plasma in supersonic airflow. *J. Appl. Phys.* **2009**, *106*, 073307. [CrossRef]

23. Benard, N.; Moreau, E. Electrical and mechanical characteristics of surface AC dielectric barrier discharge plasma actuators applied to airflow control. *Exp. Fluids* **2014**, *55*, 1846. [CrossRef]

24. Roupassov, D.V.; Nikipelov, A.A.; Nudnova, M.M.; Starikovskii, A.Y. Flow Separation Control by Plasma Actuator with Nanosecond Pulsed-Periodic Discharge. In Proceedings of the 46th AIAA Aerospace Sciences Meeting and Exhibit, Reno, NV, USA, 7–10 January 2008.

25. Durasiewicz, C.; Singh, A.; Little, J. A Comparative Flow Physics Study of Ns-DBD vs. Ac-DBD Plasma Actuators for Transient Separation Control on a NACA 0012 Airfoil. In Proceedings of the 2018 AIAA Aerospace Sciences Meeting, Kissimmee, FL, USA, 8–12 January 2018.

26. Sevant, N.E.; Bloor, M.I.G.; Wilson, M.J. Aerodynamic Design of a Flying Wing Using Response Surface Methodology. *J. Aircr.* **2015**, *37*, 562–569. [CrossRef]

27. Huber, K.; Schutte, A.; Rein, M. Numerical Investigation of the Aerodynamic Properties of a Flying Wing Configuration. In Proceedings of the 30th AIAA Applied Aerodynamics Conference, New Orleans, LA, USA, 25–28 June 2012.

28. Xu, X.; Zhou, Z. Analytical study on the synthetic jet control of asymmetric flow field of flying wing unmanned aerial vehicle. *Aerosp. Sci. Technol.* **2016**, *56*, 90–99. [CrossRef]

29. Han, M.; Li, J.; Niu, Z.; Liang, H.; Zhao, G.; Hua, W. Aerodynamic performance enhancement of a flying wing using nanosecond pulsed DBD plasma actuator. *Chin. J. Aeronaut.* **2015**, *28*, 377–384. [CrossRef]

30. Yao, J.K.; Zhou, D.J.; He, H.B.; He, C.J.; Shi, Z.W.; Du, H. Experimental investigation of lift enhancement for flying wing aircraft using nanosecond DBD plasma actuators. *Plasma Sci. Technol.* **2017**, *19*, 11–18. [CrossRef]

Article

Hydrodynamic-Interaction Analysis of an Autonomous Underwater Hovering Vehicle and Ship with Wave Effects

Chen-Wei Chen, Ying Chen * and Qian-Wen Cai

Ocean College, Zhejiang University, Zhoushan 316000, China; cwchen@zju.edu.cn (C.-W.C.);
21634097@zju.edu.cn (Q.-W.C.)
* Correspondence: ychen@zju.edu.cn

Received: 17 August 2019; Accepted: 25 September 2019; Published: 29 September 2019

Abstract: A new vertical axis-symmetrical dish-shaped autonomous underwater vehicle (AUV) with excellent maneuverability, known as the autonomous underwater hovering vehicle (AUH), is proposed. This study investigates an important working model of the AUH approaching a host ship in waves. The working model of AUH–Ship interactions deals with hydrodynamic interaction, seakeeping performance for communication, launch, and recovery near a free surface. The AUH is able to navigate and implement homing automation through acoustic positioning equipment, a depth sensor, a heading compass, and a Doppler velocity log (DVL) in the working area based on numerical analysis of AUH–Ship hydrodynamic performance in this study. The hydrodynamic-interaction performance of the AUH and ship near free surfaces is analyzed in the frequency and time domains using a potential-based surface-panel method based on a commercial computational fluid dynamics (CFD) solver (ANSYS-AQWA), i.e., a 3D panel code of seakeeping performance module in the ANSYS platform where the fluid is assumed to be irrotational, inviscid, and incompressible. The motion performance of the AUH approaching the host ship, with a dynamic positioning system in waves, is studied by estimating interactive response-amplitude operators (RAOs) of the AUH and host ship in 6-DOF that were estimated and analyzed at different wave amplitudes and frequencies. In the ship and AUH interaction simulations, the host ship is assumed to be a well-posed station keeping in waves with zero service speed. The AUH and ship interference effect is studied at different distances to appropriate recovery and launch positions for the AUH at the following sea and beam sea, i.e., wave-encounter angles 0° and 90°, respectively. In addition, the hydrodynamic interaction of the AUH and ship in yaw and roll at different AUH velocities is estimated. The AUH motion performance approaching the ship in long-crested irregular seas is analyzed in the time domain using the Pierson–Moskowitz wave spectrum model. Viscid hydrodynamic force on AUH motion in roll near a free surface was significant. A damping model was adopted to formulate the viscid effect to enhance the effectiveness of the ANSYS-AQWA inviscid potential-based solver. Numerical analysis of motion RAO of the AUH in roll with the damping effect was compared to the experimental data in wave-frequency range 0.2–1.0 Hz, resulting in the average error being reduced from 21.03% to 9.95% to verify the method's accuracy. The proposed method conveniently and accurately predicted hydrodynamic-interaction characteristics and motion RAO for a dish-type AUH and host ship for the precise use of mounted sensors in waves. The results of these simulations can be used to analyze the homing automation and adaptive controllability to advance the AUV development and design.

Keywords: Autonomous Underwater Vehicle (AUV); Autonomous Underwater Hovering Vehicle (AUH); hydrodynamic interaction; response amplitude operator (RAO); wave effects

1. Introduction

From oceanographic investigations to military applications, new axisymmetric types of autonomous underwater vehicles (AUVs) have been developed to conduct missions in complex marine environments in recent years [1,2]. The highly autonomous, flexible, maneuverable, and controllable overall design of AUVs plays an important role in ocean exploration and development.

Development of a launch-and-recovery system for AUVs from a research vessel is significant to the overall design. Submarine analysis of launch-and-recovery underwater-vehicle models was carried out by Chen and Su in 2011 [3]. An automated launch-and-recovery system for AUVs from an unmanned surface vehicle was proposed by Sarda and Dhanak in 2017 [4]. However, hydrodynamic-interaction effects of wave-induced disturbance on the surface vehicle and axisymmetric AUVs were ignored in these studies.

Research on the hydrodynamic-interaction performance of wave-induced disturbance on a hovering AUV and a host ship is significant to the overall design process of the AUV and/or scientific-research ships to smoothly guarantee the successful deployment, operation, and homing automation of AUVs from the deep sea to the free surface, or launching from host ships into waves. However, quickly estimating wave-induced hydrodynamic-interaction forces between a vertical axisymmetric AUV hull form and large-scale host ship is difficult and challenging. The viscous computational fluid dynamic solver is a time-consuming method to solve a wave-induced and multibody hydrodynamic-interaction problem. Thus, an inviscid-potential-based surface-panel method is an efficient and effective method for solving fluid and rigid-body interaction problems. The potential flow modelling in the surface-panel method was assumed to be irrotational, inviscid, and incompressible.

Hydrodynamic-interaction performance of AUVs in waves is critical to attitude-controller design and has recently been receiving intensive attention. A few works have been done on the prediction of hydrodynamic-interaction forces and moments between surface and underwater vessels [5–8]. The hydrodynamic-interaction issues associated with AUVs approaching a cone-shaped dock in ocean currents were studied in the literature [5]. Xie et al. [6] presented an experimental study of wave loads on a small vehicle in close proximity to a large vessel. The results of this investigation could be adopted to inform planning marine operations, such as the recovery of an AUV from a mothership in low-to-moderate sea states.

Skomal et al. [7] presented a depth-limited AUV mounted with a viable tool for directly observing the behavior of marine animals in order to investigate the behavior, habitat, and feeding ecology of white sharks *Carcharodon carcharias* near Guadalupe Island, off the coast of Mexico. However, the hydrodynamic interactions between the AUV and waves were not considered in this study. In fact, the effect of hydrodynamic interaction on wave loads on the AUV model was substantial enough to cause inefficiently controlled vessel motions during such operations. Zhi et al. [8] presented quasistatic analysis of hydrodynamic interactions between an AUV and a host submarine for controlling the AUV recovered to the submarine well.

In the literature of seakeeping analysis of surface vessels or underwater vehicles, Tian et al. [9] presented research on the influence of surface waves on the hydrodynamic performance of an axis-symmetrical slender-body AUV when the AUV sailed near the free surface. The results showed that the variation of AUV hydrodynamic forces was significantly affected by the waves. Mansoorzadeh and Javanmard [10] adopted a two-phase flow CFD solver with the Volume of Fluid (VOF) model to investigate the free-surface effects on drag-and-lift coefficients of an AUV with an axisymmetric hull form. Much CPU time is needed when adopting this method. Therefore, Malik et al. [11] adopted the 3D panel method using the commercial CFD code, ANSYS-AWQA, to carry out the time-saving simulations of wave-force prediction on a slender underwater vehicle in the frequency and time domains.

Lighthill [12] studied second-order wave radiation and diffraction problems by using the 3D frequency-domain method, and proposed that second-order wave force can be obtained from virtual radiation potential. Chen and Fang [13] analyzed the hydrodynamic interaction between two vessels traveling in close parallel waves. Choi and Hong [14], Masashi et al. [15], and Hong et al. [16] adopted the high-order boundary-element method in the frequency domain to evaluate hydrodynamic disturbance between multiple buoys. Their numerical simulation results were verified by physical-model tests. The literature review reveals that researchers have investigated hydrodynamic disturbances between large floating bodies, but few have focused on hydrodynamic disturbances between floating bodies with large size differences (e.g., an AUV and a host ship).

This study is focused on the prediction of the hydrodynamic performance of wave-induced disturbances on an AUV and a host ship for solving problems of motion and control of launching and recovery of AUVs from host ships in advance. A new dish-shaped AUV with a vertical axisymmetric hull form, the autonomous underwater hovering vehicle (AUH), is proposed and was developed by the Ocean College of Zhejiang University (ZJU) in 2016. The AUH was designed with a flying-saucer shape for enhanced maneuverability, to be efficient and effective and able to perform full circumferential steering, fixed-point hovering, and free take-off and landing on the sea bed. The designed speed of the AUH was 1 knot in surge. In low design speed, the horizontal maneuverability and vertical-motion stability of the disk-shaped AUH would be better than conventional torpedo-shaped AUVs [17]. Preliminary related research on the configurational design and system integration of the AUH prototype was implemented, including structural design with streamline-hull form, hybrid propulsion system with a buoyancy engine and four vectoring electric propellers, controller design in calm seas [18], analysis of seakeeping performance of second-order wave forces on the AUH [19], and prediction of antisea-current ability [20], and of added mass for the AUH near the sea bottom [21].

In addition, the AUH has motion-control, power-transmission, and optical-communication systems for the presentation of special homing automation functions in deep-sea and near-free-surface work. Excellent autonomous automation homing-navigation models have been published in the literature [22–25]. Miller et al. [22] proposed a navigation system integrating high-rate inertial-measurement-unit (IMU) accelerometers and gyros in allowable time propagation while multisensors provide measurement corrections. Bo et al. [23] proposed a SEIF-SLAM navigation approach to improve the accuracy and navigation of a slender AUV. Recently, a vision-based underwater-vehicle navigation system was proposed in the literature [24,25]. These vision-based navigation approaches could improve conventional acoustic-navigation approaches to enhance the efficiency and effectiveness of AUV recovery in clean sea water. Part of the mentioned navigation system was adopted and successfully extended by the AUH design in our navigation model. The overview of the AUH structure and module system encapsulated in the pressure hull is shown in Figure 1a,b, respectively. The ZJU AUH prototype is shown in Figure 1, and its main parameters (hull form, hybrid propulsion, and multidisciplinary systems) are shown in Table 1.

For AUH–Ship working conditions, hydrodynamic analysis should be done to understand AUH behavior in hydrodynamic interactions, allowing the advanced implementation of control strategies. Additionally, behavioral predictions for emerging and recovering an AUH from a ship in waves are important to advance AUH adaptive controllability, homing automation, and overall design.

Hydrodynamic modeling of AUH–Ship interactions near or on a free surface is challenging [26,27]. Intrinsically, waves are irregular, unsteady, nonuniform fluids with frequent breaking and vertexing effects. Hydrodynamic-wave forces can, theoretically, be divided into discrete forces: First-order wave-induced, low-varying disturbance (second-order wave excitation, wave mean drift, current, and wind forces), radiation, viscosity, etc.

Due to the significant size difference between AUH and scientific research ship (host ship), the AUH moving toward or away from the host ship is disturbed by distinct, ship-induced wave interference. If an appropriate hydrodynamic-interaction theory model is used to predict AUH–Ship motion performance, appropriate seakeeping locations can be determined for the AUH communicating with a nearby host ship mounted with active sonar.

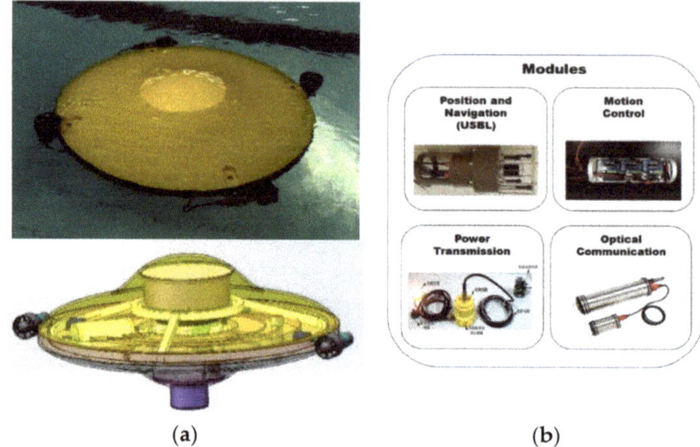

(a)   (b)

Figure 1. Zhejiang University (ZJU) autonomous underwater hovering vehicle (AUH) prototype: (**a**) Structural design; (**b**) module system.

Table 1. Main parameters of Zhejiang University (ZJU) autonomous underwater hovering vehicle (AUH).

Parameter	Symbol	Unit	Value
Diameter	L	m	1.0
Height	T	m	0.44
Design depth	H	m	1000
Seawater density	ρ	$kg \cdot m^{-3}$	1020–1030
Design speed	U	$m \cdot s^{-1}$	0.5144–1.5432
Mass/payload	m	kg	122.6/15
Vertical distance between CG and CB	BG	mm	37
Battery	null	kwh/kg	2 * 12 V–30 Ah
Number of thrusters/thrust	null	kg	2/50 kgf
Buoyancy engines/volume	null	ml	2/500
USBL/DVL	null	null	AT: 50, 150 mm/20 kHz
Pressure hull	null	null	1

Seakeeping analysis of AUH–Ship interactions can improve AUH launch and recovery efficiency, staff safety at work, and can be conducted by studying the AUH–Ship Response Amplitude Operator (RAO). The RAO is used to study six-degrees-of-freedom (6-DOF) motion performance of marine vehicles maneuvering in waves that can be divided into two types: Motion RAO, i.e., the ratio of vehicle-motion amplitude to the amplitude of incident regular waves in the frequency domain (motion RAO utilizing the frequency-domain approach to evaluate radiation forces); and the force RAO, combined with the wave spectrum (obtaining loads for the sea state and sailing conditions) [19].

In the overall design of the AUH, a hydrodynamic-interaction study of the AUH–Ship relationship could significantly enhance the effectiveness and efficiency of launch and recovery operations from a vessel onto a free surface. This study thus focused on researching the hydrodynamic-interaction performance of the AUH and host ship. Considering computing efficiency, this study adopted a potential-based panel method based on an inviscid CFD solver, namely, ANSYS-AQWA software. In the seakeeping characteristics of the AQWA module, the hydrodynamic-interaction performance (wave radiation force and diffraction force) between two buoys can be efficient in calculations, and the system can be effectively integrated with a comprehensive consideration of random environmental loads: Wind, current, wave, surge load, and propulsion effect.

In this study, this ANSYS-AQWA panel method was used to study the hydrodynamic interactions between AUH and host ship considering the effect of random wave loads for the purpose of optimizing AUV positions for communication with, and recovery and launch from, a surface vessel in regular and long-crested irregular seas. Two aspects of AUH–Ship interaction were studied: Motion RAO variations in AUH–Ship interactions with 6-DOF, and AUH–Ship interaction motions in irregular waves. The relationship between AUH and ship distances, and hydrodynamic disturbances was analyzed to determine suitable positions for launch and recovery. Cases studies were conducted evaluating different near-ship AUH locations, motion RAOs in 6-DOF, and AUH motion performance in the time domain. In addition, viscous hydrodynamic forces on the vertical axis-symmetrical AUH motion in roll or pitch near a free surface is significant. A damping model was adopted to formulate viscous effect to enhance effectiveness of the ANSYS-AQWA inviscid potential-based CFD solver. Numerical analysis of motion RAO of the AUH in roll with the damping effect were compared to experimental data in the range of wave frequency 0.2–1.0 Hz, resulting in average errors reduced from 21.03% to 9.95% to verify the method accuracy.

In the following sections, a mathematical model of a surface-panel method describing rigid-body hydrodynamic interactions in potential flow with wave effects is presented. In Section 3, simulation results of hydrodynamic interactions of AUH and host ship are introduced. The AUH and ship interference effect was studied at different distances for appropriate recovery and launch positions for the AUH at following, beam, and irregular seas. The motion RAO of the AUH near the ship in waves was estimated to determine the optimal positions to recover the AUH. In Section 4, the experiments of AUH motion RAO in rolling degrees of freedom are presented. The experiments were implemented in the large-scale cross-sectional wave flume of Zhejiang University Ocean College in Zhoushan in December 2017. Finally, the concluding remarks are presented in Section 5.

2. Mathematical Model

Modeling multibody hydrodynamic interference using a potential surface-panel method is presented and the definition of the coordinate system and the wave-encounter angle are presented in this section.

2.1. Coordinate System and Wave-Encounter Angle

The AUH–Ship interaction simulations contained a free surface, host ship, and AUH in the coordinate system, as shown in Figure 2a. Coordinate system O-XYZ is an inertial coordinate system. Coordinate systems O_1-$X_1Y_1Z_1$ and O_2-$X_2Y_2Z_2$ are the body-fixed coordinate systems of the host ship and the AUH, respectively (translating-rotation coordinates). Their hulls oscillated with the body-fixed coordinate systems. All coordinate-system origins were located on the hydrostatic surface using the right-hand coordinate system. Origin point O_1 was located at the center of gravity of the host ship, the positive OX_1 axis coincided with the navigation direction of the ship, and the OZ_1 axis was vertical (positive in the upward direction).

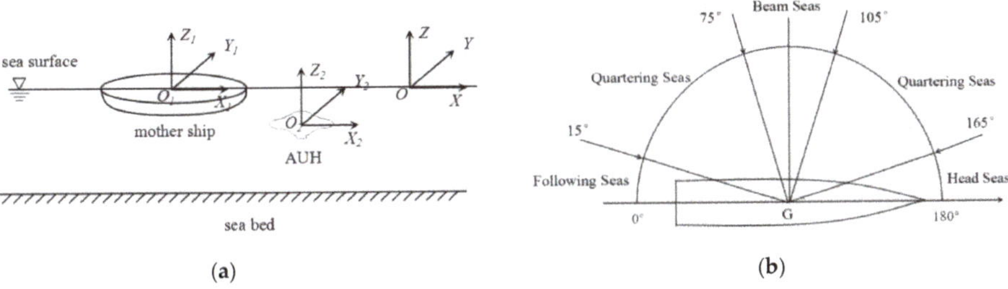

Figure 2. (a) Definition of coordinate system between the AUH and host ship; (b) wave-encounter angle.

When the host ship and AUH navigated in waves, they were subject to wave impact from all directions. These wave angles could be divided into four categories [28]: Following sea, quartering sea, beam wave, and head sea. The host ship and AUH both had 6-DOF motion in the waves, as shown in Figure 2b.

2.2. Potential Surface-Panel Method

Fluid was assumed to be irrotational, inviscid, and incompressible based on potential-flow theory. The floating-body system in the computational domain behaved as a simple harmonic with slight wave motion. The following velocity-potential function satisfied the Laplace equation [29]:

$$\nabla^2 \Phi = 0 \tag{1}$$

where Φ is the velocity potential

$$\Phi = -Ux + \phi \tag{2}$$

in which the term $-Ux$ denotes AUH sailing at constant speed U, and ϕ denotes perturbation potential induced by the presence of the AUH.

Two floating bodies were present near the flow field, making the total velocity of the flow field [30]

$$\Phi(x,y,z,t) = \Phi_I(x,y,z,t) + \Phi_D(x,y,z,t) + \Phi_R^1(x,y,z,t) + \Phi_R^2(x,y,z,t) \tag{3}$$

where Φ_I represents the velocity potential of the incident wave, Φ_D represents the diffraction power of multifloating system, Φ_R^1 represents the radiation potential of Floating Body 1 in the multifloating system, and Φ_R^2 represents the radiation potential of Floating Body 2 in the multifloating system.

The steady-state response of Φ_D can be expressed as

$$\Phi_D(x,y,z,t) = \text{Re}\left\{\phi_D(x,y,z)e^{-iwt}\right\} \tag{4}$$

where Re{} represents the real portion of the amount in {} and ω represents the wave frequency.

The radiation potential of the floating body m in a floating body system is

$$\Phi_R^{(m)}(x,y,z,t) = \text{Re}\left\{\phi_R^{(m)}(x,y,z)e^{-iwt}\right\} \tag{5}$$

where $\phi_R^m = i\omega \sum_{j=1,6} \xi_j \phi_j$, ξ_j denotes the 6-DOF motion amplitudes and m denotes multi-marine buoys.

The Laplace equation and free surface, object surfaces, bottom, and far-field radiation conditions for each of these buoys should be satisfied by these value calculations constrained by boundary conditions. These boundary conditions, including free surface, sea bed, body boundary, radiation, and initial conditions were considered. The Laplace equation and free surface, object surface, bottom, and far-field radiation conditions for each of these buoys should be satisfied as following:

1. *Body Boundary Condition*

The Neumann condition was obtained based on a zero-flux condition for fluid across the boundary:

$$\left.\frac{\partial \phi}{\partial n}\right|_{S_B} = \mathbf{U} \cdot \mathbf{n} \tag{6}$$

where $\mathbf{U} \cdot \mathbf{n}$ denotes normal velocity on AUH boundary S_B due to its motion.

2. *Sea Bottom Boundary Condition*

The zero-flux condition for fluid across the sea-bottom boundary was calculated as

$$\left.\frac{\partial \phi}{\partial n}\right|_{S_H} = 0 \tag{7}$$

where S_H denotes the sea-bottom boundary.

    3. *Combined Kinematic and Dynamic Free Surface Conditions*

$$\left[\left(i\omega_e - U\frac{\partial}{\partial x}\right)^2 + g\frac{\partial}{\partial z}\right]\phi = 0 \text{ on } z = 0 \tag{8}$$

where ω_e denotes encounter frequency, z denotes depth in sea.

    4. *Boundary Condition at Infinity*

The disturbance generated by the AUH body vanished as distance r approached infinity:

$$\nabla\phi \to 0 \text{ as } r \to \infty \tag{9}$$

In concluding remarks, the problem-saving process can be illustrated: To solving the Laplace equation constrained by the above boundary conditions, we can get the velocity potential. Taking gradient of the velocity potential, the velocity distribution on the surfaces of the AUH and host ship can be obtained. Then, the AUH–Ship–Wave hydrodynamic interaction loads on the AUH can be obtained by adopting well-known Bernoulli equation of motion.

3. Simulation Results

In this section, numerical analysis that was carried out of AUH–Ship–Wave interactions is outlined. These case studies include: (1) AUH and host ship traveling at 0 knot speed in regular waves; (2) AUH traveling at 0.5–3.0 knot speed approaching the host ship; (3) AUH and host ship interacting in irregular waves with significant wave height (1 m), the Pierson–Moskowitz(P–M) wave-spectrum model adopted. In these cases, it is assumed that the mother ship has zero speed due to executing the dynamic positioning system which enables the ship to be fixed on the homing location of the AUH in waves. The initial set and layout of these simulations are shown in Figure 3.

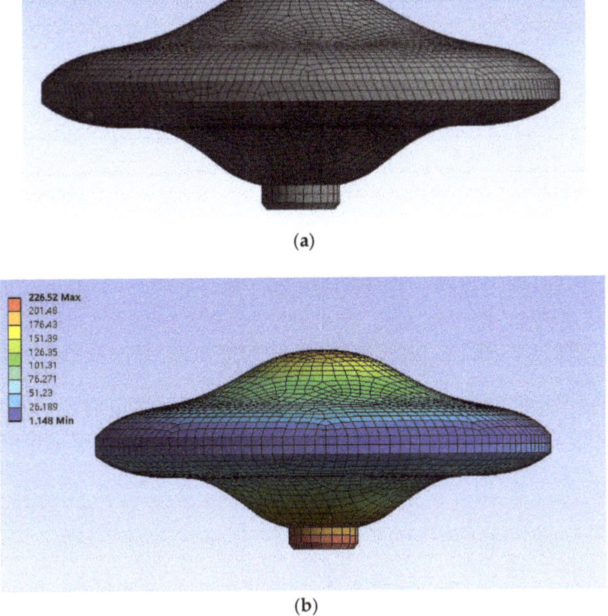

(a)

(b)

Figure 3. *Cont.*

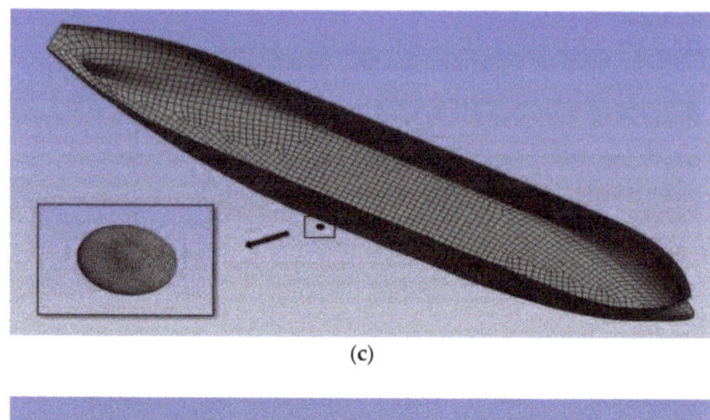

(c)

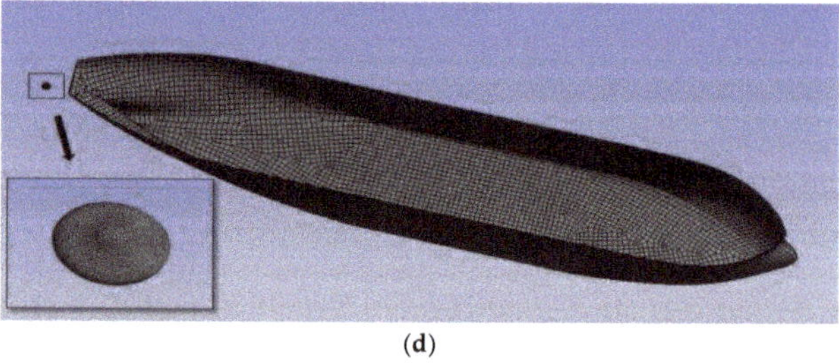

(d)

Figure 3. (**a**) Grids on the AUH surface with numbers of 6219 panels; (**b**) pressure distribution on the AUH panel surface with wave frequency of 1.48 rad/s; (**c**) the AUH approaching the midship of the host ship; (**d**) the AUH approaching the stern of the host ship.

3.1. Variations of Motion RAOs in AUH–Ship Interactions

The study calculated the 6-DOF RAOs [31,32] for AUHs located in three different positions along the X_1-axis of the host ship. RAOs are a concept in engineering statistics used to describe the motion responses of a floating buoy moving in the field of a ship or buoy design. This is performed by a transfer function that transforms wave excitation to floating-body movement while incorporating the effects of diffraction and radiation. Motion RAOs in 6-DOF motions for different sailing conditions can be expressed as

$$RAO(\omega_E, \chi) = \eta_i(\omega_E, \chi)/\zeta(\omega_E, \chi), \tag{10}$$

where η_i represents the magnitude of the i-DOF of the floating body, ζ represents the amplitude of a certain wave height in random wave conditions, ω_E represents encounter frequency, and χ represents encounter angle. When the amplitude is constant, larger RAO values correspond with greater floating-body-movement amplitudes in the i-DOF. Additionally, force RAOs can be directly derived using these determined motion RAOs as a base.

The degree of interference the host ship exerts on the AUH changes with the distance between the two. Identifying the position where the host ship influences the AUH could improve the feasibility and safety of AUH recovery, saving related resources. This study used the following parameters: Ship length L = 20 m, beam B = 4 m, draft t = 0.85 m, and AUH sizes shown in Tables 2 and 3 for three case studies (AUHs located at different host-ship X_1-coordinates). The host ship and AUH were both simulated to be travelling at speeds of 0 knots. The degree of interference the host ship exerted upon the AUH was investigated when the two floating bodies were at different distances in following sea (wave direction was 0°). The AUH was located in the three different X_1-axis positions shown in Table 2.

Table 2. Case study of AUH located in different X_1 coordinates.

Position	X_1 (m)	Y_1 (m)	Z_1 (m)
1	10	−2.4	−2
2	7	−2.4	−2
3	3	−2.4	−2

Table 3. Case study of AUH located in different Y_1 coordinates.

Position	X_1 (m)	Y_1 (m)	Z_1 (m)
1	10	−2.4	−2
2	10	−5	−2
3	10	−10	−2

The 6-DOF RAOs of the AUH were calculated for this study as shown in Figure 4a–d, where the AUH experienced disturbances (interference) from the host ship. Four degrees of freedom (surge, heave, roll, and pitch) were selected for the AUH–Ship interaction case studies, and the three curves shown for each of these represent the RAO variations versus wave frequencies in the three X_1-axis positions. $X_1 - Y_1$ plane positioning remained constant in these calculations. The red curve with triangles represents the AUH located near the center of the ship, the curve with black squares indicates a position near the rear midship, and the blue curve with dots indicates an AUH located near the stern of the ship.

For surge, shown in Figure 4a, AUH oscillations were significantly higher in Positions 1 and 2 than Position 3. For heave, in Figure 4b, all three curves reached a small peak when wave frequency reached 1 rad/s (indicating the AUH reached its resonance frequency), and Position 2 was unstable at this frequency. In practice, this wave frequency should be avoided. For roll, in Figure 3c it can be seen that the AUH oscillated more intensely than in Position 1. For pitch, in Figure 4d, AUH oscillations were most pronounced in Position 2. The results of the above analysis showed that oscillation amplitude in Position 3 was the smallest, indicating this position could be selected as an appropriate launch or recovery location. Three cases study were conducted to select three different Y_1 position in the same $X_1 - Z_1$ plane, as shown in Table 3.

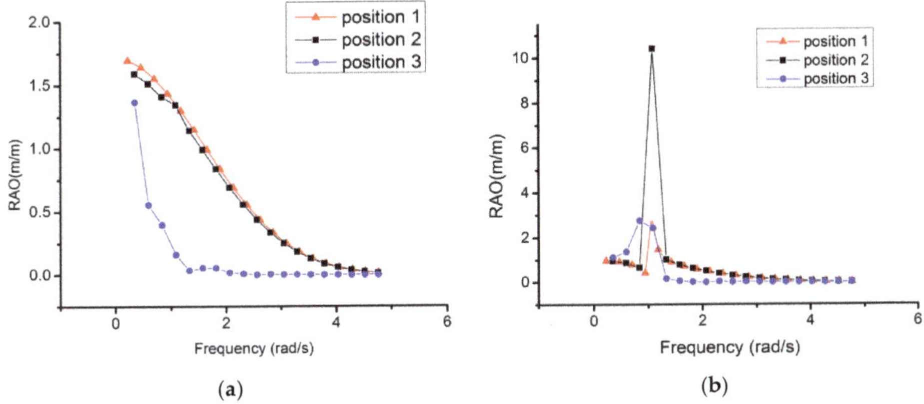

(a)

(b)

Figure 4. *Cont.*

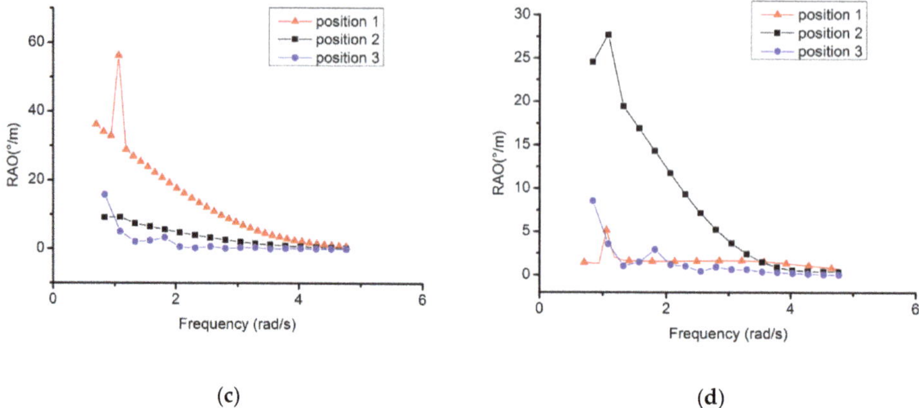

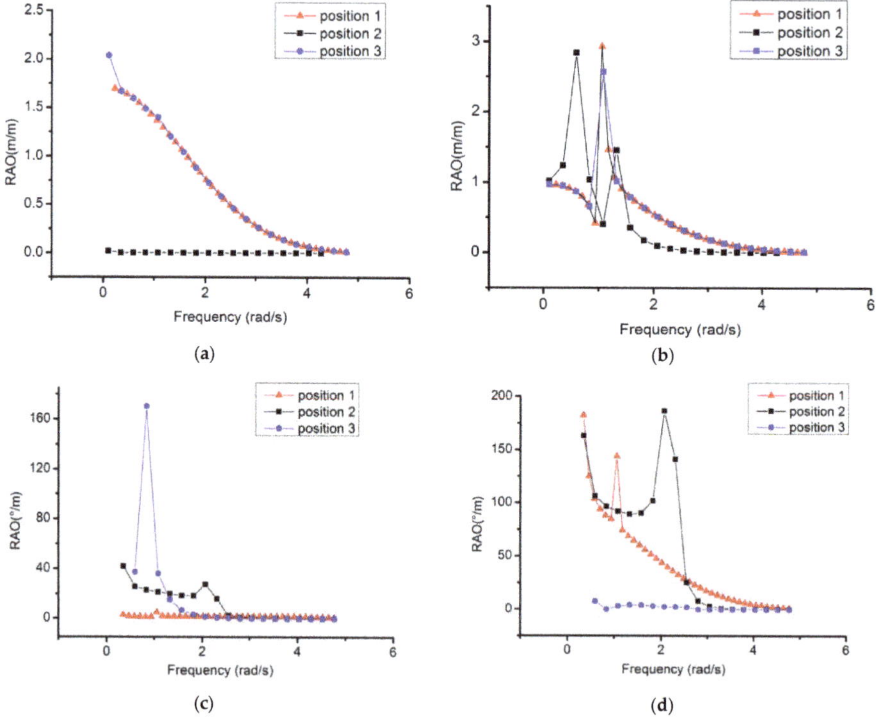

Figure 4. Six-degrees-of-freedom (6-DOF) motion response amplitude operators (RAOs) of AUH located in three different positions in X-axis. (**a**) Surge; (**b**) heave; (**c**) roll; (**d**) pitch.

For surge, as shown in Figure 5a, the AUH oscillates more slowly in Position 2. For heave, shown in Figure 5b, the three curves reach a peak when wave frequency is 1 rad/s (indicating a resonance phenomenon), implying work with AUHs should avoid these conditions. For pitch, seen in Figure 5c, the AUH is unstable at Position 3 with a wave frequency of 1 rad/s. For yaw, seen in Figure 5d, AUH oscillation amplitudes at Positions 1 and 2 are larger than at Position 3. Overall, Position 1 is shown to be the favorable location for AUH recovery and launch.

Figure 5. AUH 6-DOF RAOs in different position in Y-axis. (**a**) Surge; (**b**) heave; (**c**) pitch; (**d**) yaw.

3.2. Comparing AUH and AUH–Ship RAOs in Optimal Positions

Position $X_1 = 3$ m, $Y_1 = -2.4$ m, $Z_1 = -2$ m in the host-ship coordinate system was selected as the launch and recovery position due to its smaller oscillation amplitude in the above calculation results, indicating significant workforce and material-resource savings. This study examined the interference of the host ship with the AUH during recovery at the stern of the host ship. Motion RAO variations in the 6-DOF of AUH–Ship interactions are shown in Figure 6a–f for AUHs in the three investigated locations. Figure 6 shows that the host ship exerts significant influence on the AUH RAOs, especially sway and pitch. In sway, as shown in Figure 6b, the presence of the host ship greatly increases AUH oscillations, deteriorating the seakeeping performance of the AUH.

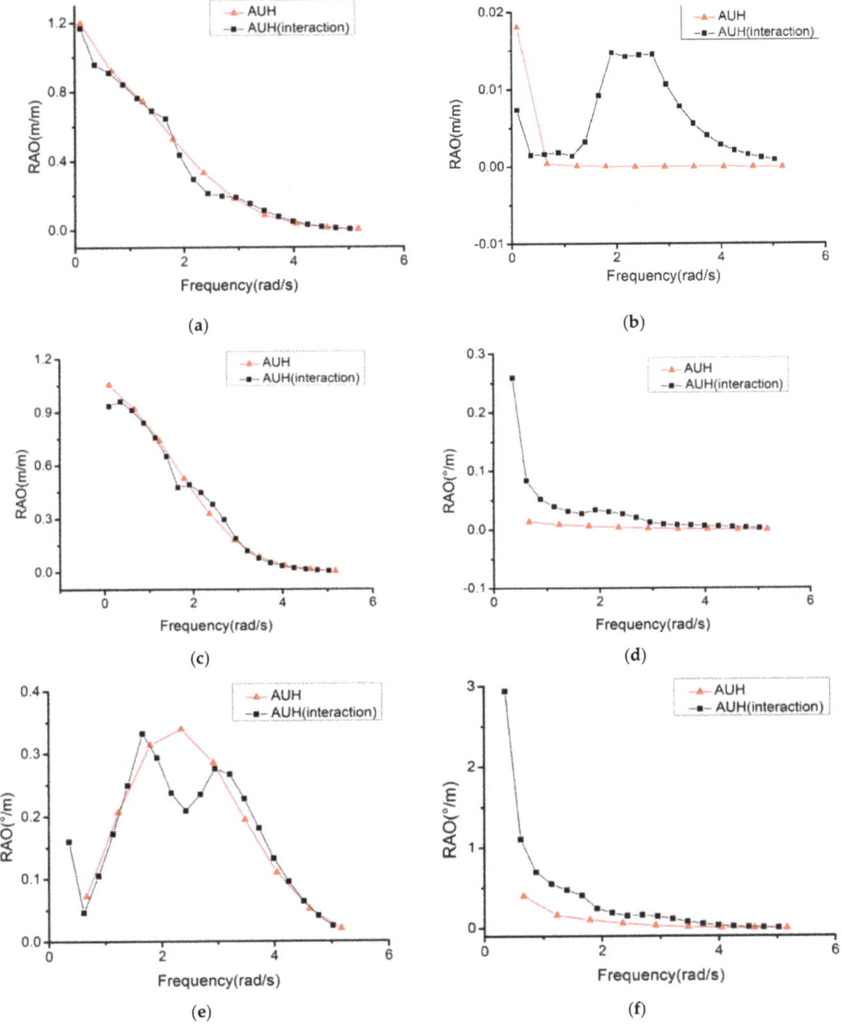

Figure 6. Six-DOF RAOs of AUH in special position. (**a**) Surge; (**b**) sway; (**c**) heave; (**d**) roll; (**e**) pitch; (**f**) yaw.

3.3. Hydrodynamic Interaction at Different Inflow Velocities

The hydrodynamic interaction of the AUH at different inflow velocities was calculated in the same conditions as Section 3.2 to observe their influence. Figure 7 shows the 2-DOF RAOs of the AUH in roll and yaw under different inflow velocities of 0.5–3.0 knots when disturbed by the host ship.

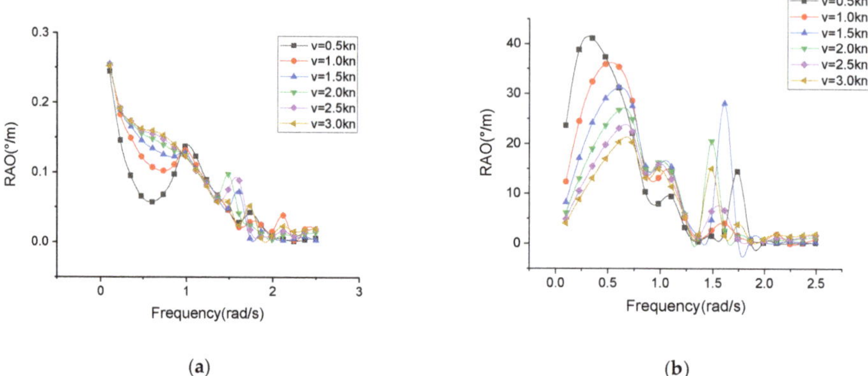

Figure 7. Two-DOF RAOs of AUH in different inflow velocities. (**a**) Roll and (**b**) yaw.

Figure 7 shows that AUH motion responses in roll and yaw are different when the AUH is at different velocities. At low wave frequencies, impact on AUH motion can be experienced. When controlling AUH maneuverability, special attention should be paid to the movement amplitude of the AUH at low frequency in order to prevent it from being in dangerous conditions.

3.4. AUH–Ship Interaction Motion in Irregular Waves

Time-domain analysis for an AUH at a specific location was conducted to further analyze the hydrodynamic interactions of AUHs influenced by disturbances from the host ship in irregular waves. Long-crested irregular waves were superimposed over a number of regular waves of different amplitudes and wavelengths, which could be expressed by a frequency spectrum [32]. According to analysis by Soares [32], wave-load response depends on spectral-model selection, including long- and short-crested irregular sea models. The former model is referred to as fully developed in the dominant wind direction. The latter represents wave crests formed at the right angles to the wind direction with waves propagating in different directions with a dominant spreading direction (directional spectrum). A motion-response spectrum [33] is given by

$$S_r(\omega_E, \chi) = \left| RAO(\omega_E, \chi) \right|^2 S_\zeta(\omega_E, \chi) \tag{11}$$

where $S_\zeta(\omega_E, \chi)$ represents a known seaway spectrum incorporating encounter frequency ω_E and encounter angle χ; $S_r(\omega_E, \chi)$ represents the response spectrum of r in 6-DOF (surge, sway, heave, roll, pitch, and yaw) within the distribution of the amplitude squared over ω_E and χ; elementary-wave-motion RAOs for the response spectrum of r were calculated as in Equation (9).

For long-crested irregular seas, this study adopted the Pierson–Moskowitz (P–M) wave-spectrum model to study AUH–Ship interaction-performance limitations in time series, as this spectrum is sufficient to cover many oceanic observations [34]. In this study, the significant wave height of 1 m of the P–M spectrum was translated into a time series, as shown in Figure 8, for time-variation simulations of RAO-based AUH motion in 6-DOF, as shown in Figure 9. The P–M wave spectrum is expressed as

$$S_X(\omega) = \frac{0.78}{\omega^5} \exp\left\{\frac{-0.78g_4}{U'\omega_4}\right\} (m^2 s) \tag{12}$$

where U' is wind speed at a height of 19.5 m. This wind speed can be determined using

$$U' = 6.85 \sqrt{h_{1/3}} \tag{13}$$

where $h_{1/3}$ is the significant wave height (1 m) and zero crossing period of 10 s. AUH and host-ship speeds are 0 knots for both, the encounter angle is 0° (following sea), and the resonant frequency value is

$$\omega_0 = 0.4 \sqrt{g/h_{1/3}} \tag{14}$$

The time series for wave-elevation numerical simulations, as shown in Figure 8, can be generated by Equation (15).

$$\zeta(x, y, t) = \sum_{n=1}^{N} \sqrt{2S_\zeta(\omega^*)\Delta\omega} \cos(\omega^* t + \varepsilon_n - k_n(x \cos \chi - y \sin \chi)) \tag{15}$$

where wave height $\zeta(x, y, t)$ incorporates space and time with ω^* taken randomly in each interval $[\omega_n - \Delta\omega/2, \omega_n + \Delta\omega/2]$ [35].

The 6-DOF RAO–based motion response of the AUH interacting with the ship in a time series with different wave directions is shown in Figures 9a–f and 10a–f, depicting AUH motions in surge (a), sway (b), heave (c), roll (d), pitch (e), and yaw (f). The speed of the AUH and the host ship is 0 knots and the wave-encounter angles are 0° and 90° (following sea and beam sea, respectively). In Figures 9 and 10, the oscillation amplitude of each DOF at different wave angles in irregular waves is not large, and the yaw amplitude, the largest of the DOFs, is up to 1 m. The simulation results of the RAO-based AUH responses show that the AUH moves gently in this position, making it a viable recovery and launch position.

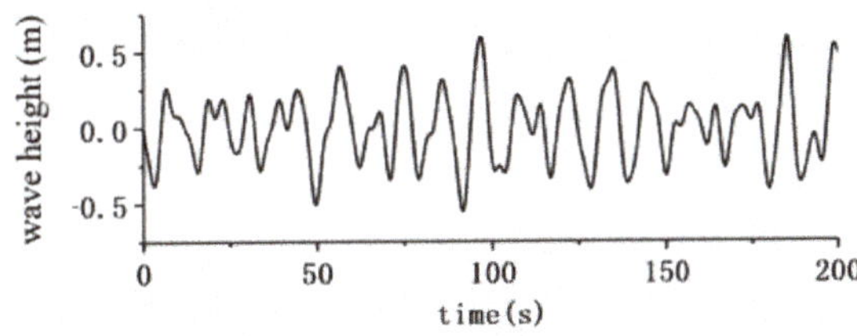

Figure 8. Time series at origin for Pierson–Moskowitz (P–M) spectrum for significant wave height of 1 m.

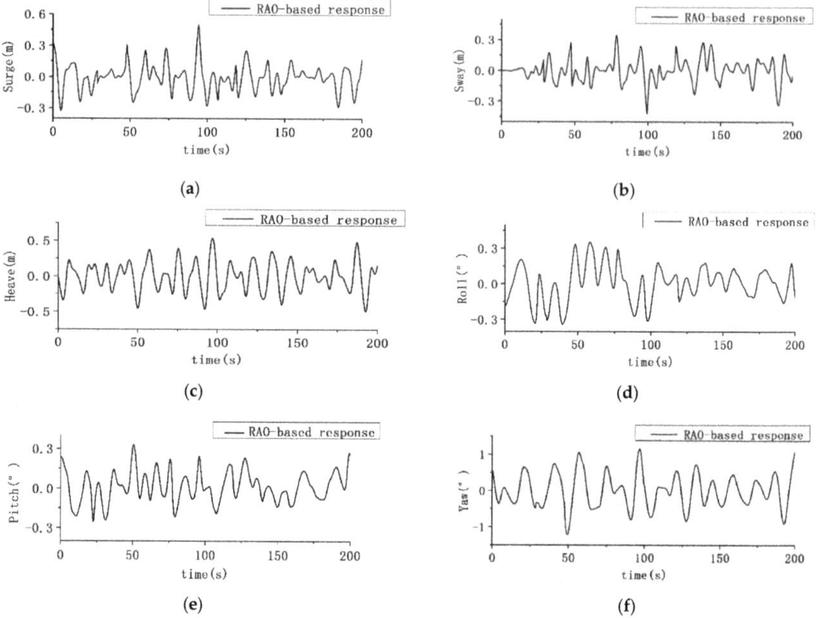

Figure 9. Time-series of RAO-based AUH motion in 6-DOF in following sea. (**a**) Surge; (**b**) sway; (**c**) heave; (**d**) roll; (**e**) pitch; (**f**) yaw.

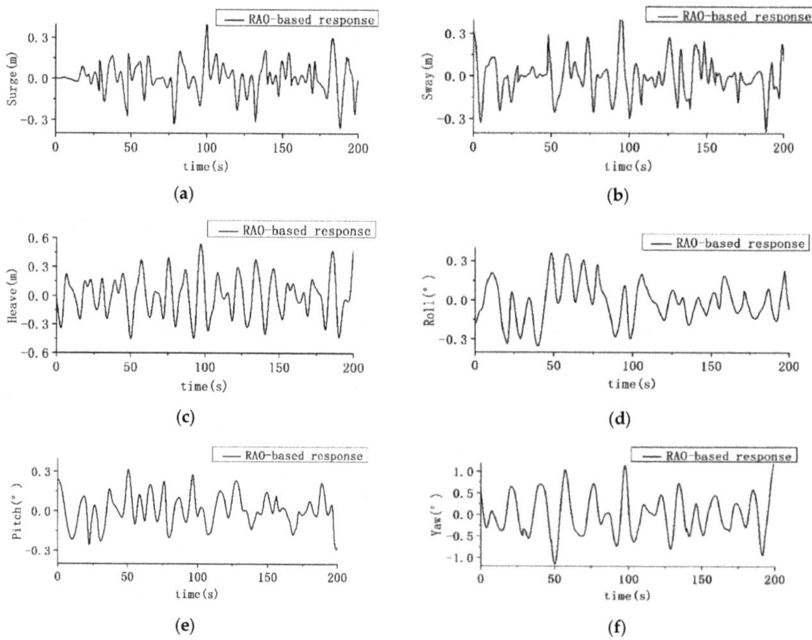

Figure 10. Time-series of RAO-based AUH motion in 6-DOF in beam sea. (**a**) Surge; (**b**) sway; (**c**) heave; (**d**) roll; (**e**) pitch; (**f**) yaw.

4. Experiment Validation

The experiments were carried out in the large-scale cross-sectional wave flume of Zhejiang University Ocean College in Zhoushan in December 2017. The flume was 75 m in length, 2 m in height, and 1.8 m in width, as shown in Figure 11a; the AUH was preadjusted to neutral buoyancy and then placed in the flume. A diagrammatic sketch of the wave-flume cross-section within the AUH is shown in Figure 11b. There was a wave maker at the end of the flume that could generate regular and irregular waves within a 0.5–5.0 s wave-period range and 0.02–0.6 m wave-height range. The waves in frequency range of 0.2–1.0 rad/s were made in the experiment to simulate a moderate sea state for recovering AUH in waves. The perspective view of the wave flume is shown in Figure 11c. The simulations of the wave flume and AUH with discrete panels, boundary conditions, and wave propagation direction for validation study is as shown in Figure 11d.

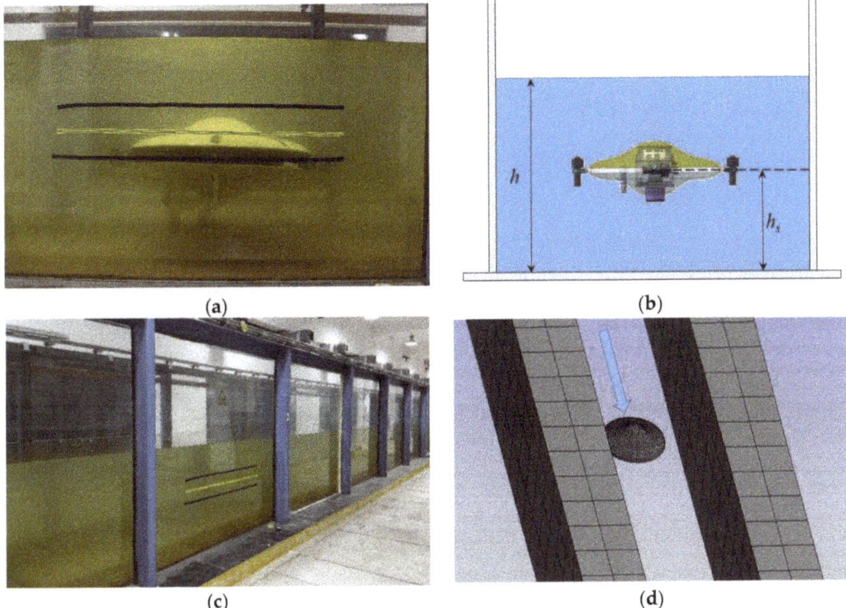

Figure 11. (**a**) AUH set in experiment wave flume in lateral view; (**b**) diagrammatic sketch of AUH in wave flume in frontal view; (**c**) wave flume with total length of 75 m; (**d**) simulation model of the wave flume and AUH.

The purpose of this experiment was to verify the accuracy of the numerical calculation results of the AUH wave force. Water depth h was 1.2 m, and distance h_S from bottom to the center of gravity was 0.6 m. Wave period T_w was set as 1, 1.5, 2, 2.5, 3, 3.5, 4, 4.5, and 5 s, and wave height H_w was set as 0.1 and 0.2 m. A WT901BLE-type attitude sensor was used to record AUH motion angle and angular-velocity information. Sensor measurement stability was 0.05°, and output frequency was 10 Hz.

The rolling angle of the AUH for different wave periods and heights was measured. Figure 12a–c shows the rolling angles of AUH change over time in three conditions: (1) $T_w = 2$s, $H_w = 0.2$m, (2) $T_w = 2.5$s, $H_w = 0.2$m, and (3) $T_w = 5$s, $H_w = 0.1$m. The rolling-angle response was more regular in high-frequency waves when compared to that in low-frequency waves.

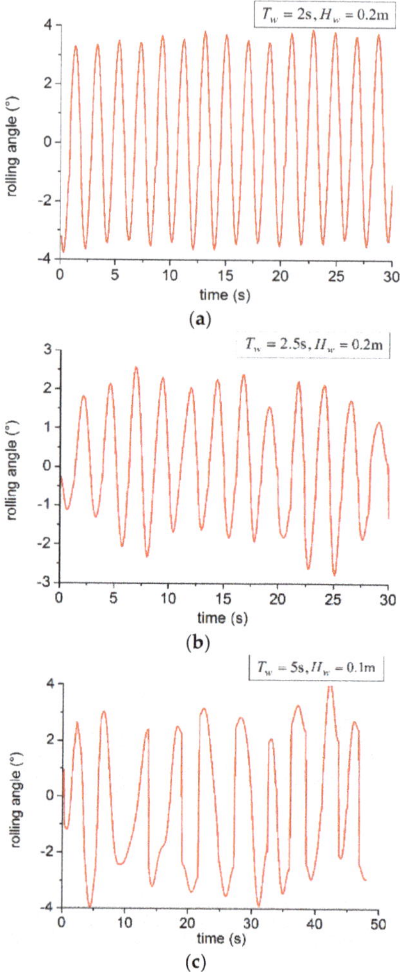

Figure 12. AUH rolling response in regular waves.

At the same time, AUH Numerical Simulation 1 in the restricted flume was conducted as shown in Figure 13. The boundary conditions were set in accordance with the experiment. Due to ANSYS-AQWA being potential-based hydrodynamic-simulation software, damping influence was neglected. Damping has great influence on the rolling motion of the AUH, i.e., viscous damping forces playing an important role while the AUH is in the condition where the lower wave frequency implies the greater wavelength encountered. The factor of damping was taken into account in Numerical Simulation 2, which increased the accuracy of the numerical simulation of the rolling direction. The improved effectiveness is obvious especially at low frequencies, as in Table 4. The proposed damping term for the vertical axisymmetrical AUH can be expressed as

$$D_{critical} = 2\sqrt{mK} = 2\sqrt{(I_{xx} + \Delta I_{xx})K_{Roll}} \tag{16}$$

where $D_{critical}$ represents critical damping; I_{xx} represents rolling inertia mass; K_{Roll} represents rolling stiffness matrix; and K and ΔI_{xx} represent the stiffness of corresponding degrees of freedom and added mass inertia mass, respectively.

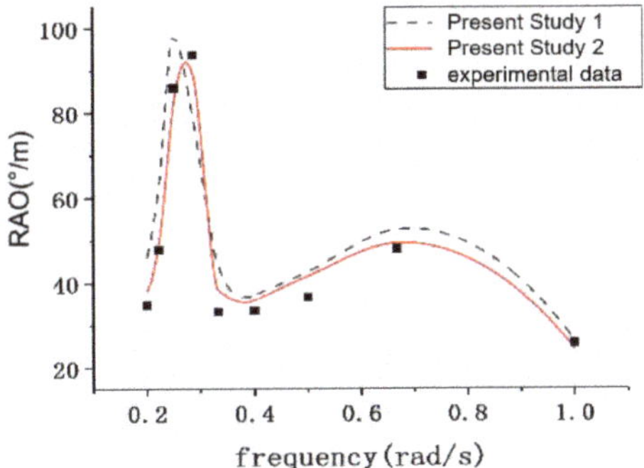

Figure 13. Comparison of RAO variations in roll versus wave frequency using the panel methods and experiment data.

Table 4. Comparison between experiment data and numerical results.

Frequency (Hz)	Experiment Data (°/m)	Study 1 (°/m)	Error 1	Study 2 (°/m)	Error 2
1	32.3	26.3	18.6%	24.3	14.13%
0.67	45.6	52.5	15.1%	49.3	8.11%
0.5	36.8	42.6	15.8%	41.6	13.04%
0.4	34.3	37.1	8.2%	35.9	4.67%
0.33	34.1	44.1	29.3%	38.3	12.32%
0.29	82.2	79.4	3.4%	89.3	9.97%
0.25	75.2	98.2	30.6%	82.3	9.44%
0.22	45.5	62.7	37.8%	49.9	9.67%
0.2	35.4	46.2	30.5%	38.3	8.19%
Average error			21.03%		9.95%

A comparison between numerical results of the AUH rolling RAO and the experiment data is shown in Figure 13 and Table 4. The overall trend of the AUH motion RAO in rolling was relatively consistent. From Table 4 we know that in wave-frequency range 0.2–1.0 Hz, the average error of the AUH rolling RAO was 21% in different wave frequencies, and minimum error was 3.4% in 2.9 Hz. Within the limits of experimental error and sensor precision, viscous force was not considered in the numerical calculation. Thus, the tolerance error was acceptable to prove the numerical results as a valid and rational. Study 1 was further optimized, and the results of Study 2 were more accurate by adding the damping term, as shown in Equation (16). This process reduced the error from 21.03% to 9.95%.

5. Conclusions

This study analyzed the hydrodynamic interaction of a new type of underwater robot, the AUH, in locations near a host ship using the computational fluid dynamic ANSYS-AQWA solver, a potential–based panel method. The simulation results of the investigated AUH–Ship interactions are significant to practical works, including research on automated ship recovery and launch systems to be used in advance for AUH operations in waves. This research can significantly enhance the efficiency of AUH homing automation to host ships in waves. The 6-DOF RAOs of an AUH at various distances from the host ship were predicted, and a suitable location area for AUH launch and recovery was determined.

The host ship exerted significant influence on AUH motion performance in sway and pitch. The influence of the host ship on interactive AUH hydrodynamic disturbances was determined in terms of 6-DOF RAOs. The RAOs of the AUH at different velocities were successfully predicted for enhanced adaptive controllability of homing automation. A specific location for optimizing AUH seakeeping performance during recovery and launch operations in long-crested irregular waves was analyzed in the time domain using the P–M wave-spectrum model to observe good positions. Simulation results showed that the sea condition with a wave frequency of 1 rad/s should be avoided due to the resonance phenomenon. AUH oscillation amplitude was the smallest at X = 3 m, Y = −2.4 m, Z = −2 m, and X = 10 m, Y = −2.4 m, Z = −2 m; these positions should be selected as an appropriate location for launch or recovery.

Experiment validation of the dish-type AUH motion RAO during roll and pitch was carried out. Numerical analysis of motion RAO in roll with a proposed damping model was compared to experiment data in the wave-frequency range of 0.2–1.0 Hz, resulting in average errors being reduced from 21.03% to 9.95%, verifying the accuracy of the adopted method. Comparisons between numerical results and experiment data showed that the tolerance error was acceptable to prove that the numerical results were valid and rational. In the future, AUH behavior induced by sensors and controllability in short-crested irregular waves will be further studied. An experimental study of the designed AUH configurations mounted with different aquatic transducer lengths, to achieve an acceptable trade-off between hydrodynamic optimization and acoustic navigation, will be carried out in future research and for practical applications.

Author Contributions: Conceptualization, C.-W.C.; Data curation, Y.C.; Formal analysis, C.-W.C. and Q.-W.C.; Funding acquisition, C.-W.C. and Y.C.; Investigation, C.-W.C. and Y.C.; Methodology, C.-W.C., Y.C. and Q.-W.C.; Project administration, Y.C.; Resources, Q.-W.C.; Software, C.-W.C., Y.C. and Q.-W.C.; Visualization, C.-W.C., Y.C. and Q.-W.C.; Writing—original draft, C.-W.C. and Q.-W.C.

Acknowledgments: The authors wish to thank the National Key Research and Development Program of China (No. 2017YFC0306100), and the National Natural Science Foundation of China (No. 51409230) for financially supporting the research on the hydrodynamic behavior of the Autonomous Underwater Hovering Vehicle.

Conflicts of Interest: The authors declare no conflict of interest.

Nomenclature

\	**Velocity potential**
U	AUH speed
−Ux	AUH sailing at a constant speed U
ϕ	Perturbation potential induced by AUH presence
S_B	Body boundary
U · n	Normal velocity on the AUH boundary
S_H	Sea-bottom boundary
ω_E	Wave-encounter frequency
ϕ_I	Velocity potential of incident wave
ϕ_D	Diffraction power of multifloating system
ϕ_{R1}	Radiation potential of floating body 1 in multifloating system
ϕ_{R2}	Radiation potential of floating body 2 in multifloating system
Re{}	Real part of amount in {}
H_w	Wave height
$X_2Y_2Z_2$	Position of AUH in body-fixed coordinates
h_S	Distance from bottom to center of gravity
CG, CB	Center of gravity, center of buoyancy
AUH	Autonomous Underwater Helicopter
AUH–Ship	Study on hydrodynamic interaction between AUH and Ship
ζ	**Wave height**
ξ	Bottom clearance coefficient
T	AUH height
RAO	Response amplitude operator
χ	Wave encounter angle
S_ζ	Seaway spectrum

S_r	Response spectrum of r in 6-DOF of surge, sway, heave, roll, pitch, and yaw
U'	Wind speed at height of 19.5 m
η_i	Magnitude of i-DOF of floating body
t	Nearest vertical distance between AUH and slope
T_w	Wave period
$D_{critical}$	Critical damping
I_{xx}	Rolling inertia mass
K_{Roll}	Rolling stiffness matrix
K	Stiffness of corresponding degrees of freedom
ΔI_{xx}	Added mass inertia mass
BG	Vertical distance between the CB and CG
RAO	Response Amplitude Operator

References

1. Ribas, D.; Palomeras, N.; Ridao, P.; Carreras, M.; Mallios, A. Girona 500 AUV: From Survey to Intervention. *IEEE/ASME Trans. Mechatron.* **2012**, *17*, 46–53. [CrossRef]
2. Yang, R.; Clement, B.; Mansour, A.; Li, M.; Wu, N. Modeling of a Complex-Shaped Underwater Vehicle for Robust Control Scheme. *J. Intell. Robot. Syst.* **2015**, *80*, 491–506. [CrossRef]
3. Chen, Q.; Su, R. Analysis of launch and recovery UUV model for submarine. *Ship Sci. Technol.* **2011**, *33*, 146–149.
4. Sarda, E.I.; Dhanak, M.R. A USV-Based Automated Launch and Recovery System for AUVs. *IEEE J. Ocean. Eng.* **2017**, *42*, 37–55. [CrossRef]
5. Wu, L.; Li, Y.; Su, S.; Yan, P.; Qin, Y. Hydrodynamic Analysis of AUV Underwater Docking with a Cone-Shaped Dock under Ocean Currents. *Ocean Eng.* **2014**, *85*, 110–112. [CrossRef]
6. Xie, N.; Iglesias, G.; Hann, M.; Pemberton, R.; Greaves, D. Experimental Study of Wave Loads on a Small Vehicle in Close Proximity to a Large Vessel. *Appl. Ocean Res.* **2019**, *83*, 77–87. [CrossRef]
7. Skomal, G.B.; Hoyos-Padilla, E.M.; Kukulya, A.; Stokey, R. Subsurface Observations of White Shark Carcharodon Carcharias Predatory Behaviour using an Autonomous Underwater Vehicle. *J. Fish Biol.* **2015**, *87*, 1293–1312. [CrossRef]
8. Leong, Z.Q.; Ranmuthugala, D.; Penesis, I.; Nguyen, H. Quasi-Static Analysis of the Hydrodynamic Interaction Effects on an Autonomous Underwater Vehicle Operating in Proximity to a Moving Submarine. *Ocean Eng.* **2015**, *106*, 175–188. [CrossRef]
9. Tian, W.; Song, B.; Ding, H. Numerical Research on the Influence of Surface Waves on the Hydrodynamic Performance of an AUV. *Ocean Eng.* **2019**, *183*, 40–56. [CrossRef]
10. Mansoorzadeh, S.; Javanmard, E. An Investigation of Free Surface Effects on Drag and Lift Coefficients of an Autonomous Underwater Vehicle (AUV) using Computational and Experimental Fluid Dynamics Methods. *J. Fluids Struct.* **2014**, *51*, 161–171. [CrossRef]
11. Malik, S.A.; Pan, G.; Liu, Y. Numerical Simulations for the Prediction of Wave Forces on Underwater Vehicle using 3D Panel Method Code. *Res. J. Appl. Sci. Eng. Technol.* **2013**, *5*, 5012–5021. [CrossRef]
12. Lighthill, M.J. Waves and hydrodynamic loading. In Proceedings of the Second International Conference on the Behaviour of Offshore Structures, London, UK, 7–10 July 1979.
13. Chen, G.-R.; Fang, M.-C. Hydrodynamic interactions between two ships advancing in waves. *Ocean Eng.* **2001**, *28*, 1053–1078. [CrossRef]
14. Choi, Y.R.; Hong, S.Y. An analysis of Hydrodynamic Interaction of Floating Multi-body Using High-Order Boundary Element Method. In Proceedings of the Twelfth International Offshore and Polar Engineering Conference, Kitakyushu, Japan, 26–31 May 2002; pp. 303–308.
15. Masashi, K.; Kazuaki, E.; Hiroshi, Y. Wave Drift Force and Moments on Two Ships Arranged side by side in Waves. *Ocean Eng.* **2005**, *32*, 529–555.
16. Hong, S.Y.; Kim, J.H.; Cho, S.K.; Choi, Y.R.; Kim, Y.S. Numerical and Experimental study on Hydrodynamic Interaction of side-by-side moored Multiple Vessels. *Ocean Eng.* **2005**, *32*, 783–801. [CrossRef]
17. Chen, C.W.; Jiang, Y.; Huang, C.H.; Ji, D.X.; Sun, G.Q.; Yu, Z.; Chen, Y. Computational fluid dynamics study of the motion stability of an autonomous underwater helicopter. *Ocean Eng.* **2017**, *143*, 227–239. [CrossRef]

18. Chen, C.W.; Huang, C.H.; Dai, X.K.; Huang, H.C.; Chen, Y. Motion and Control Simulation of a Dished Autonomous Underwater Helicopter. In Proceedings of the OCEANS 2017, Anchorage, AK, USA, 18–21 September 2017; pp. 1–6.
19. Chen, C.W.; Yan, N.M.; Leng, J.X.; Chen, Y. Numerical Analysis of Second-Order Wave Forces Acting on an Autonomous Underwater Helicopter using Panel Method. In Proceedings of the OCEANS 2017, Anchorage, AK, USA, 18–21 September 2017.
20. Chen, C.-W.; Jiang, Y. Computational Fluid Dynamics Study of Magnus Force on an Axis-Symmetric, Disk-Type AUV with Symmetric Propulsion. *Symmetry* **2019**, *11*, 397. [CrossRef]
21. Chen, C.W.; Yan, N.M. Prediction of Added Mass for an Autonomous Underwater Vehicle Moving Near Sea Bottom Using Panel Method. In Proceedings of the 2017 4th International Conference on Information Science and Control Engineering (ICISCE), Changsha, China, 21–23 July 2017; pp. 1094–1098.
22. Miller, P.A.; Farrell, J.A.; Zhao, Y.; Djapic, V. Autonomous Underwater Vehicle Navigation. *IEEE J. Ocean. Eng.* **2010**, *35*, 663–678. [CrossRef]
23. He, B.; Zhang, H.; Li, C.; Zhang, S.; Liang, Y.; Yan, T. Autonomous Navigation for Autonomous Underwater Vehicles Based on Information Filters and Active Sensing. *Sensors* **2011**, *11*, 10958–10980. [CrossRef]
24. Pérez-Alcocer, R.; Torres-Méndez, L.A.; Olguín-Díaz, E.; Maldonado-Ramírez, A.A. Vision-Based Autonomous Underwater Vehicle Navigation in Poor Visibility Conditions using a Model-Free Robust Control. *J. Sens.* **2016**, *2016*, 1–16. [CrossRef]
25. Ji, D.; Li, H.; Chen, C.W.; Song, W.; Zhu, S. Visual Detection and Feature Recognition of Underwater Target using a Novel Model-Based Method. *Int. J. Adv. Robot. Syst.* **2018**, *15*. [CrossRef]
26. Rauch, C.G.; Purcell, M.J.; Austin, T.; Packard, G.J. Ship of opportunity launch and recovery system for REMUS 600 AUV's. In Proceedings of the OCEANS 2008, Quebec City, QC, Canada, 15–18 September 2008.
27. Sharp, K.; Cronin, D.; Small, D.; Swanson, R.; Augustus, T. A cocoon-based shipboard launch and recovery system for large autonomous underwater vehicles. In Proceedings of the MTS/IEEE Oceans 2001, Honolulu, HI, USA, 5–8 November 2001.
28. Newman, J.N.; Landweber, L. Marine Hydrodynamics. *J. Appl. Mech.* **1977**, *45*, 457. [CrossRef]
29. Joseph, D.D. Potential flow of viscous fluids: Historical notes. *Int. J. Multiph. Flow* **2005**, *32*, 285–310. [CrossRef]
30. Seif, M.S.; Inoue, Y. Dynamic analysis of floating bridges. *Mar. Struct.* **1998**, *11*, 29–46. [CrossRef]
31. Sun, B.; Miao, Q.; Feng, X.; Jiang, H. A Numerical Method for the Calculation of the Wave Forces Acting on a Submarine Travelling near the Free Surface. *J. Ship Mech.* **1997**, *1*, 21–26.
32. Soares, C.G. Effect of spectral shape uncertainty in the short term wave-induced ship response. *Appl. Ocean Res.* **1990**, *12*, 54–69. [CrossRef]
33. Perez, T. *Ship Motion Control: Course Keeping and Roll Stabilisation Using Rudder and Fins*; Springer: London, UK, 2005.
34. Moreira, L.; Soares, C.G. H_2 and H_∞ Designs for Diving and Course Control of an Autonomous Underwater Vehicle in Presence of Waves. *IEEE J. Ocean. Eng.* **2008**, *33*, 69–88. [CrossRef]
35. Refsnes, J.E.; Sorensen, A.J.; Pettersen, K.Y. Model-Based Output Feedback Control of Slender-Body Underactuated AUVs: Theory and Experiments. *IEEE Trans. Control Syst. Technol.* **2008**, *16*, 930–946. [CrossRef]

 *symmetry*

Article

Computational Fluid Dynamics Study of Water Entry Impact Forces of an Airborne-Launched, Axisymmetric, Disk-Type Autonomous Underwater Hovering Vehicle

Chen-Wei Chen * and Yi-Fan Lu

Institute of Marine Structures and Naval Architecture, Ocean College, Zhejiang University, Zhoushan 316000, China
* Correspondence: cwchen@zju.edu.cn

Received: 2 August 2019; Accepted: 20 August 2019; Published: 2 September 2019

Abstract: An autonomous underwater hovering vehicle (AUH) is a novel, dish-shaped, axisymmetric, multi-functional, ultra-mobile submersible in the autonomous underwater vehicle (AUV) family. Numerical studies of nonlinear, asymmetric water entry impact forces on symmetrical, airborne-launched AUVs from conventional single-arm cranes on a research vessel, or helicopters or planes, is significant for the fast and safe launching of low-speed AUVs into the target sea area in the overall design. Moreover, a single-arm crane is one of the important ways to launch AUVs with high expertise and security. However, AUVs are still subject to a huge load upon impact during water entry, causing damage to the body, malfunction of electronic components, and other serious accidents. This paper analyses the water entry impact forces of an airborne-launched AUH as a feasibility study for flight- or helicopter-launched AUHs in the future. The computational fluid dynamics (CFD) analysis software STAR-CCM+ solver was adopted to simulate AUH motions with different water entry speeds and immersion angles using overlapping grid technology and user-defined functions (UDFs). In the computational domain for a steady, incompressible, two-dimensional flow of water with identified boundary conditions, two components (two-phase flow) were modeled in the flow field: Liquid water and free surface air. The variations of stress and velocity versus time of the AUH and fluid structure deformation in the whole water entry process were obtained, which provides a reference for future structural designs of an AUH and appropriate working conditions for an airborne-launched AUH. This research will be conducive to smoothly carrying out the complex tasks of AUHs on the seabed.

Keywords: autonomous underwater vehicle (AUV); airborne-launched AUV; autonomous underwater hovering vehicle (AUH); water entry impact force; computational fluid dynamics (CFD); two-phase flow

1. Introduction

Abundant mineral resources, power sources, and biological resources are conserved in the ocean, which is meaningful for economic development [1]. Because of the dark and hypoxic environment underwater, the need for an unmanned machine to replace people to complete underwater inventions is imperative. Autonomous underwater vehicles (AUVs) have a wide range of applications in oceanic geoscience, and they were created to accomplish resource exploration tasks on the seabed, including energy exchange, pipeline inspection, and roaming the deep-sea seabed [2]. AUVs have revolutionized our ability to detect and image the seabed with the real-time ability to exchange high-resolution, oceanographic, photomosaic information at abyssal depths [3–5]. However, conventional axisymmetric, torpedo-shaped AUVs have poor maneuverability when swaying and yawing because the added mass and added moment of inertia of the slender body dominates, which causes unstable AUV motions due to the occurrence of the Munk effect. In addition, motion instability in heave situations results from inefficient surface control in low-velocity conditions. Thus, the conventional horizontal, axisymmetric AUVs cannot effectively and efficiently accomplish unplanned super-mobile hovering tasks, for example, random landing and launching on the seabed, random hovering at some specified oceanic depth, providing services for a submarine mobile observation network, exploring submarine resources, connecting submarine operating point data in a specified smaller seabed area, and so on.

An autonomous underwater hovering vehicle (AUH), with a vertical, symmetric structural design, is a novel, dish-shaped, multi-functional, ultra-mobile submersible in the AUV family. AUHs are subject to high demands, for example, high-level and autonomous functionality, enhanced maneuverability in horizontal and vertical planes, adaptive landing and launching on the seabed, and the need to easily hover in the deep sea. The disk-shaped AUH was proposed at the Ocean College of Zhejiang University in 2016 to greatly enhance the motion stability during heaving, enable the AUH to hover in the deep sea, and improve the maneuverability of conventional, slender AUVs in yawing and swaying, including an enhanced anti-flow ability in the deep sea [6,7].

The structural design of the AUH is shown in Figure 1. This AUH could transmit data between base stations on the deep-sea bottom to communicate with scientific research ships near the free surface. The conceptual design of the AUH includes a symmetric, disk-shaped hull form, a pressure hull of 15 MPa, an energy power system, a control system, a navigation system, a communication system, and mission payload technology, e.g., recovering, landing, and launching models. The mission payload technology design in this study, i.e., airborne-launched technology, can enhance the effectiveness and efficiency of AUH missions. Thus, the study of the water entry impact forces on the disk-type AUH hull, with different water entry velocities and attitudes from a certain height, shall be carried out in this paper.

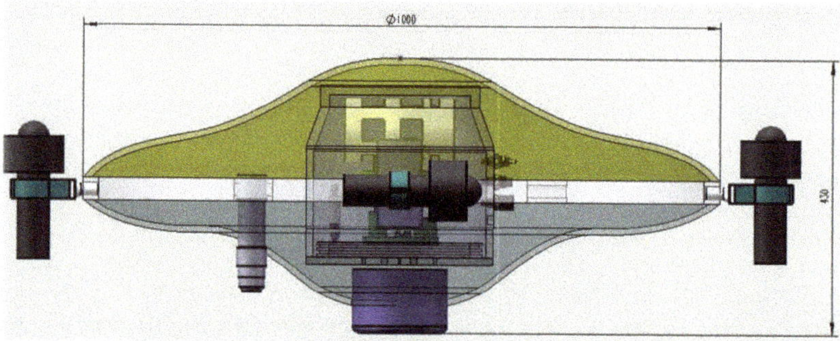

Figure 1. Conceptual design of the disk-type autonomous underwater hovering vehicle (AUH).

Research on launch and recovery systems for AUVs is significant to aid in the overall design process and guarantees successful, smooth deployment and operation of the AUH from the free surface to the deep sea [8]. The AUH hull was mounted with precise sensors, which were often damaged and lost from excessive water impact forces when the AUH was deployed and launched.

Traditionally, two main methods of AUV deployment are often adopted, i.e., a shore-based deployment or deployment by a scientific research ship [9]. Shore-based deployment technology is relatively mature; however, this needs excellent hardware support and good sea weather. Deployment by a scientific research ship is more difficult and brings about great uncertainty under atrocious sea conditions and other restrictions. Deployments by a scientific research ship include several forms: Deployment of a scientific research ship, an underwater vehicle (UV), and/or an unmanned surface vehicle (USV). An automated launch and recovery system for AUVs from an unmanned surface vehicle was proposed by Edoardo and Manhar [10].

Scientific research ships can be mounted with several launch and recovery devices, including conventional crane forms, A-shaped cranes, dedicated single-arm cranes, sliding cranes, and integrated cranes. Conventional forms are advantageous because of their simple structures and low costs; however, the operation is complex, and they are not as safe. A-shaped cranes have been widely adopted because of their simple operation. Dedicated single-arm crane systems are very safe, as shown in Figure 2. They greatly simplify operation processes and save costs [11]; however, time-consuming deployment and its inefficiency to quickly launch multiple AUVs to target sea areas are its weaknesses. Sliding-type crane systems are mainly used to continuously lay equipment or different types of remotely operated vehicle (ROV) cables. An integrated layout and recovery system operates, more or less, independently from the scientific research ship and has a high safety factor. Deployment of AUVs is better from a submarine than from a scientific research ship; however, recovering the AUVs is difficult. USV- and UV-based deployments need further improvement in order to be practical and reliable [10].

Figure 2. Conventional launching and recovery system of a single-arm crane.

In particular, airborne launch methods that use planes or helicopters to quickly launch AUVs into target sea areas have received more attention by scientists and strategists recently [11–21]. Scholars have conducted a lot of research to investigate water entry impact forces, trajectory deflection, and damage to an airborne-launched AUV, especially when the maximum impact loads occur in the initial entry stages [11].

Xia et al. [15] studied the water entry impact forces of an inclined, axisymmetric, slender body with a horizontal velocity and multiple degrees of motion freedom on the free surface. The effects of horizontal velocity, angle of attack, and inclined angle on the motion characteristics of the axisymmetric slender body were studied. A circuitous phenomenon was found when the angle of attack was greater than 22°.

Wang et al. [17] established an oblique water entry impact model, coupled with dynamic ballistic models, which was based on the theory of potential flow and the precise shape of the coupling surface between the fluid and the solid.

Qiu et al. [18] carried out simulations of the water entry impact forces on axisymmetric bodies, which was based on water entry dynamics and ballistic theories, to obtain the maximum impact load. The initial water entry conditions and the relationship between water entry impact loads were simulated. The simulation tests implemented relevant water entry processes for the revolution bodies (e.g., flat head, cone head, and round head), using commercial computational fluid dynamics (CFD) software FLUENT technologies, including dynamic mesh, user-defined functions (UDFs), and a mixture (MIXTURE) process model. The effects of velocity and head shape on the impact load and shape of cavitation were studied.

Qi et al. [19] presented the impact load of an AUV model under various water entry conditions as well as the varied rules of axial and radial forces during water infiltration through experiments and viscous CFD simulation methods. Reference data on the structural design and projection conditions of the AUV were provided.

Shi et al. [20] designed an inlet cap for an AUV and analyzed the influence of the buffer cap's structural design, material density, buffer distance, water flow velocity, and buffer effect of initial buffer on the water entry angles.

Ma et al. [21] implemented experimental investigations and analyzed the vertical water entry of a sphere. During water entry, the velocities, accelerations, and drag coefficients of the spheres were studied. The investigated results showed that the motion trajectory of the spheres presented highly nonlinear characteristics and notable fluctuations of the motion parameters, which were proportional to the entry speed.

In this paper, research on simulated water entry impact forces of an airborne-launched disk-type AUH based on the CFD method was implemented. The creative AUH hull form was different from the above water entry geometric shapes in literature. The STAR-CCM+ CFD Reynolds-averaged Navier–Stokes (RANS) solver was adopted to simulate air-launched AUH dynamic motions with different water entry speeds and immersion angles using the STAR-CCM+ volume of fluid (VOF) method, overlapping grid technology, and user-defined functions (UDFs). The simulation analysis was carried out under different water entry speeds and angles of the launched AUH in calm sea conditions. The variations of load and velocity of the disk-type AUH versus different states were obtained, i.e., in different initial free-fall velocities and water entry immersion angles. This study can provide an important reference for the disk-type, vertical, axisymmetric body of the AUHs to improve the structural design and adapt to the launching conditions, and it can enhance the effectiveness and efficiency of AUV deployments in order to smoothly carry out more complex tasks on the seabed.

2. Configuration of the AUH

In this paper, research on simulated water entry impact forces of the hung, air-launched AUH was carried out. The main parameters of the AUH are shown in Table 1, and the disk-shaped AUH prototype in the test pool is shown in Figure 3. Both Earth-fixed and body-fixed coordinates were established to describe the water impact loads and motion of the AUH, as shown in Figure 4. In the body-fixed coordinates, the hydrodynamic forces (surge, sway, and heave) and moments (roll, pitch, and yaw) exerted on the AUH in six-degrees of freedom can be designated as X, Y, Z, K, M, and N, respectively. The impact forces on the AUH were estimated in the body-fixed coordinate in this study, including the surge force (expressed in terms of X-force) and heave force (expressed as Z-force).

Table 1. Main parameters of the disk-shaped AUH.

Main Parameters	Symbol	Unit	Quantitative Values
Diameter of the hull	L	m	1.0
Height of the hull	d	m	0.45
Weight	W	Kg	136.45
Service speed	v	$m \cdot s^{-1}$	0.5144~1.5432
Density of fresh water	ρ_{water}	$kg \cdot m^{-3}$	1000.00
The viscosity coefficient of freshwater movement	γ_r	Pa·s	8.887×10^{-4}
Density of air	ρ_{air}	$kg \cdot m^{-3}$	1.00
Aerodynamic viscosity	γ_a	Pa·s	1.85508×10^{-5}
Run length	L_r	m	0.40
Reynolds number	Re	/	3.46×10^5

Figure 3. ZJU (Zhejiang University) AUH physical prototype.

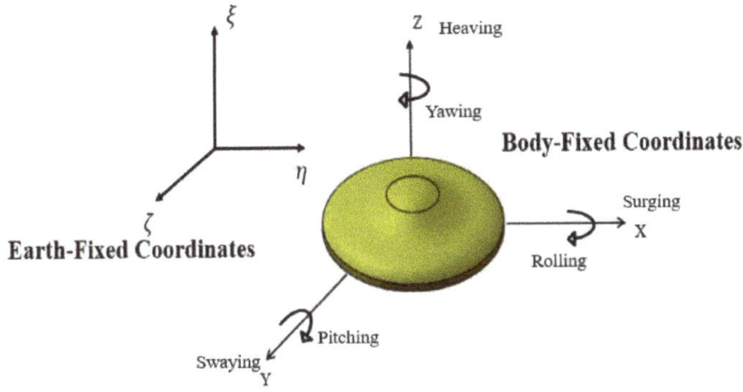

Figure 4. Earth-fixed and body-fixed coordinates.

3. Numerical Simulation of the Impact on Water Entry

This section describes the basic principles of solving the Navier–Stokes (N-S) equation by using the $k - \varepsilon$ model. It also introduces the setting to calculate the AUH water domain and boundary conditions (BC).

3.1. N-S Governing Equations and Turbulence Models

Reynolds-averaged Navier–Stokes (RANS) equations were solved by a numerical method using the software STAR-CCM+ as follows [22]:

$$\frac{\partial(\overline{u_i})}{\partial x_i} = 0, \tag{1}$$

$$\frac{\partial(\rho\overline{u_i})}{\partial t} + \rho\overline{u_j}\frac{\partial \overline{u_i}}{\partial x_j} = \rho\overline{F_i} - \frac{\partial \overline{P}}{\partial x_i} + \frac{\partial}{\partial x_j}\left(\mu\frac{\partial \overline{u_i}}{\partial x_j} - \rho\overline{u_i'u_j'}\right), \tag{2}$$

where u_i denotes the component of the average speed, u_i' denotes the turbulent fluctuation velocity component relative to the hourly average flow velocity, $\overline{F_i}$ denotes the component of the mass force, $\overline{P}$ denotes the pressure, μ is the fluid dynamic viscosity coefficient, and $\rho\overline{u_i'u_j'}$ denotes the Reynolds average stress.

The RANS equations are the current focus of computational fluid dynamics research. This method introduces fewer assumptions and is a method to calculate the viscous flow field with higher accuracy. Since RANS equations are not closed by themselves, a supplementary equation needs to be introduced to close it. Three of the most popular two-equation turbulence models, i.e., $k - \varepsilon$ model, $k - \omega$ model, and $k - \tau$ model, have been introduced in the literature [23]. In this paper, the $k - \varepsilon$ turbulence model and the STAR-CCM+ CFD solver were integrated with the volume of fluid (VOF) method, overlapping grid technology, and user-defined functions (UDFs) to simulate different situations of the disk-type AUH immersing into water, including varying the water entry velocities and angles of the AUH.

The two-equation turbulence model $k - \varepsilon$, i.e., turbulent kinetic energy k and turbulent dissipation rate ε, is expressed as follows [24]:

$$k = \frac{\overline{u_i'u_j'}}{2} = \frac{1}{2}\left(\overline{u'^2} + \overline{v'^2} + \overline{w'^2}\right) \tag{3}$$

$$\varepsilon = \frac{u}{\rho}\overline{\left(\frac{\partial u_j'}{\partial x_k}\right)\left(\frac{\partial u_i'}{\partial x_k}\right)}. \tag{4}$$

In the two-equation turbulence model $k - \varepsilon$, the corresponding transport equation can be expressed as follows:

$$\begin{cases} \frac{\partial(\rho k)}{\partial t} + \frac{\partial(\rho k u_i)}{\partial x_i} = \frac{\partial}{\partial x_j}\left[\left(\mu + \frac{\mu_i}{\sigma_k}\right)\frac{\partial k}{\partial x_j}\right] + G_k + G_b - \rho\varepsilon - Y_M + S_k \\ \frac{\partial(\rho\varepsilon)}{\partial t} + \frac{\partial(\rho\varepsilon u_i)}{\partial x_i} = \frac{\partial}{\partial x_j}\left[\left(\mu + \frac{\mu_i}{\sigma_\varepsilon}\right)\frac{\partial\varepsilon}{\partial x_j}\right] + G_{1\varepsilon}\frac{\varepsilon}{k}(G_k + G_{3\varepsilon}G_b) - G_{2\varepsilon}\rho\frac{\varepsilon^2}{k} + S_\varepsilon \end{cases} \tag{5}$$

where G_k denotes generation of turbulent kinetic energy; k is caused by the average speed gradient; G_b denotes generation of turbulent kinetic energy caused by buoyancy; Y_M is the contribution of pulsatile expansion in compressible turbulence; $G_{1\varepsilon}$, $G_{2\varepsilon}$, and $G_{3\varepsilon}$ denote the empirical constants; σ_k and σ_ε denote the Prandtl numbers corresponding to the turbulent kinetic energy k and dissipation rate ε, respectively; and S_k and S_ε denote the user-defined source items.

The VOF method was adopted to tackle the problem of free surfaces. Both the liquid phase (water) and the gas phase (air) above the free surfaces are treated clearly by putting forward a (liquid) volume fraction α_1 and gas volume fraction α_2. The combined volume fraction of both phases should satisfy the conservation property. A conservation equation was solved to transport the volume fraction of one of the phases in this study. The density, ρ, and viscosity, μ, at any point are acquired by averaging volume phases as follows [25]:

$$\begin{cases} \rho = \alpha\rho_{water} + (1-\alpha)\rho_{air} \\ \mu = \alpha\mu_{water} + (1-\alpha)\mu_{air} \end{cases}. \tag{6}$$

A single momentum equation was solved for the whole domain, resulting in a shared velocity field for both phases. The VOF defines a step function, α, equal to unity at any point occupied by water, and zero elsewhere, such that for volume fraction α, three conditions are considered as follows:

$$\alpha = \begin{cases} 1 & \text{if} \quad \text{cell} \quad \text{is} \quad \text{full} \quad \text{of} \quad \text{water} \\ 0 & \text{if} \quad \text{cell} \quad \text{is} \quad \text{full} \quad \text{of} \quad \text{air} \\ 0 < \alpha < 1 & \text{if} \quad \text{cell} \quad \text{contains} \quad \text{both} \quad \text{water} \quad \text{and} \quad \text{air} \end{cases} . \tag{7}$$

The VOF method adopted to treat two-phase flows has been given in detail in Hirt and Nichols (1981) [26]. Tracking the interface between air and water is completed by solving a volume fraction continuity equation as follows:

$$\frac{\partial \alpha}{\partial t} + U_j \frac{\partial \alpha}{\partial x_j} = 0. \tag{8}$$

3.2. Boundary Conditions

As shown in Figure 5, the computational domain for a steady, incompressible, two-dimensional flow of water with physically specified boundary conditions (BC) was identified. Two compounds were modeled in the flow field: Liquid water and air above the free surface of the calm sea. The AUH model was 1.0 m in diameter and 0.45 m in height. The coordinate origin was arranged at the center of gravity (CG) of the model. The computational domain was rectangular with the dimensions of 25.0 m × 3.0 m × 10.0 m, and the overset area near the AUH, where STAR-CCM+ overlapping grid technology was implemented, had the dimensions of 2.0 m × 1.0 m × 1.0 m. The velocity inlet was specified at 5.5 m upstream of the AUH bow. The pressure outlet was specified at 9.5 m downstream of the AUH. The initial height of the AUH model was specified as 0.5 m above the free surface, and the water depth was set as 6.0 m.

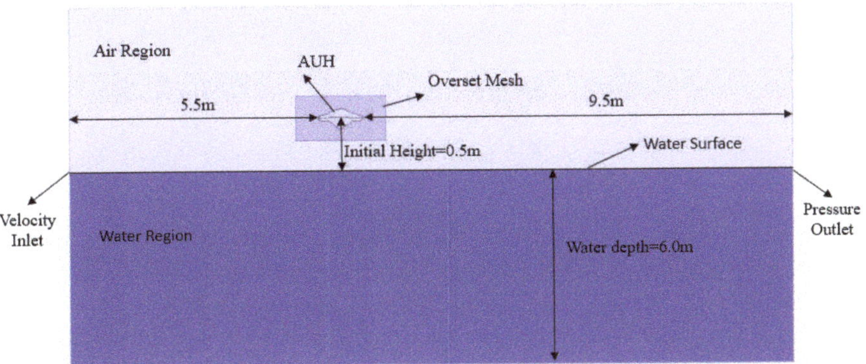

Figure 5. Computational domain for a steady, incompressible, two-phase flow of water and air with a free surface.

3.3. Meshing the Computational Domain

The CFD meshes could be divided into two categories: Structured meshes and unstructured meshes [27]. A hybrid mesh integrates structured and unstructured meshes to increase the mesh density near walls. In this paper, six hexahedron meshes were adopted by using a cutting mesh generator, and the total number of cells was 1.0×10^6. The encryption process of meshing the AUH required a minimum mesh size of 7.5×10^{-3} m to ensure an accurate solution for the turbulence model that could meet y+ values greater than 30 [28]. It is worth mentioning that, to clearly simulate the variable process of AUH immersing into the water, meshing encryption at the air–water interface needs to be arranged, and connatural and miscellaneous dimensions need to be activated in the cutting mesh generator. The generated meshes of the two-phase flow and AUH are shown in Figure 6.

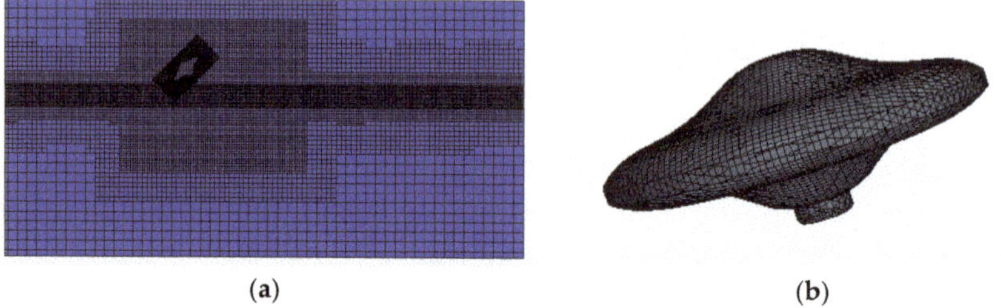

(a)   (b)

Figure 6. Mesh generations of the two-phase flow and AUH; (**a**) mesh of computational domain; (**b**) mesh partition of the AUH surface.

4. Simulation Results

This chapter introduces the water inflow process of AUH. It analyzes the changes to load and velocity in all directions during the AUH water inflow, and it analyzes the maximum load of the AUH at different velocities and at different water inlet angles.

4.1. Impact Force Load of the Water Entry Process

In this paper, the $k - \varepsilon$ turbulence model and the VOF method simulation technique in STAR-CCM+ software were adopted to simulate different situations of the AUH immersing into water from 0.5 m above the surface at different initial velocities. Figure 7a,b shows the two-phase flow simulation of the AUH water entry at an initial velocity of 3 m/s with different immersion angles: 30° and 60°, respectively. Distribution of the volume fraction of water to air is shown in the CFD simulation results (Figure 7). A time-varying, deformable cavity formed, and free surfaces were captured while the AUH was immersed into the water with immersion angles of 30°, 45°, 60°, and 90°. The initial velocities of the AUH were set to 3–8 m/s. In summary, since the forces acting on the symmetric, disk-type, non-spinning, inclined AUH after the impact are dictated by the cavity's dynamics, they are also affected by free surface conditions in cases where the cavity forms asymmetrically. The asymmetric degree of the formed cavity is highly correlated with the water entry velocity and immersion angle of the AUH.

Figure 8a–d shows that the surge force (X-force) changes under different immersion conditions during the initial AUH free fall, where the surge impact force on the AUH was measured with the body-fixed coordinates. When the AUH immersed into the water, the surge force reached the maximum value and then slowly decreased to a stable value. At the same immersion angle, the greater the initial velocity was, the greater the impact on the AUH. At the same initial speed, with an increase in the immersion angle of the AUH, the impact force on the AUH decreased. Notably, at the highest initial velocity of 8 m/s, the impact momentum of the surge on the AUH was smaller when the immersion angle was 90°, whereas a lower immersion angle of 30° caused a prominent impulsive force on the disk-type AUH.

Figure 9a–d shows the variations of the heave force (Z-force) in body-fixed coordinates versus different immersion angles over time. These results were similar to those of the surge force changes during the initial free-falling period; thus, the heave force is negligible. After the value reaches its peak, it decreases gradually over time until steady-state conditions are reached. In the case of the same immersion angle, a greater the initial velocity causes a greater impact force on the AUH. For different immersion angles with the same velocity as in Figure 9, a greater immersion angle causes a smaller impact force on the AUH. Therefore, a greater immersion angle can decrease the force of the impact on the AUH in surge and heave, which can be conducive to the floating state of the AUH when the AUH immerses into water.

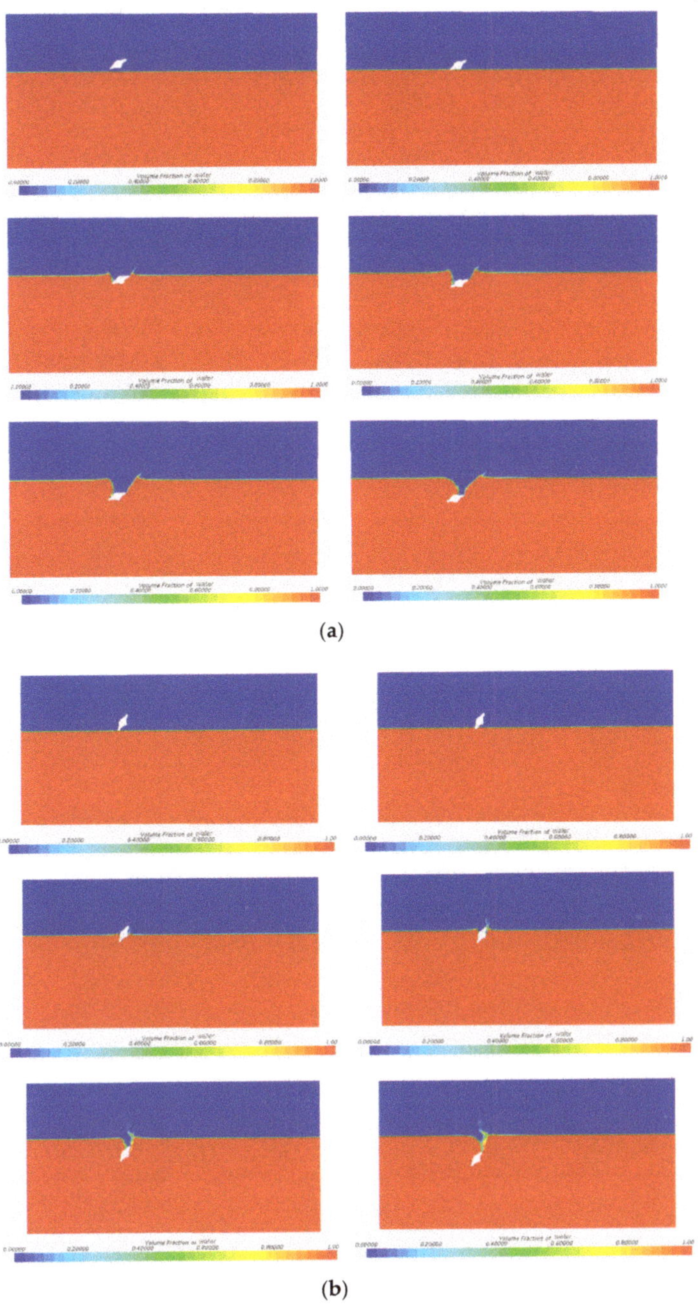

Figure 7. Two-phase flow simulation of the AUH at an initial velocity of 3 m/s with different immersion angles of 30° and 60°. (**a**) Immersion angle: 30°, initial velocity: 3 m/s; (**b**) immersion angle: 60°, initial velocity: 3 m/s.

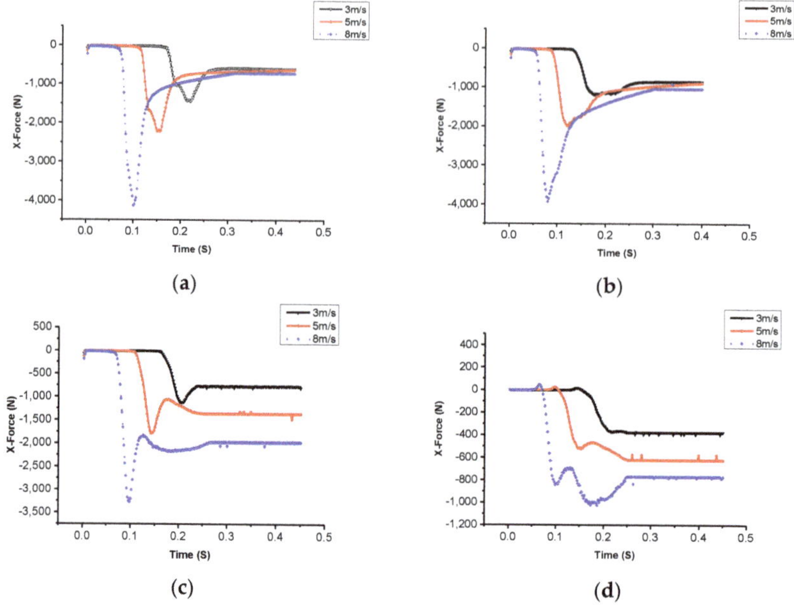

Figure 8. The variations of surge loads versus initial velocities of 3, 5, and 8 m/s at different immersion angles of 30°, 45°, 60°, and 90° over time. (**a**) Immersion angle: 30°; (**b**) immersion angle: 45°; (**c**) immersion angle: 60°; and (**d**) immersion angle: 90°.

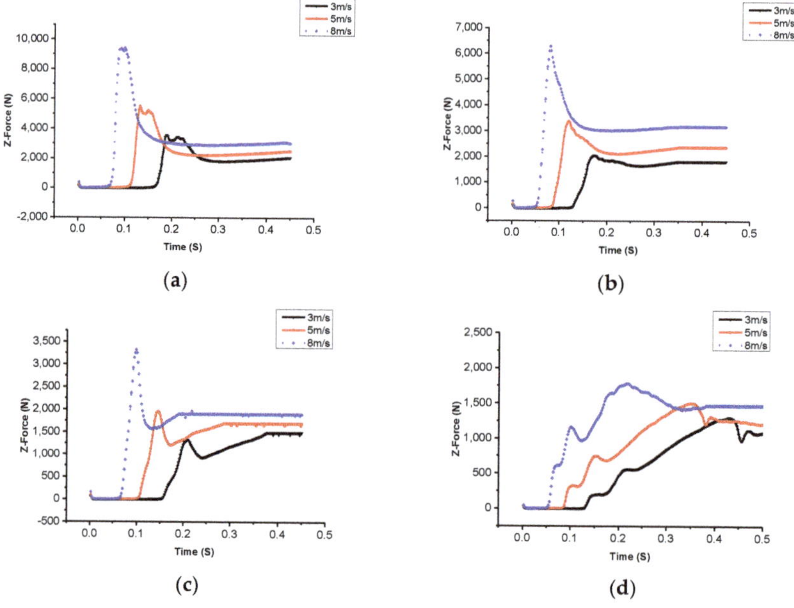

Figure 9. The variations of heave loads versus initial velocities of 3, 5, and 8 m/s at different immersion angles of 30°, 45°, 60°, and 90° over time. (**a**) Immersion angle: 30°; (**b**) immersion angle: 45°; (**c**) immersion angle: 60°; and (**d**) immersion angle: 90°.

4.2. Variations of the AUH Water Entry Velocity

Figure 10a–d shows the speed changes under different immersion conditions. There was an initial acceleration process for a very short period because this is the free-falling motion of the AUH. The greater the initial velocity, the shorter the free-fall duration was; the greater the immersion angle, the shorter the free-fall duration was. Subsequently, as the AUH head entered the water, the velocity of the AUH decreased over time, and the greater the initial velocity, the greater the speed reduction. For instance, when the immersion angle of the AUH was at 30° and the AUH hull was completely immersed in water, the velocity reduced by 7.0, 4.3, and 2.7 m/s at 8, 5, and 3 m/s, respectively. The smaller the immersion angle, the greater the maximum vertical velocity that could be achieved. The larger the immersion angle, the smaller the vertical velocity change was, and the AUH quickly stabilized.

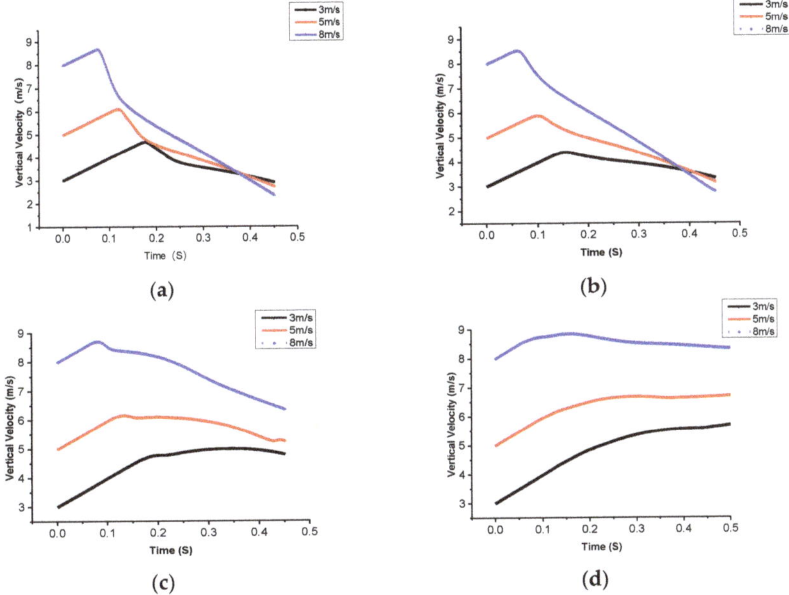

Figure 10. The variations of falling velocity in heave versus initial velocity (water entry velocity: 4.34, 5.90, and 8.59m/s) and immersion angles over time. (**a**) Immersion angle: 30°; (**b**) immersion angle: 45°; (**c**) immersion angle: 60°; and (**d**) immersion angle: 90°.

4.3. Maximum Load Analysis

Figure 11 shows the peak values of the impact surge loads on the AUH at different immersion angles. At the same angle, the initial velocity was directly proportional to the surge load. When the AUH immersion angle approached 30° and the initial velocity was 8 m/s, the load was up to 4000 N, which would seriously impact the safety of the AUH and would even cause electronic components to fail. At the same initial velocity, the smaller immersion angle caused a greater surge load to be experienced. Therefore, the results suggest that, to avoid a small immersion angle and a greater speed to decreasing the damage, the AUH structure should be mounted with precise sensors.

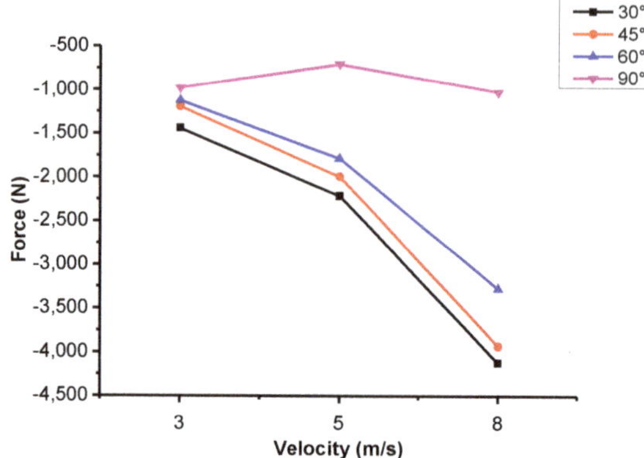

Figure 11. Variations of surge loads versus velocity at different immersion angles.

5. Conclusions

In this paper, viscous CFD two-phase flow simulations of a novel, vertical, axisymmetric AUH launched into water were carried out based on STAR-CCM+ VOF, UDF, and overlapping grid technologies. Numerical analyses of the axisymmetric impact forces on the disk-type AUH in the vertical plane were implemented with different water entry velocities and immersion angles. The deformable free surface and flow structure, from entry of the inclined AUH structure, were captured in a numerical tank, and the cavity around the AUH was found to form in a highly asymmetric way. The asymmetric degree of the formed cavity is highly correlated with the water entry velocity and immersion angle of the AUH.

In summary, the following conclusions are drawn:

(1) The total process of the AUH, with a traditional single-arm suspension or air-launched into the water, is accompanied by some significant phenomena such as surface uplift, splashing, asymmetrical cavity formation, and so on. The numerical stress analysis and maximum peak impact loads of the vertical axisymmetric hull body is emphasized in this paper. The variations of surge and heave loads versus initial velocities of 3, 5, and 8 m/s at different immersion angles of 30°, 45°, 60°, and 90° over time was studied.

(2) While the AUH with an improved single-arm suspension was immersed into the water, a peak impact load occurred. As the water depth increased, the impact decreased until a stable value was reached. The impact load of the AUH decreased with an increasing water entry angle, and the greater the initial falling velocity of the AUH was, the greater was the load that was experienced. In these simulations, the maximum peak impact loads on the AUH in surge and heave appeared at the initial velocity 8 m/s and immersion angle 30°, approaching 4000 N and 10^4 N, respectively. These quantities shall be used as a reference for the structural strength design of the AUHs.

(3) While the immersion angle of the air-launched AUH with a single-arm suspension immersed into the water decreased, the surge peak load increased. Thus, the dish-shaped AUH hull form should not have a small immersion angle into the water. Nevertheless, the silent phenomenon of acoustic transducers may appear when the water entry angle is 90°. In the trade-off study, these results suggest that the appropriate water entry immersion angle for the AUH should be 45°. The desired immersion angle will provide further design reference for air-launching transmitters of the AUHs.

Improvements will be made in future studies, including experiments with the same conditions as the numerical simulation. Based on a fabricated AUH, water entry impact force data on airborne-launched AUH, or on the air-launched AUHs using an improved single-arm suspension, will be collected in a sea trial to compare with the numerical simulation results. In addition, current and wave effects on launched AUHs will be considered in future work.

Author Contributions: Conceptualization, C.-W.C.; Data curation, Y.-F.L.; Formal analysis, Y.-F.L.; Funding acquisition, C.-W.C. and Y.-F.L.; Resources, C.-W.C.; Writing—original draft, C.-W.C. and Y.-F.L.; Writing—review & editing, C.-W.C. and Y.-F.L. The revision was done by C.-W.C. under the direction and supervision by C.-W.C.

Funding: This research was funded by the National Key R&D Program of China grant number 2017YFC0306100. And this research was funded by the National Science Foundation of China grant number 51409230. And this research was funded by the Zhejiang Zhoushan Science and Technology Project grant number 2018C81041.

Acknowledgments: The authors wish to thank the National Key R&D Program of China (Project Title: Autonomous Underwater Hovering-Vehicle, Project Number: 2017YFC0306100), the National Science Foundation of China (Project Title: Research on Vortex-Based Boundary Integral Method Estimating AUV Maneuverability Based on Euler–Rodriguez Quaternion Method, Project Number: 51409230), and the Zhejiang Zhoushan Science and Technology Project (Project Number: 2018C81041) for financially supporting the research on the hydrodynamic behavior of the AUHs.

Conflicts of Interest: The authors declare no conflicts of interest.

References

1. Yamazaki, T. Status of Development for Deep-sea Mineral Resources. *J. Jpn. Inst. Energy* **2009**, *88*, 569–576.
2. Leonard, J.J.; Bahr, A. Autonomous Underwater Vehicle Navigation. *J. Ocean. Eng. Technol.* **2010**, *35*, 663–678.
3. Yi, R.; Hu, Z.; Lin, Y.; Gu, H.; Ji, D.; Liu, J.; Wang, C. Maneuverability design and analysis of an autonomous underwater vehicle for deep-sea hydrothermal plume survey. In *2013 OCEANS—San Diego*; IEEE: San Diego, CA, USA, 2013; pp. 1–5.
4. Fossen, T.I. *Marine Control System, Guidance, Navigation and Control of Ships, Rigs and Underwater*; Marine Cybernetics: Trondheim, Norway, 2002; p. 586.
5. Phillips, A.B.; Turnock, S.R.; Furlong, M. The use of computational fluid dynamics to aid cost-effective hydrodynamic design of autonomous underwater vehicles. *Proc. Inst. Mech. Eng. Part M J. Eng. Marit. Environ.* **2010**, *224*, 239–254. [CrossRef]
6. Chen, C.-W.; Jiang, Y.; Huang, H.-C.; Ji, D.-X.; Sun, G.-Q.; Yu, Z.; Chen, Y. Computational fluid dynamics study of the motion stability of an autonomous underwater helicopter. *Ocean Eng.* **2017**, *143*, 227–239. [CrossRef]
7. Chen, C.-W.; Jiang, Y. Computational Fluid Dynamics Study of Magnus Force on an Axis-Symmetric, Disk-Type AUV with Symmetric Propulsion. *Symmetry* **2019**, *11*, 397. [CrossRef]
8. Chen, Q.; Sun, R. Analysis of launch and recovery UUV model for submarine. *Ship Sci. Technol.* **2011**, *33*, 146–149.
9. Zhang, H.L.; Deng, Z.Y.; Luo, Y.G. Development actuality of submarine launch and recovery system. *Ship Sci. Technol.* **2012**, *34*, 3–6.
10. Sarda, E.I.; Dhanak, M.R. A USV-Based Automated Launch and Recovery System for AUVs. *J. Ocean. Eng. Technol.* **2017**, *42*, 37–55. [CrossRef]
11. Yan, G.-X.; Pan, G.; Shi, Y.; Chao, L.-M.; Zhang, D. Experimental and numerical investigation of water impact on air-launched AUVs. *Ocean Eng.* **2018**, *167*, 156–168. [CrossRef]
12. Techet, A.; Truscott, T. Water entry of spinning hydrophobic and hydrophilic spheres. *J. Fluids Struct.* **2011**, *27*, 716–726. [CrossRef]
13. Truscott, T.T.; Epps, B.P.; Belden, J. Water Entry of Projectiles. *Annu. Rev. Fluid Mech.* **2014**, *46*, 355–378. [CrossRef]
14. Zhao, C.; Wang, C.; Wei, Y.; Zhang, X.; Sun, T. Experimental study on oblique water entry of projectiles. *Mod. Phys. Lett. B* **2016**, *30*, 1650348. [CrossRef]
15. Xia, W.; Wang, C.; Wei, Y.; Li, J. Experimental Study on Water Entry of Inclined Circular Cylinders with Horizontal Velocities. *Int. J. Multiph. Flow* **2019**, *118*, 37–49. [CrossRef]
16. Nair, V.V.; Bhattacharyya, S. Water entry and exit of axisymmetric bodies by CFD approach. *J. Ocean Eng. Sci.* **2018**, *3*, 156–174. [CrossRef]
17. WANG, Y.H.; SHI, X.H.; WANG, P.; WANG, S.W.; ZHAO, J.R. Modeling and Simulation of Oblique Water-Entry of Disk Ogive. *Torpedo Technol.* **2008**, *16*, 14–17.

18. QIU, H.Q.; YUAN, X.L.; WANG, Y.D.; LIU, C.L. Simulation on Impact Load and Cavity Shape in High Speed Vertical Water Entry for an Axisymmetric Body. *Torpedo Technol.* **2013**, *21*, 161–164.

19. Qi, D.; Feng, J.; Xu, B.; Zhang, J.; Li, Y. Investigation of water entry impact forces on airborne-launched AUVs. *Eng. Appl. Comput. Fluid Mech.* **2016**, *10*, 475–486. [CrossRef]

20. Shi, Y.; Pan, G.; Huang, Q. Water entry impact cushioning performance of mitigator for AUV. In *Oceans 2017–Aberdeen*; IEEE: Aberdeen, UK, 2017.

21. Ma, Q.P.; He, C.T.; Wang, C.; Wei, Y.J.; Lu, Z.L.; Sun, J. Experimental investigation on vertical water-entry cavity of sphere. *Explos. Shock. Waves* **2014**, *34*, 174–180.

22. Nouri, N.M.; Zeinali, M.; Jahangardy, Y. AUV hull shape design based on desired pressure distribution. *J. Mar. Sci. Technol.* **2016**, *21*, 203–215. [CrossRef]

23. Sodja, J. *Turbulence Models in CFD*; University of Ljubljana: Ljubljana, Slovenia, 2007; pp. 1–18.

24. Sakthivel, R.; Vengadesan, S.; Bhattacharyya, S.K. Application of non-linear k-e turbulence model in flow simulation over underwater axisymmetric hull at higher angle of attack. *J. Nav. Arch. Mar. Eng.* **2011**, *8*, 149–163. [CrossRef]

25. Nematollahi, A.; Dadvand, A.; Dawoodian, M. An axisymmetric underwater vehicle-free surface interaction: A numerical study. *Ocean Eng.* **2015**, *96*, 205–214. [CrossRef]

26. Hirt, C.W.; Nichols, B.D. Volume of fluid (VOF) method for the dynamics of free boundaries. *J. Comput. Phys.* **1981**, *39*, 201–225. [CrossRef]

27. Chow, P.; Cross, M.; Pericleous, K. A natural extension of the conventional finite volume method into polygonal unstructured meshes for CFD application. *Appl. Math. Model.* **1996**, *20*, 170–183. [CrossRef]

28. Jinxin, Z.; Yumin, S.; Lei, J.; Jian, C. Hydrodynamic performance calculation and motion simulation of an AUV with appendages. In Proceedings of the 2011 International Conference on Electronic & Mechanical Engineering and Information Technology, Harbin, China, 12–14 August 2011; pp. 657–660.

 *symmetry*

Article

Numerical Simulation and Mathematical Modeling of Electro-Osmotic Couette–Poiseuille Flow of MHD Power-Law Nanofluid with Entropy Generation

Rahmat Ellahi [1,2,*], Sadiq M. Sait [3], N. Shehzad [2] and N. Mobin [2]

[1] Center for Modeling & Computer Simulation, Research Institute, King Fahd University of Petroleum & Minerals, Dhahran 31261, Saudi Arabia

[2] Department of Mathematics & Statistics, FBAS, IIUI, Islamabad 44000, Pakistan

[3] Center for Communications and IT Research, Research Institute, King Fahd University of Petroleum & Minerals, Dhahran 31261, Saudi Arabia

* Correspondence: rellahi@alumni.ucr.edu

Received: 18 July 2019; Accepted: 8 August 2019; Published: 12 August 2019

Abstract: The basic motivation of this investigation is to develop an innovative mathematical model for electro-osmotic flow of Couette–Poiseuille nanofluids. The power-law model is treated as the base fluid suspended with nano-sized particles of aluminum oxide (Al_2O_3). The uniform speed of the upper wall in the axial path generates flow, whereas the lower wall is kept fixed. An analytic solution for nonlinear flow dynamics is obtained. The ramifications of entropy generation, magnetic field, and a constant pressure gradient are appraised. Moreover, the physical features of most noteworthy substantial factors such as the electro-osmotic parameter, magnetic parameter, power law fluid parameter, skin friction, Nusselt number, Brinkman number, volume fraction, and concentration are adequately delineated through various graphs and tables. The convergence analysis of the obtained solutions has been discussed explicitly. Recurrence formulae in each case are also presented.

Keywords: electroosmotic flow; power law fluid; nanoparticles; MHD; entropy generation; convergence analysis; residual error

---

1. Introduction

Conventional fluids like glycol, acetone, kerosene and water have low heat transfer characteristics and play an important role in laboratories and industrial engineering applications. Several experiments and procedures have been demonstrated to improve heat transfer characteristics. A significant contribution of boosting the heat transfer ability of fluids in an industrial process by the insertion of nanoparticles into the fluids was found. The pioneering work of Choi [1] led to the discovery that heat transfer rate and higher thermal conductivity can be enriched by a combination of base fluids and nanoparticles. Later, Xuan [2] experimentally investigated the enrichment of thermal conductivity in nanofluids. Some core developments have been collated in references [3–9].

Further, power-law fluids have attracted the attention of many scientists as the description of power-law fluids happens to be the best description of fluid behavior (with right choice of power-law index) of shear-dependent fluids. There are several more models that better describe the range of shear rates, but they do so at the cost of simplicity. For this reason, the power law is used to describe fluid behavior as compared to other fluid models. Recently, Ouyang et al. [10] proposed a two-dimensional squirmer model for power law fluids. They concluded that the selection of power law relation is the best choice to demonstrate the rheological properties of three sorts of fluids, namely: (i) Newtonian, (ii) shear-thickening, and (iii) shear-thinning fluids. Also, the exact velocity at the bottom region of shear rate cannot augur the basis of the power law model. After overcoming the insufficiency of power law, it cannot be used to find the exact velocity profile for the shear rate of the bottom and zero regions [11]. Also, power law fluid in some cases is used to describe non-Newtonian fluids [12,13]

Electro-osmotic flow, also known as electro-osmosis flow, is used in microfluidic devices and electronically controlled fluid flow or any other fluid conduit. Zhao et al. [14] have invoked electro-osmotic flow in a channel with the power law model. Das and Chakraborty [15] investigated the impact of non-Newtonian fluid on electro-osmotic flow. Several attempts regarding electro-osmosis flow can be found in references [16–19].

It is well known that magnetic fluids have significant utilization in applications such as heat transfer, cancer therapy, opto-electronic devices, sensors, cooling of nuclear reactors, liquid metal flow control, high temperature plasmas, solidification of binary alloys, drying processes etc. [20–25]. In addition, a number of researchers have conducted promising investigations on nanoparticles. Examples include the work of Malvandi et al. [26] who have studied MHD mixed convection with nanoparticle migration and Yousif et al. [27] who publicized magnetohydrodynamics (MHD) Carreau nanofluids with internal heat source/sink radiation. Further, magnetic nanofluids also exhibit interesting properties that are used in various microfluidic applications, dipolar nanoparticles, magnetic fluid hyperthermia and magnetic resonance imaging [28–31]. The existing literature bears witness that magnetic fluids with nanoparticles exhibit several interesting structural characteristics depending on the applied magnetic field strength. As a result, several studies have also been undertaken on MHD rheological fluids [32–34].

In recent years, prominent authors have been inspired to study entropy generation by various applications, such as diffusion and Joule heating. Entropy is caused by irreversible processes of a system and can be reduced only when it interacts with some other system whose entropy increases in the process. Bejan [35] studied the entropy generation for a flow system and discussed the main ideas to control the energy loss in flow problems and enhanced the ability of the system. Zeeshan et al. [36] discussed the radiative and electro-magnetohydrodynamic effects of a titanium dioxide/water based nanofluid on entropy generation. Ranjit and Shit [37] presented the effects of entropy generation with MHD on electro-osmotic flow. Numerical analysis for entropy generation on nanofluids with the suspension of nanoparticles (such as copper, Al_2O_3 and TiO_3) in water as a base fluid, which passes through wavy walls, was conducted by Cho et al. [38]. Different researchers have devoted their efforts to explore the impact of entropy generation to real world problems [39–43].

Thus far, the simultaneous effects of entropy generation, MHD and electro-osmosis on power law nanofluid flow has not been studied. In order to fill this gap, the next section is devoted to developing a mathematical formulation of the problem, which comprises mass, momentum, and energy equations. The situation becomes difficult as the resulting equations are not only nonlinear, but also coupled. The analytical findings are developed by a homotopic tactic [44] that has been used effectively for the last two decades [45–50].

The acquired results satisfied the governing equations and boundary conditions. The physical features involve parameters which are adequately determined through various graphs and tables.

2. Problem Design

2.1. Physical Considerations

Incompressible and steady nanofluids pass through horizontal parallel plates as illustrated in Figure 1. The wall at $\bar{y} = a$ moves with constant velocity U^* while the second wall remains stationary at $\bar{y} = -a$. B_0 is uniform transverse magnetic field, ξ_1 is zeta potential at lower wall, ξ_2 is zeta potential at upper wall, ψ is electric double layer potential and $2a$ is the total width of the channel.

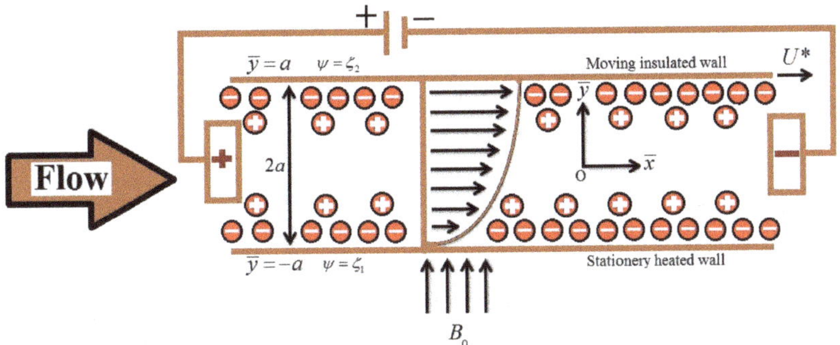

Figure 1. Schematic representation of electro-osmotic flow of the physical problem.

2.2. Electrical Potential Distribution

During the process of electro-osmotic flow (EOF), the separation of ions takes place and an electrical double layer (EDL) forms adjacent to the channel walls, thereby developing electric potential distribution. The electric potential $\overline{\psi}$ within the channel, described by the Poisson–Boltzmann equation [51] in the Cartesian co-ordinate system, can be calculated as:

$$\frac{\partial^2\overline{\psi}}{\partial \overline{x}^2} + \frac{\partial^2\overline{\psi}}{\partial \overline{y}^2} = -\frac{\overline{\rho}_e(\overline{y})}{\varepsilon} \tag{1}$$

For small values of electrical potential $\overline{\psi}$ of the EDL, the Debye–Hückel approximation can be applied and Equation (1) becomes:

$$\nabla^2\overline{\psi} = \overline{\kappa}^2\overline{\psi}, \ \overline{\kappa}^2 = \frac{2n_0z_v^2e^2}{\varepsilon k_B \hat{T}}. \tag{2}$$

The plates of the channel are made of different materials and have different zeta potentials.

$$\overline{\psi} = \zeta_1 \text{ at } \overline{y} = -a \text{ and } \overline{\psi} = \zeta_2 \text{ at } \overline{y} = a. \tag{3}$$

The electrical potential under the action of Equation (3) can be explored as:

$$\overline{\psi}(\overline{y}) = \frac{\zeta_1 \text{Sinh}(\overline{\kappa}(a-\overline{y})) + \zeta_2 \text{Sinh}(\overline{\kappa}(a+\overline{y}))}{\text{Sin}(2a\overline{\kappa})} \tag{4}$$

The electric double layer effects are produced by the external field relation $\mathbf{E} = (E_x, 0, 0)$ and charge density of nanoparticles. The external electric force $\overline{\rho}_e\mathbf{E}$, also called the electro-kinetic force [52,53], is generated outside of the charge particle. Along with $\overline{\rho}_e(\overline{y})$, electric charge density is defined as:

$$\overline{\rho}_e(\overline{y}) = -\varepsilon\overline{\kappa}^2\left(\frac{\zeta_1 \text{Sinh}(\overline{\kappa}(a-\overline{y})) + \zeta_2 \text{Sinh}(\overline{\kappa}(a+\overline{y}))}{\text{Sin}(2a\overline{\kappa})}\right) \tag{5}$$

where $\varepsilon, \overline{\kappa}, n_0, z_v, e, k_B$ and $\hat{T}$ are the relative permittivity of the medium, Debye parameter, ion density of bulk liquid, valence of ions, electron charge, Boltzmann constant and absolute temperature, respectively.

2.3. Power Law Model

In the current study, the following power law fluid model [54] is used:

$$\tau = \mu_{nf}\left(\frac{\partial \overline{u}}{\partial \overline{y}}\right)^{n-1}\left(\frac{\partial \overline{u}}{\partial \overline{y}}\right) = \begin{cases} \mu_{nf}\left(\frac{\partial \overline{u}}{\partial \overline{y}}\right)^{n} & \text{for } \frac{\partial \overline{u}}{\partial \overline{y}} > 0 \\ -\mu_{nf}\left(-\frac{\partial \overline{u}}{\partial \overline{y}}\right)^{n} & \text{for } \frac{\partial \overline{u}}{\partial \overline{y}} < 0. \end{cases} \tag{6}$$

where τ is shear stresses, n is the power-law or flow behavior index of the fluid and μ_{nf} is the viscosity of the nanofluid [55] and is defined along the consistency index δ as:

$$\mu_{nf} = \left(123\phi^2 + 7.3\phi + 1\right)\mu_f \tag{7}$$

and the viscosity of the base fluid in the current discussion is taken as:

$$\mu_f = \delta\left(\frac{\partial \overline{u}}{\partial \overline{y}}\right)^{n-1}. \tag{8}$$

One signifies a Newtonian fluid for $n = 1$ whereas $n < 1$ and $n > 1$ respectively denote the shear-thinning and shear-thickening of fluids.

To estimate the shear stresses of a fluid, we will illustrate further investigations with shear-thinning properties. Thus, shear stress for the non-Newtonian power-law model can be written as:

$$\tau = \delta\left(123\phi^2 + 7.3\phi + 1\right)\left(\frac{\partial \overline{u}}{\partial \overline{y}}\right)\left(\frac{\partial \overline{u}}{\partial \overline{y}}\right)^{n-1}. \tag{9}$$

2.4. Governing Equations

The electro-osmotic flow occurs in the channel due to the movement of an upper wall, whereas the flow around the channel of the walls is generated because of an applied electric field. The governing equations for the current flow phenomenon are:

$$\nabla.\mathbf{V} = 0 \tag{10}$$

$$(\mathbf{V}.\nabla)\mathbf{V} = \frac{1}{\rho_{nf}}\left(-\nabla\overline{p} + \nabla.\tau + Body\ force\right) \tag{11}$$

$$(\mathbf{V}.\nabla)T = \alpha_{nf}\nabla^2 T + \frac{1}{\left(\rho C_p\right)_{nf}\sigma_{nf}}J^2 + \frac{1}{\left(\rho C_p\right)_{nf}}\Gamma \tag{12}$$

where $\mathbf{V}$, T and Γ correspondingly represent velocity, temperature and viscous dissipation. The body force contains magnetic, electrical and buoyancy effects.

$$Body\ force = \underbrace{(\mathbf{J}\times\mathbf{B})}_{\text{Lorentz force}} + \underbrace{\overline{\rho}_e\mathbf{E}}_{\text{External Electric force}} + \underbrace{(\rho\beta)_{nf}\mathbf{g}(T - T_w)}_{\text{Buoyancy force}} \tag{13}$$

and

$$\Gamma = \mu_{nf}\left(\frac{\partial \overline{u}}{\partial \overline{y}}\right)^2. \tag{14}$$

Under the application of Lorentz force, it produces the following Joules heating effects.

$$\frac{1}{\sigma_{nf}}J^2 = \sigma_{nf}B_0^2\overline{u}^2 \text{ and } \mathbf{J}\times\mathbf{B} = \left(-\sigma_{nf}B_0\overline{u}, 0, 0\right) \tag{15}$$

The governing Equations (10) and (12) in components form after omitting the axial heat conductions at the walls and in the fluid [54] and become:

$$\frac{\partial \bar{p}}{\partial \bar{x}} = \mu_{nf}\frac{\partial^2 \bar{u}}{\partial \bar{y}^2} - \sigma_{nf}B_0{}^2\bar{u} + (\rho\beta)_{nf}(T - T_w)g + \bar{P}_e(\bar{y})E_x \tag{16}$$

$$\bar{u}\frac{\partial T}{\partial \bar{x}} = \alpha_{nf}\frac{\partial^2 T}{\partial \bar{y}^2} + \frac{\sigma_{nf}B_0{}^2}{(\rho C_p)_{nf}}\bar{u}^2 + \frac{\mu_{nf}}{(\rho C_p)_{nf}}\left(\frac{\partial \bar{u}}{\partial \bar{y}}\right)^2. \tag{17}$$

The associated boundary conditions are:

$$\left.\begin{array}{l}(\text{At upper wall}): \bar{u} = U^*, k_f\frac{\partial T}{\partial \bar{y}} = 0 \text{ at } \bar{y} = a \\[4pt] (\text{At lower wall}): \bar{u} = 0, -k_f\frac{\partial T}{\partial \bar{y}} = q_w \text{ at } \bar{y} = -a\end{array}\right\} \tag{18}$$

The effective density, heat capability, and thermal and electrical conductivities of a nanofluid [55] are respectively given by:

$$\frac{\rho_{nf}}{\rho_p} = \left[(1-\phi)\frac{\rho_f}{\rho_p} + \phi\right] \tag{19}$$

$$\frac{(\rho C_p)_{nf}}{(\rho C_p)_p} = \left[(1-\phi)\frac{(\rho C_p)_f}{(\rho C_p)_p} + \phi\right] \tag{20}$$

$$\frac{k_{nf}}{k_f} = \left(4.97\phi^2 + 2.72\phi + 1\right) \tag{21}$$

$$\frac{\sigma_{nf}}{\sigma_f} = \left[1 + \frac{3\left(\frac{\sigma_p}{\sigma_f} - 1\right)\phi}{\left(\frac{\sigma_p}{\sigma_f} + 2\right) - \left(\frac{\sigma_p}{\sigma_f} - 1\right)\phi}\right]. \tag{22}$$

The following dimensionless transformations

$$\left.\begin{array}{l}\bar{y} = ay, \ \bar{u} = u_m u, \ U^* = u_m U, \ \bar{p} = \rho_f u_m{}^2 p(a/u_m)^n, \ \theta = \frac{T - T_w}{q_w a/k_f}, \\[4pt] \bar{P}_e = -\left(\varepsilon\zeta_1/a^2\right)\rho_e, \ \bar{\kappa} = \kappa/a, \ \psi = \bar{\psi}/\zeta_1\end{array}\right\} \tag{23}$$

Transform Equations (16) and (17) by using (23) in the dimensionless form as:

$$\left(123\phi^2 + 7.3\phi + 1\right)n\left(\frac{\partial u}{\partial y}\right)^{n-1}\frac{\partial^2 u}{\partial y^2} - A_4 M^2 u + A_3 Gr\theta + \beta_u \rho_e - ReP = 0, \tag{24}$$

$$\left(4.97\phi^2 + 2.72\phi + 1\right)\frac{\partial^2 \theta}{\partial y^2} + Br\left(123\phi^2 + 7.3\phi + 1\right)\left(\frac{\partial u}{\partial y}\right)^{n+1} - B_1\gamma u A_4 + BrM^2 u^2 = 0. \tag{25}$$

$$\left.\begin{array}{llll}u = U, & \frac{\partial \theta}{\partial y} = 0 & \text{at} & y = 1 \quad (\text{Upper wall}) \\[4pt] u = 0, & \frac{\partial \theta}{\partial y} = -1 & \text{at} & y = -1 \quad (\text{Lower wall})\end{array}\right\}. \tag{26}$$

In which

$$\left.\begin{array}{l} Gr = \frac{(\rho\beta)_f g q_w a^3}{\delta k_f u_m}\left(\frac{a}{u_m}\right)^{n-1}, \ Re = \frac{\rho_f u_m a}{\delta}\left(\frac{a}{u_m}\right)^{n-1}, \ M^2 = \frac{\sigma_f B_0{}^2 a^2}{\delta}\left(\frac{a}{u_m}\right)^{n-1} \\[2mm] Br = \frac{\delta u_m{}^2}{q_w a}\left(\frac{a}{u_m}\right)^{1-n}, \ U_{Hs} = -\frac{\varepsilon \zeta_1 E_x}{\delta}\left(\frac{a}{u_m}\right)^{n-1}, \ \beta_u = \frac{U_{Hs}}{u_m}, \ \gamma = \frac{k_f u_m a}{\alpha_f q_w}\frac{\partial T}{\partial x}, \\[2mm] \rho_e(y) = \kappa^2\left(\frac{\mathrm{Sinh}(\kappa(1-y))+R_\zeta \mathrm{Sinh}(\kappa(1+y))}{\mathrm{Sin}(2\kappa)}\right), \ \alpha_f = \frac{(\rho C_p)_f}{k_f} \\[2mm] A_4 = \frac{\sigma_{nf}}{\sigma_f}, \ A_3 = \frac{(\rho\beta)_{nf}}{(\rho\beta)_f}, \ B_1 = \frac{(\rho C_p)_{nf}}{(\rho C_p)_f}, \ R_\zeta = \zeta_2/\zeta_1. \end{array}\right\} \tag{27}$$

where κ is the electro-osmotic parameter, ρ is density, μ is dynamic viscosity, β is volumetric volume expansion, Gr is Grashof number and C_p is specific heat. Assume that $u_m = -(a^2/2\mu_f)\partial p/\partial x$ is the maximum velocity between two plates, and β_u is the ratio between electro-osmotic velocity U_{Hs} and maximum velocity of the fluid. Thermophysical properties of alumina and the base fluid polyvinyl chloride (PVC) are illustrated below in Table 1.

Table 1. Physical properties of PVC [56] and Al$_2$O$_3$ [52].

Properties		C_p (J kg^{-1} K^{-1})	$\beta \times 10^{-5}$ (K^{-1})	ρ (kg m^{-3})	k (W m^{-1} K^{-1})	μ (Ns m^{-2})
Al$_2$O$_3$		765	8.5×10^{-1}	3970	40	- - - - -
	2%	4117.56	21.9	1006.24	0.586	0.0015
	3%	4085.34	21.8	1010.25	0.579	0.00107
PVC	4%	4053.12	21.8	1014.27	0.572	0.00114
	5%	4020.9	21.8	1018.29	0.718	0.00114
	6%	3988.68	21.8	1022.31	0.559	0.00116
	7%	3956.46	21.7	1026.33	0.552	0.00119

2.5. Thermophysical Relations

Skin friction is defined as:

$$C_f = \frac{2\tau_w}{\rho_f u_m^2} \ \text{and} \ \tau_w = \delta\left(123\phi^2 + 7.3\phi + 1\right)\left(\frac{\partial \overline{u}}{\partial \overline{y}}\right)\left(\frac{\partial \overline{u}}{\partial \overline{y}}\right)^{n-1}; \ \overline{y} = \pm a \tag{28}$$

Moreover, in the present study, different concentration of alumina particles (Al$_2$O$_3$) are used in the polymer solution of polyvinyl alcohol in water; consequently, different values of τ in correlation with nanoparticle volume fraction ϕ for different concentrations of polyvinyl alcohol are listed in Table 2.

Table 2. Properties of the power-law equation and PVC solutions [56].

PVC (%)	Consistency Index δ	Power Index n	Shear Stress
2	0.00494	0.790	$\tau = 0.00494\left(123\phi^2 + 7.3\phi + 1\right)(\partial\overline{u}/\partial\overline{y})^{0.790}$
3	0.00925	0.764	$\tau = 0.00925\left(123\phi^2 + 7.3\phi + 1\right)(\partial\overline{u}/\partial\overline{y})^{0.764}$
4	0.01557	0.734	$\tau = 0.01557\left(123\phi^2 + 7.3\phi + 1\right)(\partial\overline{u}/\partial\overline{y})^{0.734}$
5	0.02170	0.718	$\tau = 0.02170\left(123\phi^2 + 7.3\phi + 1\right)(\partial\overline{u}/\partial\overline{y})^{0.718}$
6	0.02616	0.691	$\tau = 0.02616\left(123\phi^2 + 7.3\phi + 1\right)(\partial\overline{u}/\partial\overline{y})^{0.691}$
7	0.03033	0.663	$\tau = 0.03033\left(123\phi^2 + 7.3\phi + 1\right)(\partial\overline{u}/\partial\overline{y})^{0.663}$

Hence, the coefficient of skin friction in the dimensionless form for both walls is:

$$C_f = \frac{2}{Re}\left(123\phi^2 + 7.3\phi + 1\right)\left(\frac{\partial u}{\partial y}\right)^n \ \text{at} \ y = \pm 1. \tag{29}$$

2.6. Heat Transfer Rate

The Nusselt number is defined as:

$$Nu = \frac{ah}{k_f} \text{ and } h = \frac{q_w}{T_w - T_m}$$

(30)

Here, h is heat transfer coefficient, q_w is wall heat flux and T_m is bulk mean temperature, which is given by:

$$T_m = \frac{\int \rho_f \overline{u} T dA}{\int \rho_f \overline{u} dA}.$$

(31)

The mean temperature in the dimensionless form becomes:

$$\theta_m = \frac{k_f(T_m - T_w)}{q_w a}.$$

(32)

Therefore, the Nusselt number is:

$$Nu = -\frac{1}{\theta_m}.$$

(33)

2.7. Entropy Generation

The entropy generation of local volumetric rate can be defined as:

$$S_G = \frac{k_{nf}}{T_w^2}\left(\frac{\partial \overline{T}}{\partial \overline{y}}\right)^2 + \frac{\mu_{nf}}{T_w}\left(\frac{\partial \overline{u}}{\partial \overline{y}}\right)^2 + \frac{\sigma_{nf}B_0^2\overline{u}^2}{T_w} + \frac{\rho_e(\overline{y})E_x}{T_w}.$$

(34)

The characteristic entropy generation can be expressed as:

$$S_0 = \frac{q_w^2}{k_f T_w^2},$$

(35)

By using the transformation given in Equation (23), the non-dimensional total entropy generation may be expressed as:

$$Ns = \frac{S_G}{S_0} = \left(4.97\phi^2 + 2.72\phi + 1\right)\left(\frac{\partial \theta}{\partial y}\right)^2 + \frac{Br}{\Omega}\left((123\phi^2 + 7.3\phi + 1)\left(\frac{\partial u}{\partial y}\right)^{n+1} + A_4 M^2 u^2 + \beta_u \rho_e u\right),$$

(36)

where M, Br, β_u, ρ_e and Ω are respectively the magnetic field, Brinkman number, volumetric volume expansion, dimensional electric charge density, and dimensionless temperature difference. These parameters are defined as:

$$M^2 = \frac{\sigma_f B_0^2 a^2}{K}\left(\frac{a}{u_m}\right)^{n-1}, \; Br = \frac{K u_m^2}{q_w a}\left(\frac{a}{u_m}\right)^{1-n}, \; \beta_u = \frac{U_{Hs}}{u_m},$$
$$\rho_e = -(\varepsilon \zeta_1/a^2)\rho_e, \; \Omega = \frac{q_w a}{k_f T_w}$$

(37)

$$Be = \frac{\text{Entropy due to heat transfer}}{\text{Total entropy geeration}}$$

(38)

$$Be = \frac{\left(4.97\phi^2 + 2.72\phi + 1\right)\left(\frac{\partial \theta}{\partial y}\right)^2}{(4.97\phi^2 + 2.72\phi + 1)\left(\frac{\partial \theta}{\partial y}\right)^2 + \frac{Br}{\Omega}\left((123\phi^2 + 7.3\phi + 1)\left(\frac{\partial u}{\partial y}\right)^{n+1} + A_4 M^2 u^2 + \beta_u \rho_e u\right)},$$

(39)

3. Discussion of Results

3.1. Analytic Solution

In this section, we intend to find the analytical solutions by means of the homotopy analysis method [57]. We chose the initial guesses u_0, θ_0 and linear operators $\pounds_1, \pounds_2$, [58] as:

$$u_0(y) = \frac{1}{2}(1+y)U, \ \theta_0(y) = \frac{1}{4}y(y-2). \tag{40}$$

$$\pounds_1 = \frac{d}{dy}\left(\frac{du}{dy}\right), \pounds_2 = \frac{d}{dy}\left(\frac{d\theta}{dy}\right). \tag{41}$$

The deformation equations of homotopy for the zeroth-order are established as:

$$\left. \begin{array}{l} (1-\xi)\pounds_1[\theta(y, \xi) - \theta_0(y)] = \xi\hbar_u N_1[u(y, \xi), \theta(y, \xi)], \\ (1-\xi)\pounds_2[\theta(y, \xi) - \theta_0(y)] = \xi\hbar_\theta N_2[u(y, \xi), \theta(y, \xi)] \end{array} \right\} \tag{42}$$

$$\left. \begin{array}{lcc} \text{For} & \xi = 0 & \xi = 1 \\ u(y,\xi): & u_0(y) & u(y) \\ \theta(y, \xi): & \theta_0(y) & \theta(y) \end{array} \right\} \tag{43}$$

The nonlinear operators N_1, N_2 are can be written as:

$$\left. \begin{array}{ll} N_1[u(y, \xi), \theta(y, \xi)] = & \left(123\phi^2 + 7.3\phi + 1\right)n\left(\frac{\partial u(y,\xi)}{\partial y}\right)^{n-1}\frac{\partial^2 u(y,\xi)}{\partial y^2} - A_4 M^2 u(y, \xi) + \\ & A_3 Gr\theta(y, \xi) + \beta_u \rho_e - ReP, \\ N_2[u(y, \xi), \theta(y, \xi)] = & \left(4.97\phi^2 + 2.72\phi + 1\right)\frac{\partial^2\theta(y,\xi)}{\partial y^2} + A_4 BrM^2(u(y, \xi))^2 + \\ & Br\left(123\phi^2 + 7.3\phi + 1\right)\left(\frac{\partial u(y,\xi)}{\partial y}\right)^{n+1} - B_1\gamma u(y, \xi) \end{array} \right\}. \tag{44}$$

The solution for velocity and temperature up to mth-order approximation can be expressed as:

$$u(y) = u_0(y) + \sum_{l=1}^{m} u_l(y), \theta(y) = \theta_0(y) + \sum_{l=1}^{m} \theta_l(y). \tag{45}$$

The best solutions for solving coupled differential equation can be presented as below at 20th order approximations:

$$u(y) = C_1 + C_2 y + C_3 y^2 + C_4 y^3 + C_5 y^4 + C_6 y^5 + C_7 y^6 + C_8 y^7 \tag{46}$$

$$\theta(y) = D_1 + D_2 y + D_3 y^2 + D_4 y^3 + D_5 y^4 + D_6 y^5 \tag{47}$$

The constants C_1 to C_8 and D_1 to D_6 are specified in Appendix A.

3.2. Convergence Inspection

As pointed out by Liao [59], the convergence of homotopic results can be controlled by the auxiliary parameters $\hbar_u$ and $\hbar_\theta$. In Figures 2 and 3, it is observed that the minimum error for the velocity and temperature profiles can be achieved at $\hbar_u = -0.5$ and $\hbar_\theta = -0.65$.

In addition, the residual error norms have been utilized to further ensure the accuracy of the obtained series solutions. The residual errors of the velocity and temperature profiles for two succeeding approximations of temperature E_θ and velocity E_u up to the 20th order iteration as given in Table 3 can be obtained by the following expressions:

$$E_u = \sqrt{\frac{1}{21}\sum_{i=0}^{20}(u(i/20))^2} \text{ and } E_\theta = \sqrt{\frac{1}{21}\sum_{j=0}^{20}(\theta(j/20))^2}. \tag{48}$$

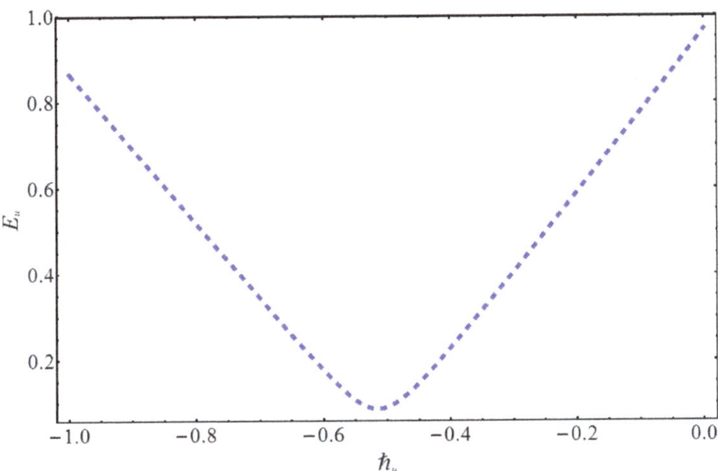

Figure 2. Graph of residual error for $\hbar_u$.

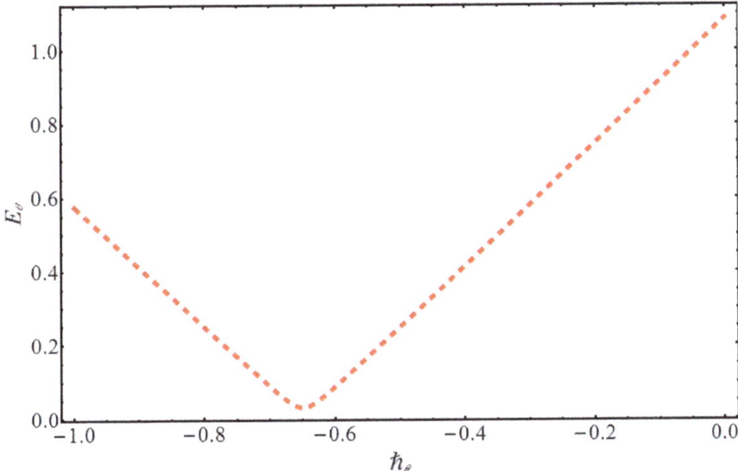

Figure 3. Graph of residual error for $\hbar_\theta$.

Table 3. Residual error of series solutions when $Gr = 3.1556$, $Br = 1$, $Re = 442.956$, $\beta = 1$, $M = 0.1$ and $pe = 1$.

Order of Approximation	Time	E_u	E_θ
05	8.2651	9.8020×10^{-3}	9.8561×10^{-3}
10	29.3761	9.3023×10^{-3}	7.6511×10^{-3}
15	62.4216	2.3452×10^{-3}	2.2438×10^{-3}
20	100.0125	1.0411×10^{-3}	1.9624×10^{-3}

3.3. Graphical Illustration

The current research is about flow and entropy on a magnetized power law nanofluid along the horizontal walls. The effects of the nanofluid can be determined at the same pressure with the electric body force and the motion of the upper plate. Flow and entropy generation on the magnetized power law of a nanofluid in the horizontal channel are studied systematically. The lower wall is heated, and the upper wall maintains the temperature. The influence of different important parameters such as Grashof number, electro-osmosis, Reynolds number, magnetic field, Brinkman number, entropy generation, nanoparticle volume fraction and Bejan number on temperature and velocity distributions were illustrated graphically in Figures 4–23. By using different parameters, we get different results which can be explained as $n = 0.764$ (PVC 3%), $\phi = 0.03$, $M = 2$, $\beta_u = 0.3$, $\gamma = 10$, $\kappa = 8$, $Gr = 2.366$, $Re = 442.956$ and $Br = 1$.

The impact of M on temperature and velocity is demonstrated in Figures 4 and 5. They show that when the magnetic parameter increases, the velocity profile decreases while the temperature profile increases. When one applies a magnetic field on electrically treated nanofluid, then it produces a force opposite to the flow direction. Consequently, the Lorentz force increases the magnetic parameter which opposes the fluid flow, and due to that, the velocity distribution decreases and the temperature distribution upsurges. The effects of volume fraction ϕ on temperature are depicted in Figure 6. The temperature profile increases by increasing the ϕ volume fraction. Various concentrations of polyvinylchloride (PVC) on temperature and velocity are illustrated in Figures 7 and 8. From Figure 7, it is observed that the velocity distribution rises with the increase of PVC. The temperature for various values of PVC are portrayed in Figure 8. It can be seen that temperature is a decreasing function between the channel. Figure 9 shows the effect of β_u on velocity profile. β_u can be expressed as the ratio of U_{Hs} and u_m of a nanofluid. Figure 9 shows that flow of a channel exceeds and gains its high value. Figure 10 reveals a minute change for increasing values of the electro-osmotic parameter for velocity and temperature. The effects of the electro-osmotic value of κ on temperature and velocity profile are shown in Figures 11 and 12. It can be seen from Figure 11 that by increasing κ, the velocity profile increases. This is due to a larger value of κ for the velocity profiles that display EDL layers. The effect of electro-osmotic κ on temperature is illustrated in Figure 12, which gives the deficiency of the Joule effect. It is also noted that if the pseudo-plastic increases in electro-osmotic parameter κ, it results in a notable increase in temperature. The effect of Br on temperature is shown in Figure 13. It is perceived that by increasing Br, the temperature decreases, from which we conclude that Br increases as compared to bulk mean temperature. The effect of Brinkman number with respect to volume friction is expressed in Figure 14. It is perceived that for a developing Brinkman number with respect to volume fraction, the Nusselt number increases. Results indicate that temperature increases for higher values of nanoparticle volume fraction. The influence of U_{Hs} and u_m of nanofluid β_u and κ on the Nusselt number are shown in Figure 15. It is seen that near the heated wall, the Nusselt number reduces; this is because first heat is moved to the fluid and then transferred into a separated plate.

The entropy generation profiles for various important parameters such as volumetric volume expansion β, magnetic field M, dimensionless temperature parameter $Br\Omega^{-1}$, and Brinkman number Br are illustrated in Figures 16–21. The impact of M on entropy generation is displayed in Figure 16. It can be observed that when M increases, the entropy generation decreases on the left side with minimum energy loss at $y = -0.25$; after this, increasing behavior is detected. The effect of β on entropy generation is depicted in Figure 17, where it can be observed that when β is increasing, the entropy generation is also increased. The energy loss on the upper wall is comparatively greater when compared to the lower wall. The significance of Br on entropy generation in Figure 18 shows that when Br increases, entropy generation decreases on the left side and has minimum energy loss at $y = 0$; after this, an increasing trend on the left side is seen. The influence of $Br\Omega^{-1}$ on entropy generation is given in Figure 19 and it is observed that entropy generation decreases on the lower wall and has minimum energy loss at $y = 0.15$. Figure 20 shows that when the electro-osmotic parameter increases, entropy generation also increases. From Figure 21, it is observed that when R_ζ increases, the entropy generation at the lower wall decreases, but after $y = 0$, it increases at the upper wall. Figure 22 shows that when the magnetic parameter increases, Bejan number also increases. It is observed that there is a dominant effect on the lower as well as the upper wall. Figure 23 shows that when $Br\Omega^{-1}$ increases, the Bejan number also increases. It is noted that the Bejan number shows a dominant role on the upper wall.

The impact of the different parameters on skin friction such as the electro-osmotic parameter, the ratio between U_{Hs} and u_m, volume concentration, Brinkman number, and magnetic field are given in Table 4. It is found that by increasing the Brinkman number and the magnetic parameter, the skin friction decreases.

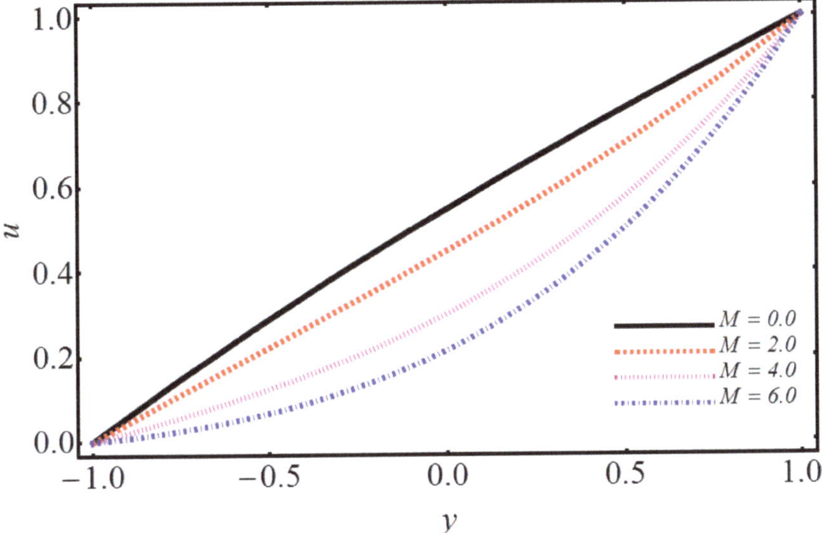

Figure 4. Performance of M on velocity.

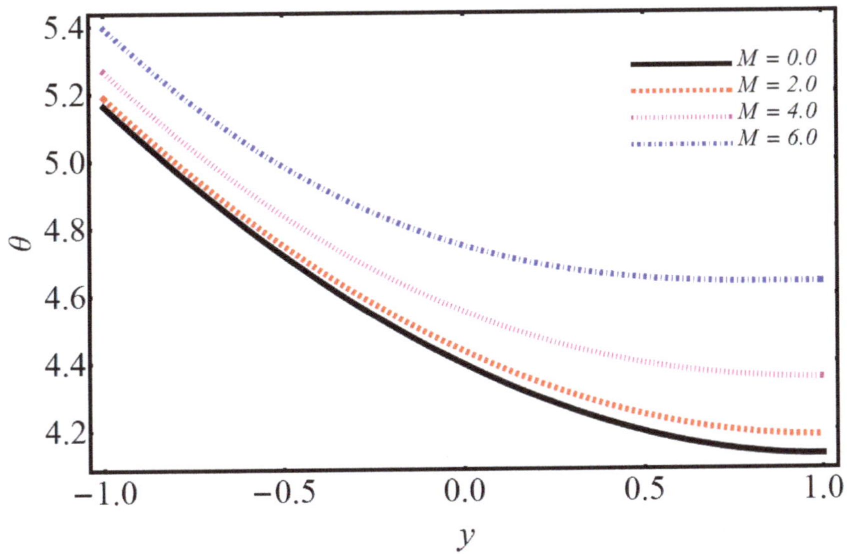

Figure 5. Performance of M on temperature.

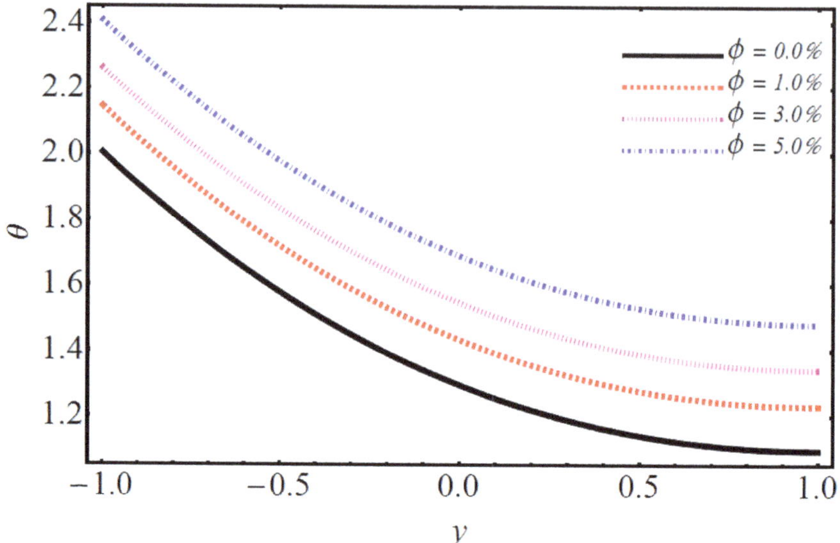

Figure 6. Performance of ϕ on temperature.

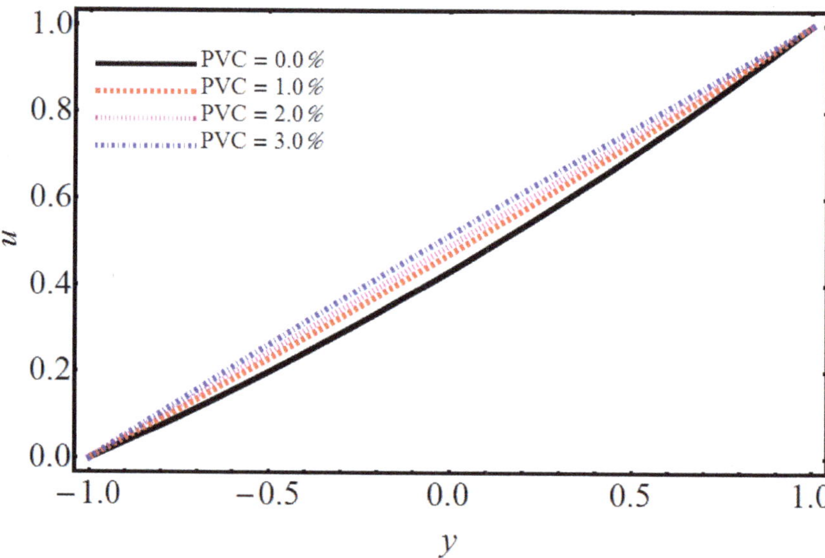

Figure 7. Performance of PVC concentration on velocity.

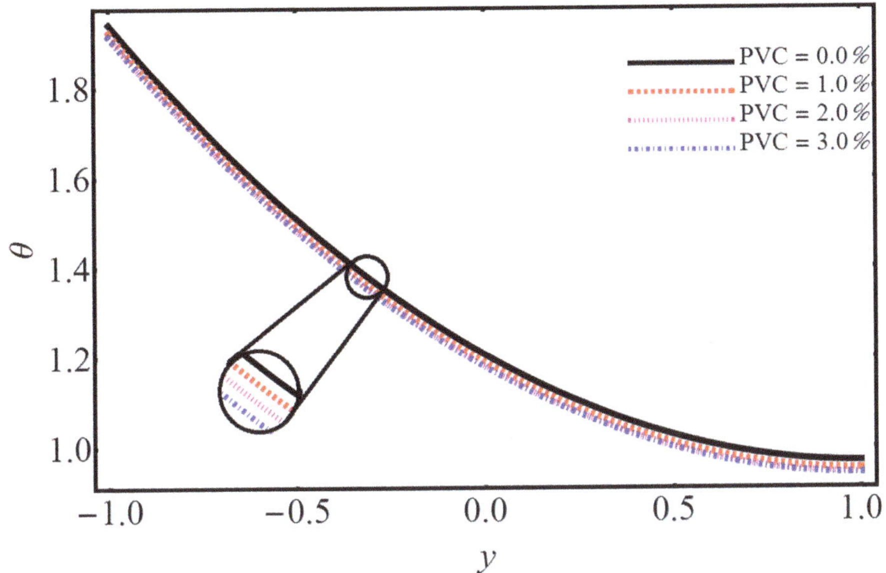

Figure 8. Performance of PVC concentration on temperature.

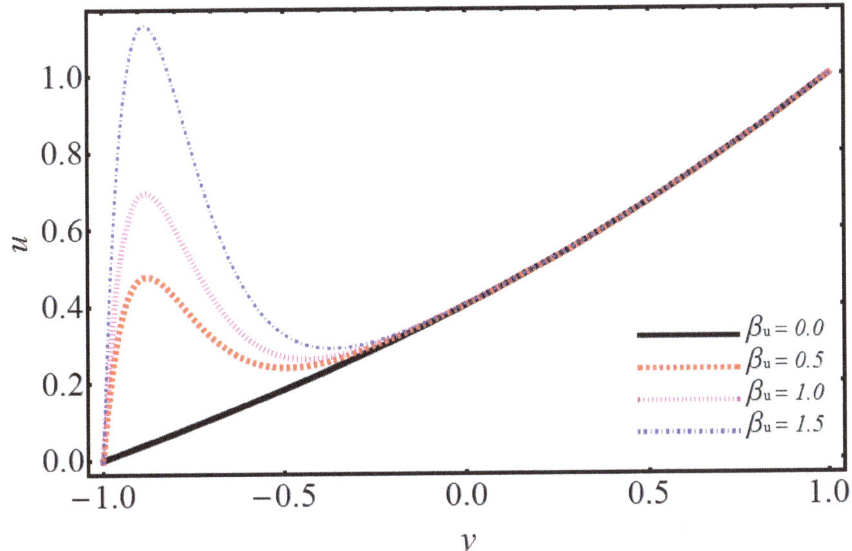

Figure 9. Performance of β_u on velocity.

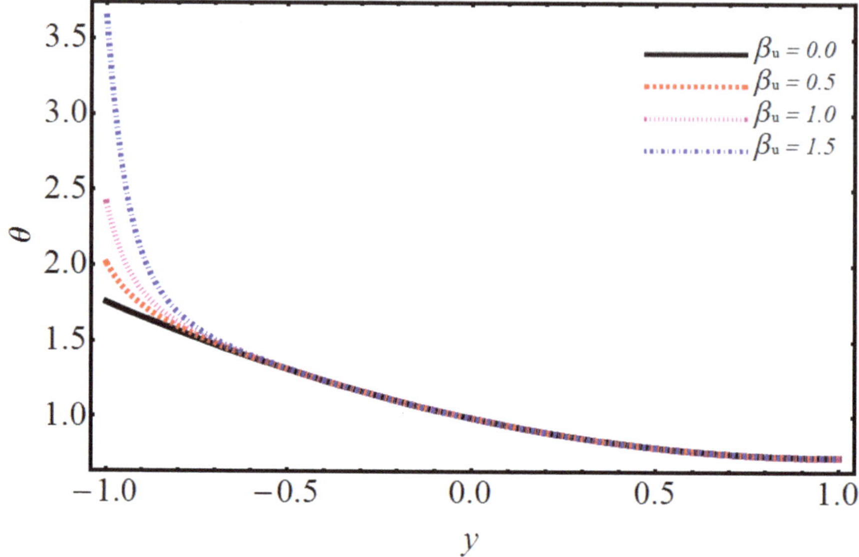

Figure 10. Performance of β_u on temperature.

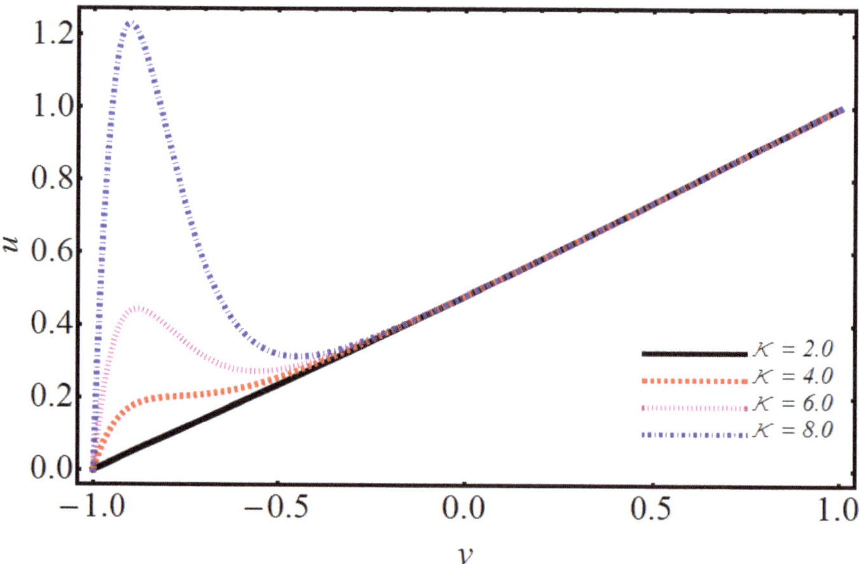

Figure 11. Effect of the electro-osmotic parameter on velocity.

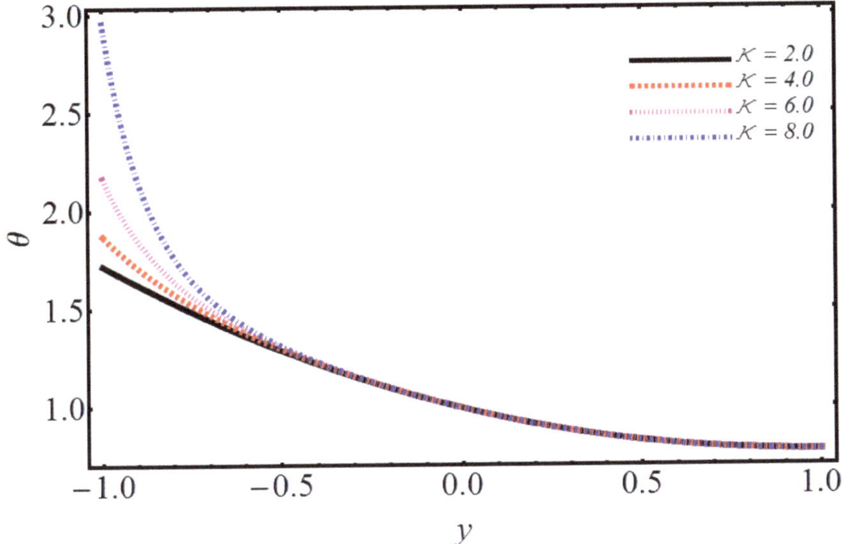

Figure 12. Performance of the electro-osmotic parameter on temperature.

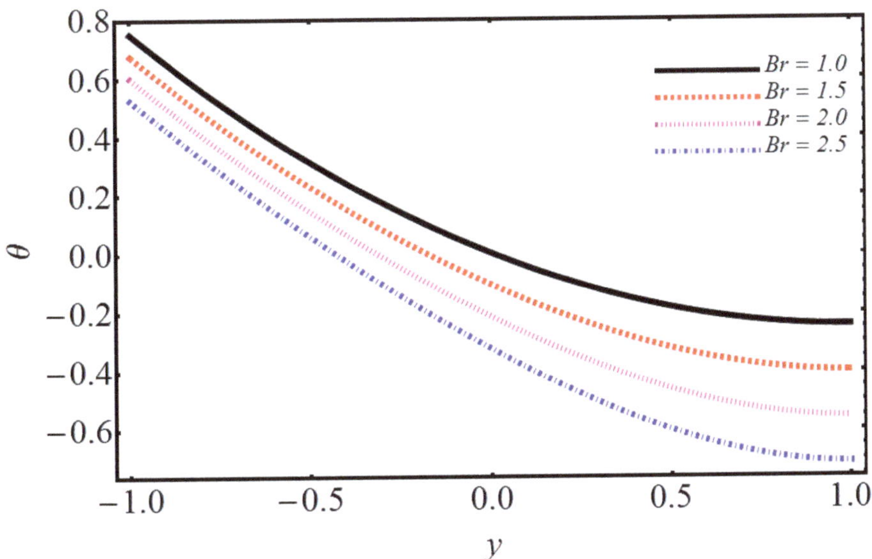

Figure 13. Performance of Brinkman number *Br* on temperature.

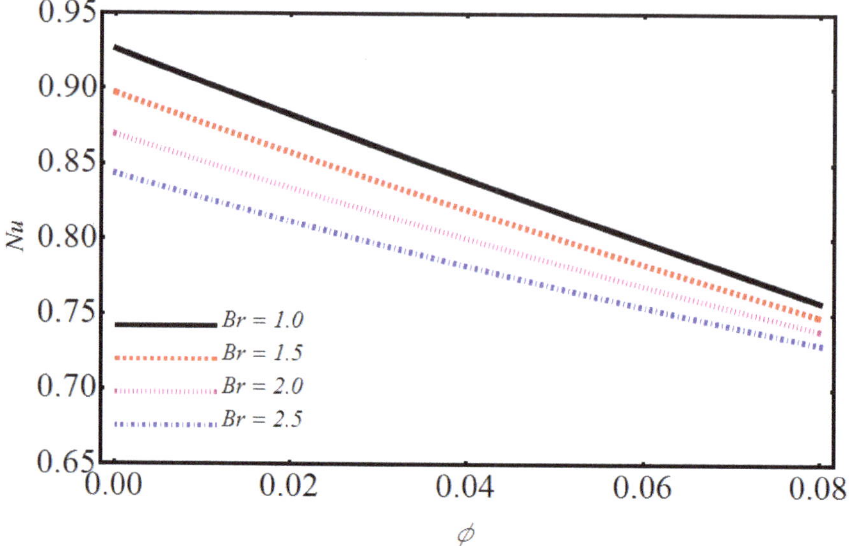

Figure 14. Performance of *Br* w.r.t ϕ on Nusselt number.

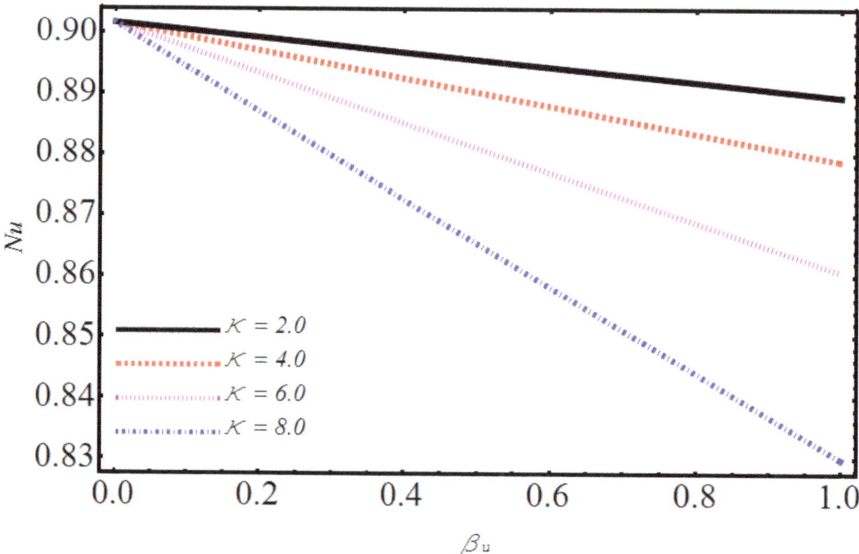

Figure 15. Performance of κ w.r.t β_u on Nusselt number.

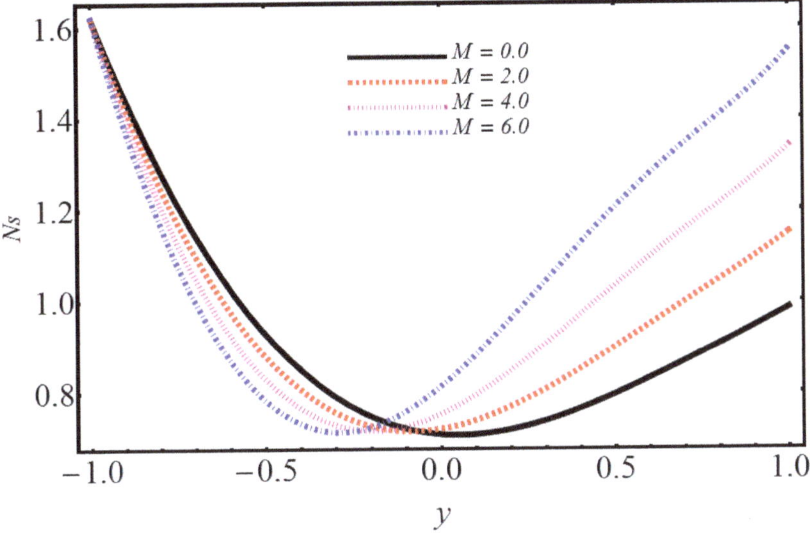

Figure 16. Performance of the magnetic parameter on entropy generation.

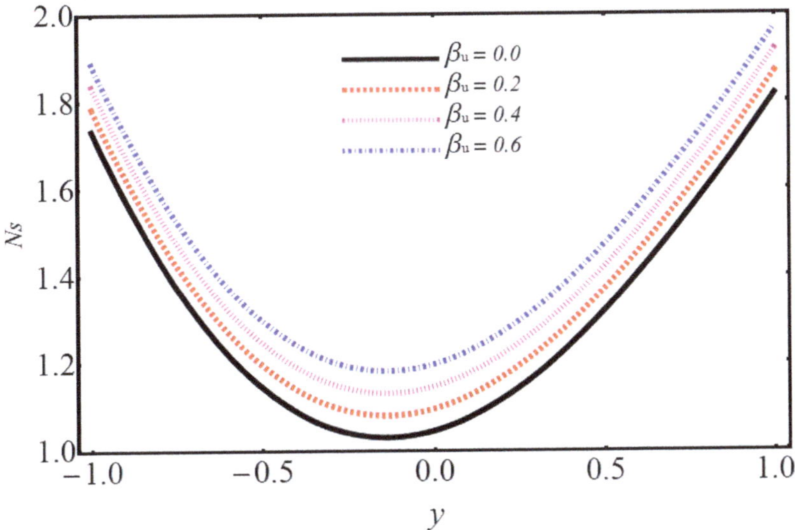

Figure 17. Performance of β for various values on entropy generation.

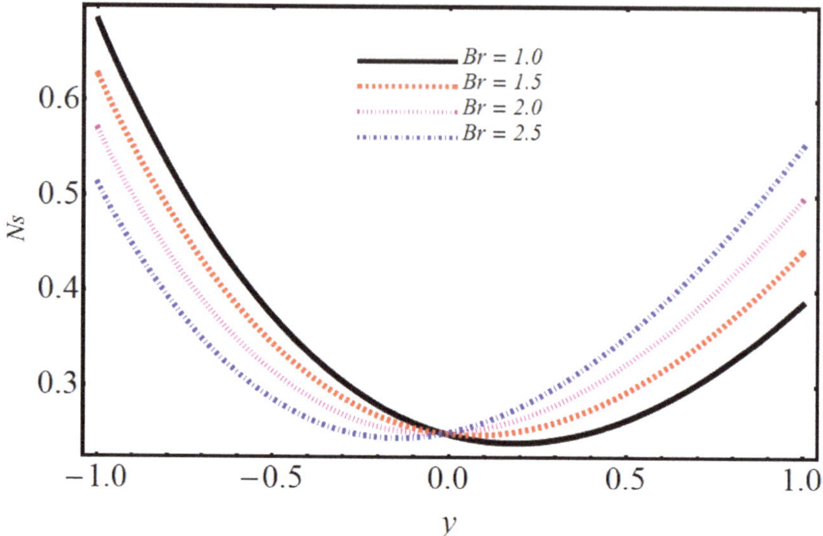

Figure 18. Performance of Brinkman number *Br* on entropy generation.

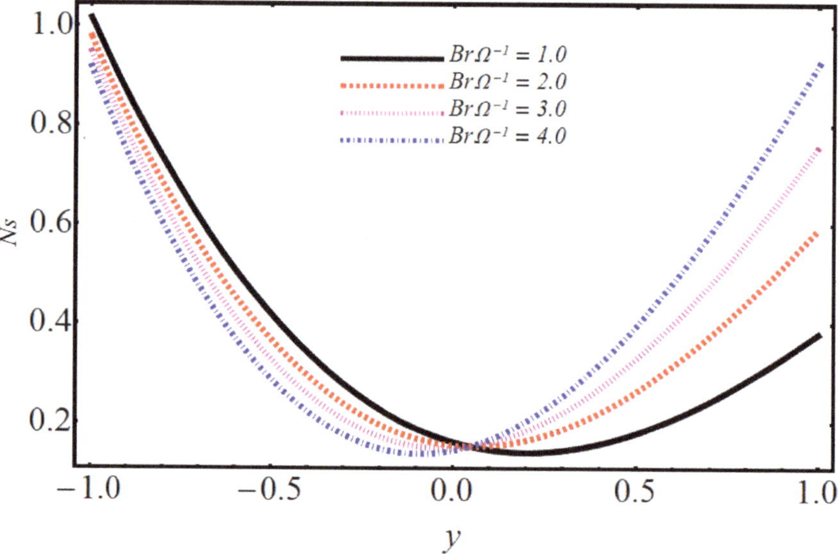

Figure 19. Performance of $Br\Omega^{-1}$ on entropy generation.

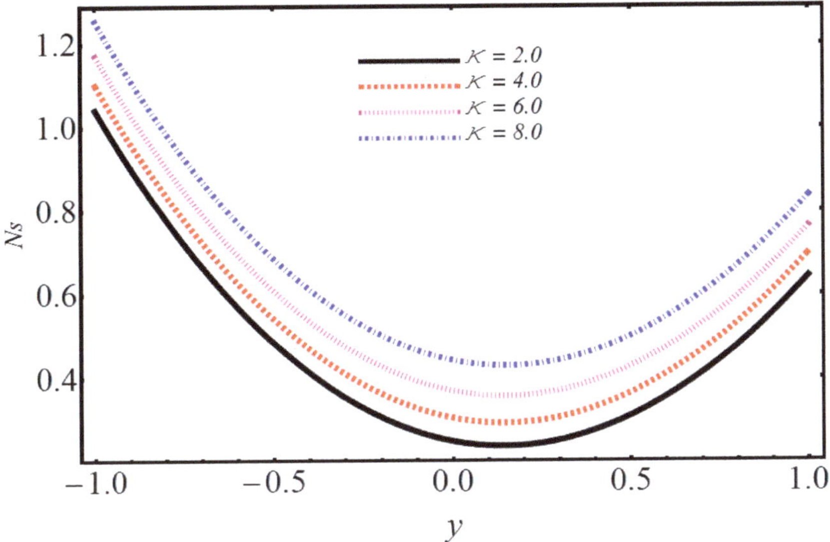

Figure 20. Performance of κ on entropy generation.

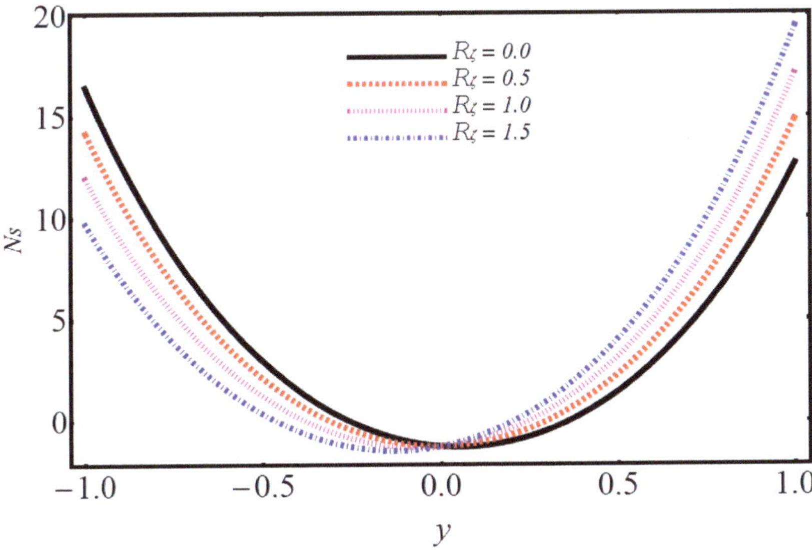

Figure 21. Performance of R_ζ on entropy generation.

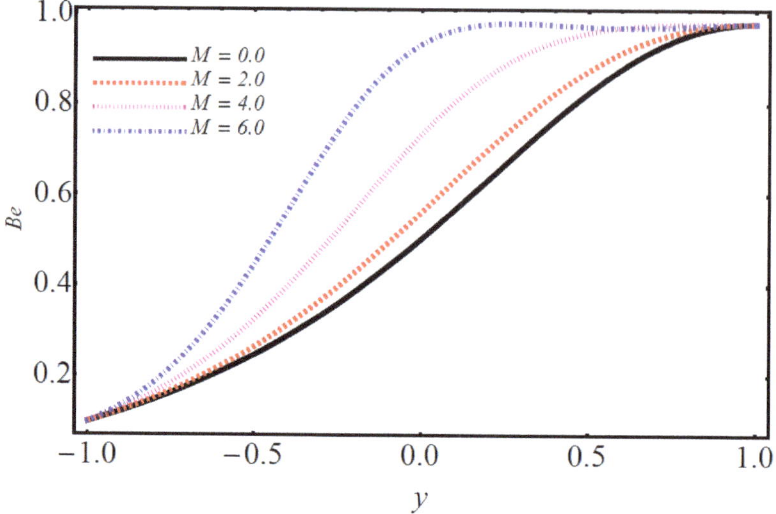

Figure 22. Performance of the magnetic parameter on Bejan number.

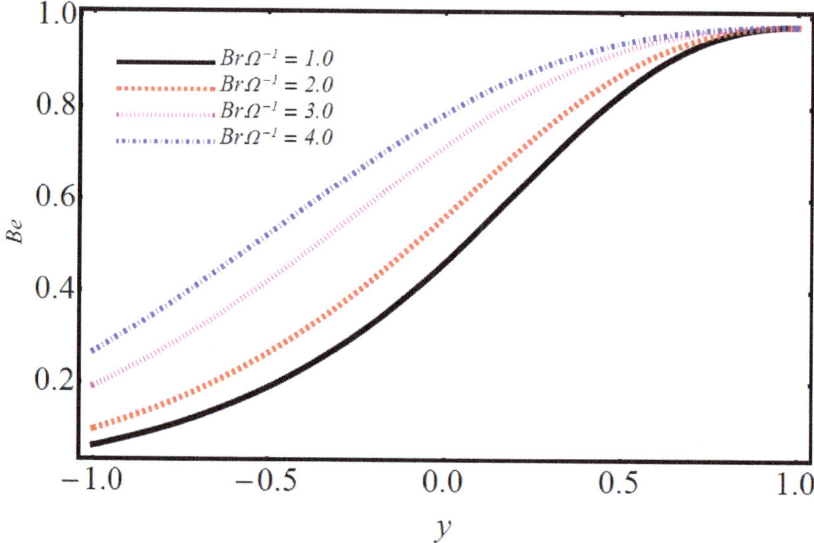

Figure 23. Performance of $Br\Omega^{-1}$ on Bejan number.

Table 4. Coefficient of skin friction C_f with M, Br and β_u along $n = 0.764$ (PVC 3%).

β	M	Br	C_{f-h1}	C_{f-h2}
0.1	0.1	0.0	0.00317506	0.00388072
0.2			0.00317497	0.00388079
0.3			0.00317489	0.00388087
0.4			0.00317481	0.00388095
	0.0		0.00317423	0.00388169
	0.5		0.00317428	0.00388157
	1.0		0.00317443	0.00388152
	1.5		0.00317468	0.00388148
		0.0	0.00317506	0.00388072
		1.0	0.00317512	0.00388066
		2.0	0.00317518	0.00388060
		3.0	0.00317524	0.00388054

4. Conclusions

In this work, the effects of entropy generation on power law nanofluid through a horizontal channel in a way that one wall is movable and the other is stationary are presented. The impact of a mixed convection magnetic field and an electrical double layer has combined for momentum conduct. Flow is produced in an axial direction by the movement of the upper wall. The important findings are:

- When the magnetic parameter and nanoparticle volume fraction are increased, then the velocity of the nanofluid decreases whereas the temperature profile is increased.
- Velocity profile is increased for increasing PVC while a decrease in temperature is detected.
- Temperature and velocity demonstrate similar behavior for increasing values of κ and the ratio between U_{Hs} and u_m.
- It is observed that C_f increases at the heated wall against higher Brinkman number and volume fraction while the reverse behavior is noted for the increasing ratio between U_{Hs} and u_m. The same phenomena are observed for the cases of electro-osmotic and magnetic factors.
- Skin friction is improved with increasing values of κ and the ratio between U_{Hs} and u_m, whereas it decreases with the increase in Brinkman number, volume fraction and the magnetic parameter.
- The Nusselt number escalates for a snowballing magnetic parameter but de-escalates with the increasing ratio between U_{Hs} and u_m, volume fraction, Brinkman number and electro-osmotic factors.
- The entropy generation increases with an increase of volumetric volume expansion β, magnetic field M and κ, while it decreases with an increase of $Br\Omega^{-1}$ and Brinkman number.
- A dual behavior of entropy generation is noted for decreasing and increasing values of R_ζ.
- The Bejan number escalates by snowballing values of both $Br\Omega^{-1}$ and magnetic elements in direct relation with each other and is depicted for both of them.

Author Contributions: Conceptualization, R.E.; Writing—Review and Editing, S.M.S.; Methodology, N.S.; Investigation, N.M.

Funding: This research received no external funding.

Acknowledgments: R. Ellahi gratefully thanks to Research Institute, King Fahd University of Petroleum & Minerals, Dhahran, Saudi Arabia to honor him with the Chair Professor at KFUPM.

Conflicts of Interest: The authors declare no conflict of interest.

Appendix A

$$C_1 = \tfrac{1}{2} - \hbar_u \tfrac{A_3Gr}{24} - \hbar_u^2 \tfrac{2^{-3-n}nA_2A_3Gr}{3} + \hbar_u\hbar_\theta \tfrac{A_3B_2Gr}{2} + 2^{-2-n}A_2A_3BrGr\hbar_u\hbar_\theta$$
$$+\hbar_u\tfrac{A_4M^2}{2} + \hbar_u^2 2^{-1-n}nA_2A_4GrM^2 - \hbar_u^2\tfrac{7A_3A_4GrM^2}{720} - \tfrac{26A_3A_4BrGrM^2}{100}\hbar_u\hbar_\theta$$
$$+\left(\tfrac{5A_3A_4BrGrM^2}{100}\right)\hbar_u\hbar_\theta + \left(\tfrac{37A_3A_4BrGrM^2}{100}\right)\left(\tfrac{27}{10}\right)^n\hbar_u\hbar_\theta + \tfrac{5A_4^2M^4}{48}\hbar_u^2$$
$$-\hbar_u Re - 2^{-n}nA_2Re\hbar_u^2 - \tfrac{5A_4M^2Re}{24}Re\hbar_u^2 + \tfrac{1}{2}y + \hbar_u\tfrac{A_3Gr}{6}y + \hbar_u^2\tfrac{2^{-1-n}nA_2A_3Gr}{3}$$
$$C_2 = \hbar_u\tfrac{A_4M^2}{6} + \hbar_u^2\tfrac{2^{-1-n}nA_2A_4M^2}{3} + \hbar_u^2\tfrac{7A_2A_4GrM^2}{720} - \hbar_u\hbar_\theta\tfrac{29A_3A_4BrGrM^2}{100}$$
$$+\hbar_u^2\tfrac{7A_4^2M^4}{720}$$
$$C_3 = -\tfrac{A_3B_2Gr}{4}\hbar_u\hbar_\theta - 2^{-n-2}A_2A_3BrGr\hbar_u\hbar_\theta - \hbar_u\tfrac{A_4M^2}{2} - \hbar_u^2 2^{-1-n}nA_2A_4GrM^2$$
$$+\hbar_u^2\tfrac{A_3A_4GrM^2}{96} - \tfrac{3A_3A_4BrGrM^2}{8}\hbar_u\hbar_\theta - \hbar_u^2\tfrac{A_4^2M^4}{8} + \hbar_u Re + 2^{-n}nA_2Re\hbar_u^2$$
$$+\tfrac{A_4M^2Re}{4}\hbar_u^2$$
$$C_4 = -\hbar_u\tfrac{A_3Gr}{6} - \tfrac{2^{-1-n}nA_2A_3Gr}{3}\hbar_u^2 - \hbar_u\tfrac{A_4M^2}{6} - \tfrac{2^{-1-n}nA_2A_4M^2}{3}\hbar_u^2$$
$$-\tfrac{A_4A_3GrM^2}{72}\hbar_u^2 - \tfrac{A_3A_4BrGrM^2}{12}\hbar_u\hbar_\theta - \tfrac{A_4^2M^4}{72}\hbar_u^2$$
$$C_5 = \hbar_u\tfrac{A_3Gr}{24} + \tfrac{2^{-3-n}nA_2A_3Gr}{3}\hbar_u^2 - \tfrac{A_3A_4BrGrM^2}{48}\hbar_u\hbar_\theta + \hbar_u^2\tfrac{A_4^2M^4}{48}$$
$$-\hbar_u^2\tfrac{A_4M^2Re}{24}$$
$$C_6 = \hbar_u^2\tfrac{A_4A_3GrM^2}{240} + \hbar_u^2\tfrac{A_4^2M^4}{240}, C_7 = -\hbar_u^2\tfrac{A_3A_4M^2Gr}{1440}$$
$$C_8 = -\hbar_u\beta_u\rho e - 2^{-n}nA_2\beta_u\rho e\hbar_u^2 - \hbar_u^2\tfrac{5A_4M^2\beta_u\rho e}{24} + \hbar_u\beta_u\rho e + \hbar_u^2 2^{-n}nA_2\beta_u\rho e$$

$$D_1 = -B_2\hbar_\theta - 2^{-n}A_2Br\hbar_\theta + \hbar_u\hbar_\theta\tfrac{5A_2A_3BrGr}{3} - \left(\tfrac{45A_3A_2BrGr}{100}\right)\left(\tfrac{27}{10}\right)^{-\frac{69n}{100}}\hbar_u\hbar_\theta$$
$$+\left(\tfrac{3A_3A_2BrGr}{100}\right)\left(\tfrac{27}{10}\right)^{-\frac{69n}{100}}\hbar_u\hbar_\theta + \tfrac{5.2^{-2-n}nA_2A_3Gr}{3}\hbar_u\hbar_\theta -$$
$$\left(\tfrac{45A_3A_2BrGr}{100}\right)\left(\tfrac{27}{10}\right)^{-\frac{69n}{100}}n\hbar_u\hbar_\theta + \left(\tfrac{3A_3A_2BrGr}{100}\right)\left(\tfrac{27}{10}\right)^{-\frac{69n}{100}}n\hbar_u\hbar_\theta - \tfrac{3A_4BrM^2}{2}\hbar_\theta$$
$$+\left(\tfrac{A_4BrM^2}{10}\right)\left(\tfrac{27}{10}\right)\hbar_\theta + \left(\tfrac{75A_4BrM^2}{100}\right)\left(\tfrac{27}{10}\right)\hbar_\theta$$
$$D_2 = -\tfrac{1}{2} - 2^{-1-n}A_2A_3BrGr\hbar_\theta\hbar_u - A_4BrM^2\hbar_\theta + 2^{-1-n}A_2A_4BrM^2\hbar_\theta\hbar_u$$
$$+2^{-1-n}A_2A_4BrM^2n\hbar_u\hbar_\theta$$
$$D_3 = \left(\tfrac{42B_1\beta_u\rho e\gamma}{100}\right)\left(\tfrac{27}{10}\right)\hbar_u\hbar_\theta + \tfrac{B_1\beta_u\rho e\gamma}{2},$$
$$D_4 = \tfrac{A_4^2BrM^2Re}{2}\hbar_u\hbar_\theta - \left(\tfrac{87A_4BrReM^2}{100}\right)\left(\tfrac{27}{10}\right)\hbar_u\hbar_\theta + \left(\tfrac{172A_4BrReM^2}{100}\right)\left(\tfrac{27}{10}\right)$$
$$D_5 = \tfrac{5.2^{-2-n}A_2A_4BrM^2}{3}\hbar_u\hbar_\theta - \left(\tfrac{A_4A_2BrM^2}{20}\right)\left(\tfrac{27}{10}\right)^{-\frac{69n}{100}}\hbar_u\hbar_\theta$$
$$+\tfrac{5.2^{-2-n}A_2A_4BrM^2}{3}n\hbar_u\hbar_\theta + \left(\tfrac{37A_4A_2BrM^2}{100}\right)\left(\tfrac{27}{10}\right)^{-\frac{69n}{100}}\hbar_u\hbar_\theta$$
$$D_6 = -\left(\tfrac{A_4A_2BrM^2}{20}\right)\left(\tfrac{27}{10}\right)^{-\frac{69n}{100}}n\hbar_u\hbar_\theta - \left(\tfrac{37A_4A_2BrM^2}{100}\right)\left(\tfrac{27}{10}\right)^{-\frac{69n}{100}}n\hbar_u\hbar_\theta$$
$$+\tfrac{65A_3A_4BrGrM^2}{48}\hbar_u\hbar_\theta - \left(\tfrac{169A_4A_3BrGrM^2}{100}\right)\left(\tfrac{27}{10}\right)\hbar_u\hbar_\theta$$

References

1. Choi, S.U.S. Enhancing thermal conductivity of fluids with nanoparticles. In Proceedings of the ASME International Mechanical Engineering Congress & Exposition, San Francisco, CA, USA, 12–17 November 1995; Volume 231, pp. 99–105.

2.	Xuan, Y.; Li, Q. Heat transfer enhancement of nanofluids. *Int. J. Heat Fluid Flow* **2000**, *21*, 58–64. [CrossRef]

3.	Karimipour, A.; Orazio, A.D.; Shadloo, M.S. The effects of different nano particles of Al_2O_3 and Ag on the MHD nanofluid flow and heat transfer in a microchannel including slip velocity and temperature jump. *Physica E* **2017**, *86*, 146–153. [CrossRef]

4.	Ellahi, R.; Zeeshan, A.; Hussain, F.; Asadollahi, A. Peristaltic blood flow of couple stress fluid suspended with nanoparticles under the influence of chemical reaction and activation energy. *Symmetry* **2019**, *11*, 276. [CrossRef]

5.	Shadloo, M.S.; Kimiaeifar, A. Application of homotopy perturbation method to find an analytical solution for magnetohydrodynamic flows of viscoelastic fluids in converging/diverging channels. *Proc. Mech. Eng. Part C J. Mech. Eng.* **2011**, *225*, 347–353. [CrossRef]

6.	Safaei, M.R.; Ahmadi, G.; Goodarzi, M.S.; Shadloo, M.S.; Goshayeshi, H.R.; Dahari, M. Heat transfer and pressure drop in fully developed turbulent flows of graphene nanoplatelets–silver/water nanofluids. *Fluids* **2016**, *1*, 20. [CrossRef]

7.	Rashidi, M.M.; Nasiri, M.; Shadloo, M.S.; Yang, Z. Entropy generation in a circular tube heat exchanger using nanofluids: Effects of different modeling approaches. *Heat Transf. Eng.* **2017**, *38*, 853–866. [CrossRef]

8.	Hosseini, S.M.; Safaei, M.R.; Goodarzi, M.; Alrashed, A.A.; Nguyen, T.K. New temperature, interfacial shell dependent dimensionless model for thermal conductivity of nanofluids. *Int. J. Heat Mass Transf.* **2017**, *114*, 207–210. [CrossRef]

9.	Hassan, M.; Marin, M.; Alsharif, A.; Ellahi, R. Convective heat transfer flow of nanofluid in a porous medium over wavy surface. *Phys. Lett. A* **2018**, *382*, 2749–2753. [CrossRef]

10.	Ouyang, Z.; Lin, J.Z.; Ku, X. The hydrodynamic behavior of a squirmer swimming in power-law fluid. *Phys. Fluids* **2018**, *30*, 083301. [CrossRef]

11.	Chhabra, R.P.; Richardson, J.F. *Non-Newtonian Flow in the Process Industries, Fundamentals and Engineering Applications*; Jordan Hill: Oxford, UK, 1999; pp. 206–392.

12.	Kakac, S.; Shah, R.K.; Aung, W. *Handbook of Single-Phase Convective HEAT Transfer*; Wiley: New York, NY, USA, 1987.

13.	Zeeshan, A.; Bhatti, M.M.; Akbar, N.S.; Sajjad, Y. Hydromagnetic blood flow of Sisko fluid in a non-uniform channel induced by a peristaltic wave. *Commun. Theor. Phys.* **2017**, *68*, 103. [CrossRef]

14.	Zhao, C.; Zholkovskij, E.; Masliyah, J.H.; Yang, C. Analysis of electroosmotic flow of power-law fluids in a slit microchannel. *J. Colloid Interface Sci.* **2008**, *326*, 503–510. [CrossRef] [PubMed]

15.	Das, S.; Chakraborty, S. Analytical solutions for velocity, temperature and concentration distribution in electroosmotic microchannel flows of a non-Newtonian bio-fluid. *Anal. Chim. Acta* **2006**, *559*, 15–24. [CrossRef]

16.	Prakash, J.; Tripathi, D. Electroosmotic flow of Williamson ionic nanoliquids in a tapered microfluidic channel in presence of thermal radiation and peristalsis. *J. Mol. Liq.* **2018**, *256*, 352–371. [CrossRef]

17.	Ali, F.; Iftikhar, M.; Khan, I.; Sheikh, N.A. Atangana–Baleanu fractional model for electro-osmotic flow of viscoelastic fluids. *Chaos Solitons Fractals* **2019**, *124*, 125–133. [CrossRef]

18.	Mondal, A.; Shit, G.C. Transport of magneto-nanoparticles during electro-osmotic flow in a micro-tube in the presence of magnetic field for drug delivery application. *J. Magn. Magn. Mater.* **2017**, *442*, 319–328. [CrossRef]

19.	Shit, G.C.; Mondal, A.; Sinha, A.; Kundu, P.K. Electro-osmotically driven MHD flow and heat transfer in micro-channel. *Physica A* **2016**, *449*, 437–454. [CrossRef]

20.	Ellahi, R. The effects of MHD and temperature dependent viscosity on the flow of non-Newtonian nanofluid in a pipe: Analytical solutions. *Appl. Math. Model.* **2013**, *37*, 1451–1457. [CrossRef]

21.	Hayat, T.; Qayyum, S.; Khan, M.I.; Alsaedi, A. Entropy generation in magnetohydrodynamic radiative flow due to rotating disk in presence of viscous dissipation and Joule heating. *Phys. Fluids* **2018**, *30*, 017101. [CrossRef]

22.	Hayat, T.; Khan, M.I.; Qayyum, S.; Alsaedi, A. Entropy generation in flow with silver and copper nanoparticles. *Colloids Surf. A* **2018**, *539*, 335–346. [CrossRef]

23.	Khan, M.I.; Waqas, M.; Hayat, T.; Alsaedi, A. A comparative study of Casson fluid with homogeneous-heterogeneous reactions. *J. Colloid Interface Sci.* **2017**, *498*, 85–90. [CrossRef]

24. Goshayeshi, H.R.; Goodarzi, M.; Safaei, M.R.; Dahari, M. Experimental study on the effect of inclination angle on heat transfer enhancement of a ferrofluid in a closed loop oscillating heat pipe under magnetic field. *Exp. Therm Fluid Sci.* **2016**, *74*, 265–270. [CrossRef]

25. Sohail, A.; Fatima, M.; Ellahi, R.; Akram, K.B. A videographic assessment of Ferrofluid during magnetic drug targeting: An application of artificial intelligence in nanomedicine. *J. Mol. Liq.* **2019**, *285*, 47–57. [CrossRef]

26. Malvandi, A.; Safaei, M.R.; Kaffash, M.H.; Ganji, D.D. MHD mixed convection in a vertical annulus filled with Al2O3–water nanofluid considering nanoparticle migration. *J. Magn. Magn. Mater.* **2015**, *382*, 296–306. [CrossRef]

27. Yousif, M.A.; Ismael, H.F.; Abbas, T.; Ellahi, R. Numerical study of momentum and heat transfer of MHD Carreau nanofluid over exponentially stretched plate with internal heat source/sink and radiation. *Heat Transf. Res.* **2019**, *50*, 649–658. [CrossRef]

28. Sheikholeslami, M.; Bhatti, M.M. Active method for nanofluid heat transfer enhancement by means of EHD. *Int. J. Heat Mass Transf.* **2017**, *109*, 115–122. [CrossRef]

29. Tripathi, D.; Jhorar, R.; Beg, O.A.; Kadir, A. Electro-magneto-hydrodynamic peristaltic pumping of couple stress biofluids through a complex wavy micro-channel. *J. Mol. Liq.* **2017**, *236*, 358–367. [CrossRef]

30. Ray, A.; Varma, V.B.; Jayaneel, P.J.; Sudharsan, N.M.; Wang, Z.P.; Ramanujan, R.V. On demand manipulation of ferrofluid droplets by magnetic fields. *Sens. Actuator B Chem.* **2017**, *242*, 760–768. [CrossRef]

31. Casula, M.F.; Corrias, A.; Arosio, P.; Lascialfari, A.; Sen, T.; Floris, P.; Bruce, I.J. Design of water-based ferrofluids as contrast agents for magnetic resonance imaging. *J. Colloid Interface Sci.* **2011**, *357*, 50–55. [CrossRef]

32. Makinde, O.D.; Chinyoka, T. MHD transient flows and heat transfer of dusty fluid in a channel with variable physical properties and Navier slip condition. *Comp. Math. Appl.* **2010**, *60*, 660–669. [CrossRef]

33. Abro, A.K.; Khan, I. Analysis of the heat and mass transfer in the MHD flow of a generalized Casson fluid in a porous space via non-integer order derivatives without a singular kernel. *Chin. J. Phys.* **2017**, *55*, 1583–1595. [CrossRef]

34. Ellahi, R.; Zeeshan, A.; Shehzad, N.; Alamri, S.Z. Structural impact of Kerosene-Al_2O_3 nanoliquid on MHD Poiseuille flow with variable thermal conductivity: Application of cooling process. *J. Mol. Liq.* **2018**, *264*, 607–615. [CrossRef]

35. Bejan, A. *Entropy Generation Minimization*; CRC: Boca Raton, NY, USA, 1996.

36. Zeeshan, A.; Shehzad, N.; Abbas, T.; Ellahi, R. Effects of radiative electro-magnetohydrodynamics diminishing internal energy of pressure-driven flow of titanium dioxide-water nanofluid due to entropy generation. *Entropy* **2019**, *21*, 236. [CrossRef]

37. Ranjit, K.; Shit, G.C. Entropy generation on electro-osmotic flow pumping by a uniform peristaltic wave under magnetic environment. *Energy* **2017**, *128*, 649–660. [CrossRef]

38. Cho, C.C.; Chen, C.L. Natural convection heat transfer and entropy generation in wavy-wall enclosure containing water-based nanofluid. *Int. J. Heat Mass Transf.* **2013**, *61*, 749–758. [CrossRef]

39. Darbari, B.; Rashidi, S.; Esfahani, A.J. Sensitivity analysis of entropy generation in nanofluid flow inside a channel by response surface methodology. *Entropy* **2016**, *18*, 52. [CrossRef]

40. Bhatti, M.M.; Abbas, T.; Rashidi, M.M.; Ali, M.E.-S. Numerical simulation of entropy generation with thermal radiation on MHD Carreau nanofluid towards a shrinking sheet. *Entropy* **2016**, *18*, 200. [CrossRef]

41. Bhatti, M.M.; Abbas, T.; Rashidi, M.M.; Ali, M.E.S.; Yang, Z. Entropy generation on MHD Eyring–Powell nanofluid through a permeable stretching surface. *Entropy* **2016**, *18*, 224. [CrossRef]

42. Hayat, T.; Javed, S.; Khan, M.I.; Khan, M.I.; Alsaedi, A. Physical aspects of irreversibility in radiative flow of viscous material with cubic autocatalysis chemical reaction. *Eur. Phys. J. Plus* **2019**, *134*, 172. [CrossRef]

43. Bhatti, M.M.; Sheikholeslami, M.; Shahid, A.; Hassan, M.; Abbas, T. Entropy generation on the interaction of nanoparticles over a stretched surface with thermal radiation. *Colloids Surf. A* **2019**, *570*, 368–376. [CrossRef]

44. Liao, S.J. *Beyond Perturbation: Introduction to Homotopy Analysis Method*; Chapman & Hall: Boca Raton, FL, USA, 2003.

45. Ellahi, R.; Hassan, M.; Zeeshan, A. Shape effects of spherical and nonspherical nanoparticles in mixed convection flow over a vertical stretching permeable sheet. *Mech. Adv. Mater. Struct.* **2017**, *24*, 1231–1238. [CrossRef]

46. Ellahi, R.; Tariq, M.H.; Hasssan, M.; Vafai, K. On boundary layer nano-ferroliquid flow under the influence of low oscillating stretchable rotating disk. *J. Mol. Liq.* **2017**, *229*, 339–345. [CrossRef]

47. Shehzad, N.; Zeeshan, A.; Ellahi, R.; Vafai, K. Convective heat transfer of a nanofluid in a wavy channel: Buongiorno's mathematical model. *J. Mol. Liq.* **2016**, *222*, 446–455. [CrossRef]
48. Hayat, T.; Khan, M.I.; Farooq, M.; Alsaedi, A.; Waqas, M.; Yasmeen, T. Impact of Cattaneo–Christov heat flux model in flow of variable thermal conductivity fluid over a variable thicked surface. *Int. J. Heat Mass Transf.* **2016**, *99*, 702–710. [CrossRef]
49. Dehghan, M.; Manafian, J.; Saadatmandi, A. Solving nonlinear fractional partial differential equations using the homotopy analysis method. Numer. *Meth. Part Differ. Equ. Int. J.* **2010**, *26*, 448–479. [CrossRef]
50. Ellahi, R.; Zeeshan, A.; Hassan, M. Particle shape effects on Marangoni convection boundary layer flow of a nanofluid. *Int. J. Numer. Methods Heat Fluid Flow* **2016**, *26*, 2160–2174. [CrossRef]
51. Goswami, P.; Chakraborty, S. Semi-analytical solutions for electroosmotic flows with interfacial slip in microchannels of complex cross-sectional shapes. *Microfluid. Nanofluid.* **2011**, *11*, 255–267. [CrossRef]
52. Shehzad, N.; Zeeshan, A.; Ellahi, R. Electroosmotic flow of MHD power law Al2O3-PVC nanofluid in a horizontal channel: Couette-Poiseuille flow model. *Commun. Theor. Phys.* **2018**, *69*, 655. [CrossRef]
53. Wang, X.; Qi, H.; Yu, B.; Xiong, Z.; Xu, H. Analytical and numerical study of electroosmotic slip flows of fractional second grade fluids. *Commun. Nonlinear Sci. Numer. Simul.* **2017**, *50*, 77–87. [CrossRef]
54. Aydın, O.; Avcı, M. Laminar forced convection with viscous dissipation in a Couette–Poiseuille flow between parallel plates. *Appl. Energy* **2016**, *83*, 856–867. [CrossRef]
55. Maiga, S.E.B.; Palm, S.J.; Nguyen, C.T.; Roy, G.; Galanis, N. Heat transfer enhancement by using nanofluids in forced convection flows. *Int. J. Heat Fluid Flow* **2005**, *26*, 530–546. [CrossRef]
56. Ellahi, R.; Hassan, M.; Zeeshan, A. A study of heat transfers in power-law nanofluid. *Therm. Sci.* **2016**, *20*, 2015–2026. [CrossRef]
57. Liao, S.J. The Proposed Homotopy Analysis Technique for the Solution of Nonlinear Problems. Ph.D. Thesis, Shanghai Jiao Tong University, Shanghai, China, 1992.
58. Van Gorder, R.A.; Vajravelu, K. On the selection of auxiliary functions, operators, and convergence control parameters in the application of the Homotopy Analysis Method to nonlinear differential equations: A general approach. *Commun. Nonlinear Sci. Numer. Simul.* **2009**, *14*, 4078–4089. [CrossRef]
59. Zhao, Y.; Liao, S.J. HAM-Based Mathematica Package BVPh 2.0 for Nonlinear Boundary Value Problems. In *Advances in the Homotopy Analysis Method*; World Scientific Press: Singapore, 2013.

Article

The Solutions of Non-Integer Order Burgers' Fluid Flowing through a Round Channel with Semi Analytical Technique

M. Imran [1], D. L.C. Ching [2], Rabia Safdar [1], Ilyas Khan [3,*], M. A. Imran [4] and K. S. Nisar [5]

1   Department of Mathematics, Government College University, Faisalabad, Punjab 38000, Pakistan
2   Fundamental and Applied Science Department, Universiti Teknologi Petronas, 32610 Perak, Malaysia
3   Faculty of Mathematics and Statistics, Ton Duc Thang University, Ho Chi Minh City 72915, Vietnam
4   Department of Mathematics, University of Management and Technology Lahore, Punjab 54770, Pakistan
5   Department of Mathematics, College of Arts and Science, Prince Sattam bin Abdulaziz University, Wadi Al-Dawaser 11991, Saudi Arabia
*   Correspondence: ilyaskhan@tdtu.edu.vn

Received: 4 April 2019; Accepted: 6 May 2019; Published: 1 August 2019

Abstract: The solutions for velocity and stress are derived by using the methods of Laplace transformation and Modified Bessel's equation for the rotational flow of Burgers' fluid flowing through an unbounded round channel. Initially, supposed that the fluid is not moving with t = 0 and afterward fluid flow is because of the circular motion of the around channel with velocity $\Omega R t^p$ with time positively grater than zero. At the point of complicated expressions of results, the inverse Laplace transform is alternately calculated by "Stehfest's algorithm" and "MATHCAD" numerically. The numerically obtained solutions in the terms of the Modified Bessel's equations of first and second kind, are satisfying all the imposed conditions of given mathematical model. The impact of the various physical and fractional parameters are also indeed and so presented by graphical demonstrations.

Keywords: Burgers' fluid; velocity field; shear stress; Laplace transform; modified Bessel function; Stehfest's algorithm; MATHCAD

---

1. Introduction

Fluids can be divided into two types i.e., Newtonian and non-Newtonian fluid. The Newtonian fluids are simple and ideal. In real life there is no exitance of Newtonian fluids but water and air consider as Newtonian fluid. However, the non-Newtonian fluids are complicated and cannot be solved easily. The fluid motion within a cylinder has a wide application in the field of physics, engineering and specially in the food industry. In (1923), Taylor presented the results of stability of fluid in rotating cylinders [1]. Stephen Childress, in 2009 talked about the lift and drag in ideal fluids in two-dimensional, Stoke's flow and gas dynamics [2]. Waters and King [3] discussed the Oldroyd-B fluid in a circular tube by taking Poiseuille flow into account. The investigation of basic unsteady pipe flow and viscoelastic upper-convected Maxwell fluid in uniform circular cross section, is available in [4]. They used the Fourier Bessel series to obtain the closed form solutions. The exact solutions of different fluids in circular cylinders (may be finite and infinite) can be obtained by applying the Laplace transformation on the system. There has been published a number of papers on this idea.

The action of circular cylinder and pressure gradient for both translational and rotational flows for viscoelastic fluids discussed in [5]. Fox and Macdonald [6] focussed on differential analysis of one dimensional and steady state flow of incompressible and non viscous fluid. He also discussed the flow of fluid through pipes and channels. Fetecau [7] worked on unidirectional unsteady flow through an infinite pipe for Oldroyd-B fluids. For analytical solutions, an expansion theorem of Steklov is used with a no slip condition. The viscoelastic fluids have many exertions in various fields of industry, also in bio engineering. Ting [8] and Srivastava [9] find out the analytical solutions of non-Newtonian fluids for second grade and Maxwell fluids respectively. Sherief et al. [10] considered an infinitely magnetic insulating circular cylinder for the steady one-dimensional flow of an incompressible MHD fluid. The slip boundary conditions are applied on velocity and they calculated the results for the micro rotation, ratio of flow and skin coefficients.

The Helical flow is the composition of translation and rotational motion, is applicable in vascular hydrodynamics and biomedical engineering. Many papers have been published on this type of flow. The general solutions of some helical flow corresponding to the second grade fluid are seems in [11]. The analytic solutions of same fluid for unsteady flow has been derived by Hayat et al. [12]. In rotating circular cylinder, Fetecau [13] obtained the analytical solutions for the helical flow of Oldroyd-B fluid. In the same domain, the solutions of Maxwell and second grade fluids are available in [14,15].

Fractional calculus plays a very important rule in the field of fluid mechanics. Podlubny [16] discussed the differential equations of fractional model. The elements of fractional model defines viscoelastic fluids of special kind. Now a days, such models frequently [17,18] commonly encounter in our daily lives. The analytical solutions of generalized second grade fluids are available in [19]. Exact solutions of generalized Burgers' fluid in annulus of circular cylinders and helical flow of Burgers' fluid, in terms of fractional derivatives, are found in [20,21] respectively. Whereas the analysis of velocity and stress field, vortex sheet of same fluid by considering the fractional anomalous diffusion by Xu et al. [22]. Song and Jiang [23] studied the five parametric constitutive equations with fractional derivatives of linear viscoelastic Jeffreys model. For the applications, they consider the Sesbania gel and Xanthan gum and get the satisfactory results. Tan et al. [24] and Xu et al. [25] applied a fractional derivative model to Maxwell and generalized second grade fluids between two parallel plates. Abdullah et al. [26] considered the fractional Maxwell fluid in a boundless circular pipe with velocity ft. To obtained the solutions for velocity and shear stress they applied the Laplace transformation and modified Bessel equation. To find out numerically inverse Laplace transformation MATLAB is used by them.

In this article, the solutions for Burgers' fluid in rotating pipe like domain are determined. Here, a Laplace transformation technique was used to study fluid motion in a circular domain. The variable of time is removed with the help of a Laplace transform and modified Bessel functions to convert the complex equations into simple algebraic equations. The derived results are complicated so that inverse Laplace transformation were difficult to apply. So "Stehfest's algorithm" [27] and "MATHCAD" software was used to find the numerical solutions for the Burgers' fluid in circular domain. Graphs were drawn to elaborate the influences of fluids on velocity against different parameters. The results and discussions of all parameters are given at the end that shows the consistency in obtained results.

2. Governing Equations

Here, the formulation of velocity $\mathbf{V}$ and the extra-stress $\mathbf{S}$ for the fluid under consideration are as [28]

$$\mathbf{V} = \mathbf{V}(r,t) = F(r,t)\mathbf{e}_\theta, \quad \mathbf{S} = \mathbf{S}(r,t), \tag{1}$$

where $\mathbf{e}_\theta$ is the unit vector of the cylindrical coordinates system and $F(r,t)$ is the component of velocity along $\mathbf{e}_\theta$. The initial conditions for the fluid at rest position are, as in [28];

$$\mathbf{V}(r,0) = \mathbf{0}, \quad \mathbf{S}(r,0) = \mathbf{0}. \tag{2}$$

The governing equations for the motion of Burgers' fluid are [29]

$$\left(1 + \lambda_1 \frac{\partial}{\partial t} + \lambda_2 \frac{\partial^2}{\partial t^2}\right) \frac{\partial F}{\partial t} = \nu \left(1 + \lambda_3 \frac{\partial}{\partial t}\right)\left(\frac{\partial^2}{\partial r^2} + \frac{1}{r}\frac{\partial}{\partial r} - \frac{1}{r^2}\right) F(r,t), \tag{3}$$

$$\left(1 + \lambda_1 \frac{\partial}{\partial t} + \lambda_2 \frac{\partial^2}{\partial t^2}\right) Q(r,t) = \mu \left(1 + \lambda_3 \frac{\partial}{\partial t}\right)\left(\frac{\partial}{\partial r} - \frac{1}{r}\right) F(r,t), \tag{4}$$

where μ is the coefficient of viscosity, $\nu = \mu/\rho$ is kinematic viscosity, ρ is constant density of the fluid, λ_i ($i = 1,2,3$) new material constants and $Q(r,t) = S_{r\theta}(r,t) \neq 0$ shear stress. By altering the inner time derivatives with the fractional derivatives in Equations (3) and (4), the governing equations for the fractional derivative of Burgers' fluid (FBF) can be obtained as

$$\left(1 + \lambda_1^\alpha D_t^\alpha + \lambda_2^{2\alpha} D_t^{2\alpha}\right) \frac{\partial F(r,t)}{\partial t} = \nu \left(1 + \lambda_3^\beta D_t^\beta\right)\left(\frac{\partial^2}{\partial r^2} + \frac{1}{r}\frac{\partial}{\partial r} - \frac{1}{r^2}\right) F(r,t), \tag{5}$$

$$\left(1 + \lambda_1^\alpha D_t^\alpha + \lambda_2^{2\alpha} D_t^{2\alpha}\right) Q(r,t) = \mu \left(1 + \lambda_3^\beta D_t^\beta\right)\left(\frac{\partial}{\partial r} - \frac{1}{r}\right) F(r,t), \tag{6}$$

where α and β are the fractional parameters such as $0 \leq \alpha \leq \beta \leq 1$. The Caputo fractional derivative is defined as [16,18]

$$D_t^\alpha f(t) = \begin{cases} \frac{1}{\Gamma(1-\alpha)} \frac{d}{dt} \int_0^t \frac{f(\tau)}{(t-\tau)^\alpha} d\tau, & 0 \leq \alpha < 1; \\[2ex] \frac{d}{dt} f(t), & \alpha = 1, \end{cases} \tag{7}$$

where $\Gamma(\cdot)$ is the gamma function. When $\alpha, \beta \to 1$, Equations (5) and (6) reduce to Equations (3) and (4), because $D_t^1 f = \frac{df}{dt}$.

3. Imposing Condition and Geometry

Initially, (when $t = 0$) the FBF is at rest in an infinite circular pipe with radius $R(> 0)$ as shown in Figure 1. After time $t = 0^+$, the pipe suddenly starts to rotate about its Z-axis having the angular velocity Ωt^p. According to mathematical situation, appropriate initial and boundary conditions are

$$F(r,0) = \frac{\partial F(r,0)}{\partial t} = 0, \ \tau(r,0) = 0; \ \ r \in [0, R], \tag{8}$$

$$F(R,t) = R\Omega t^p; \quad t > 0, \ \ p \in N, \ p > 0, \tag{9}$$

where Ω denotes the angular constant and N is the set of natural numbers.

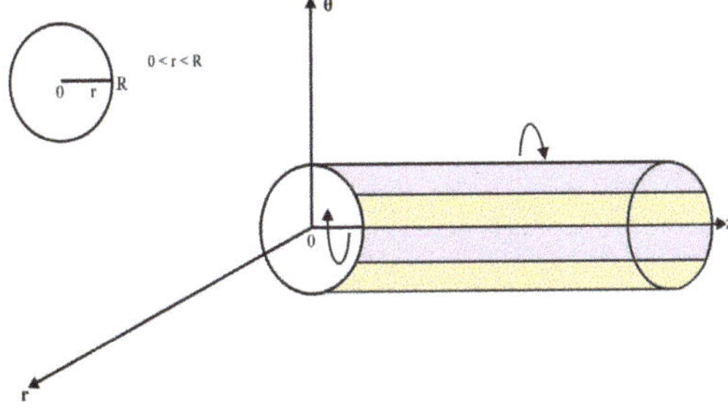

Figure 1. Geometry of the problem.

3.1. Velocity Field Due to Rotating Circular Pipe

By applying the Laplace transform to Equations (5) and (9),

$$\frac{\partial^2 \overline{F}(\mathbf{r},s)}{\partial r^2} + \frac{1}{r}\frac{\partial \overline{F}(\mathbf{r},s)}{\partial r} - \left[\frac{1}{r^2} + a(s)\right]\overline{F}(\mathbf{r},s) = 0, \tag{10}$$

and

$$\overline{F}(R,s) = \frac{R\Omega p!}{s^{p+1}}, \tag{11}$$

where

$$a(s) = \frac{s(1 + \lambda_1^{\alpha} s^{\alpha} + \lambda_2^{2\alpha} s^{2\alpha})}{\nu(1 + \lambda_3^{\beta} s^{\beta})}.$$

Also $\overline{F}(\mathbf{r},s)$ and $\overline{F}(R,s)$ are the Laplace transforms of the functions $F(\mathbf{r},t)$ and $F(R,t)$ respectively. By applying the variable transformation $z = r\sqrt{a(s)}$ in Equation (10)

$$z^2 \frac{\partial^2 \overline{F}}{\partial r^2} + z\frac{\partial \overline{F}}{\partial r} - (1^2 + z^2)\overline{F} = 0. \tag{12}$$

The above Equation (12) is the modified Bessel equation of order 1. So, the general solution of this equation is given by the successive equation

$$\overline{F}(z,s) = C_1 I_1(z) + C_2 K_1(z), \tag{13}$$

where C_1 and C_2 are constants and I_1, K_1 are the modified Bessel functions of first and second kind of order 1. For a finite solution at $r = 0 (z = 0)$, the constant C_2 should be zero i.e., $C_2 = 0$. So, Equation (13) implies

$$\overline{F}(z,s) = C_1 I_1(z). \tag{14}$$

The value of C_1 is calculated from Equation (14) by taking Equation (11) into account

$$C_1 = \frac{\Omega R p!}{s^{p+1} I_1(R\sqrt{a(s)})}. \tag{15}$$

Therefore the Equation (14) implies

$$\overline{F}(\mathbf{r},s) = \frac{\Omega R p!}{s^{p+1}} \times \frac{I_1(r\sqrt{a})}{I_1(R\sqrt{a})}. \tag{16}$$

The solution in Equation (16) is in the complex form of the first and second kind of Bessel functions. Normally, it is difficult to find out the solution of such complex expressions with an ordinary inverse Laplace transformation. However, an alternately inverse Laplace of Equation (16) was numerically calculated by using the "Stehfest's algorithm" [27]

$$f_n(x) = \ln(2)x^{-1}\sum_{i=1}^{2n} a_i(n)F(i\ln(2)x^{-1}), \qquad n \geq 1, x > 0, \tag{17}$$

where $a_i(n)$ are coefficients defined as follows

$$a_i(n) = \frac{(-1)^{n+i}}{n!}\sum_{s=\lceil\frac{(i+1)}{2}\rceil}^{\min(i,n)} s^{n+1}\, {}^nC_s\, {}^{2s}C_s\, {}^sC_{i-s}, \qquad n \geq 1, 1 \leq i \leq 2n, \tag{18}$$

and "MATHCAD" software.

3.2. Shear Stress Due to Rotating Circular Cylinder

By taking the Laplace transform of Equation (6)

$$\overline{Q}(r,s) = \sqrt{b(s)}\left[\frac{\partial \overline{F}(r,s)}{\partial r} - \frac{1}{r}\overline{F}(r,s)\right], \tag{19}$$

where

$$b(s) = \frac{\mu(1 + \lambda_3^\beta s^\beta)}{(1 + \lambda_1^\alpha s^\alpha + \lambda_2^{2\alpha} s^{2\alpha})}.$$

Putting Equation (16) into (17) and after taking the few steps of simplification,

$$\overline{Q}(r,s) = \frac{\sqrt{b}}{s^{p+1}} \times \frac{p!\Omega R}{I_1(R\sqrt{a})}\left[\sqrt{a}I_0(r\sqrt{a}) - \frac{2}{r}I_1(r\sqrt{a})\right]. \tag{20}$$

Again it is in a complicated form. So "Gaver Stehfest's algorithm" and "MATHCAD" software were used for the inverse Laplace transformation of the result given in Equation (20).

4. Results and Discussions

In this article, the main goal is to establish the numerical technique to develop the solutions of the Burgers' fluid for the rotational flow of Burgers' fluid within cylindrical domain. The expressions of velocity and shear stress are found for an incompressible non-integer order model of the Burgers' fluid in a circular pipe. The Laplace transformation is used to establish the solutions. As there are complicated expressions in Equations (16) and (20) with respect to the Laplace transformation, it was not easy to find out final results by using direct inverse Laplace on Equations (16) and (20). Although, "MATHCAD" is used along with the "Stehfest's algorithm" instead of the inverse Laplace transformation to find out the numerical results. The numerical solutions for Oldroyd-B, Maxwell, second grade and Newtonian fluids are also derived with generalization of main results i.e., Equations (16) and (20). To determine the impact of physical parameters, the graphical illustrations are made. The behaviour of time on the velocity field and stress is shown in Figures 2 and 3. These figures show that the influence of velocity and stress depending upon the parameters $r, \lambda_1, \lambda_2, \lambda_3, \alpha, \beta, \nu$ and μ. Figures 2 and 3 show that the velocity and shear stress increase with respect to the increase in the dependent variable t with fixed values of other dependent parameters. Similarly, a clear increase in velocity and stress can also be seen in Figures 4 and 5 for r with respect to the time variable t while fixing the other dependent parameters. It is also observed from Figure 4 that velocity has linear relation for r with respect to t.

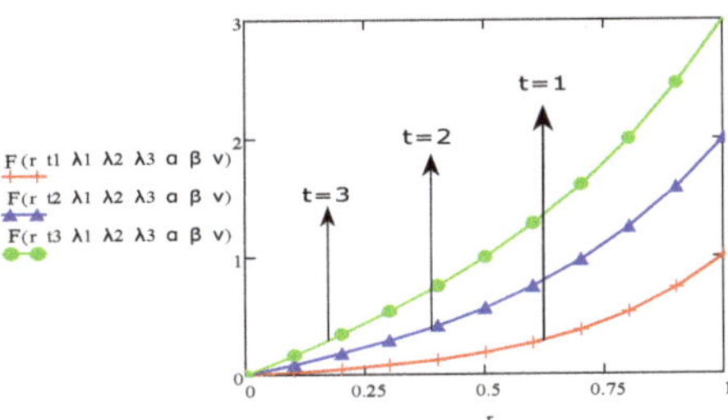

Figure 2. The aspect of velocity $F(r,t)$ for various values of t and fixed values for $\Omega = 1, R = 1, p = 1, s = 1, \alpha = 0.3, \beta = 0.7, \nu = 0.053, n = 12, \lambda_1 = 5, \lambda_2 = 20$ and $\lambda_3 = 55$.

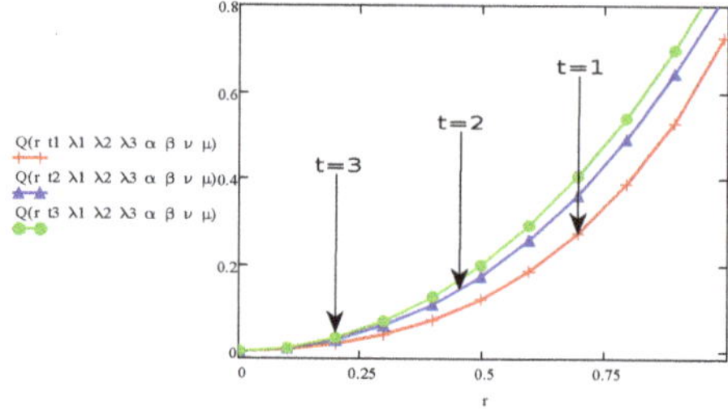

Figure 3. The aspect of stress $Q(\mathrm{r}, t)$ for various values of t and fixed values for $\mu = 0.010, \Omega = 2.5, R = 1, p = 1, s = 1, \alpha = 0.3, \beta = 0.7, v = 0.053, n = 12, \lambda_1 = 5, \lambda_2 = 20$ and $\lambda_3 = 55$.

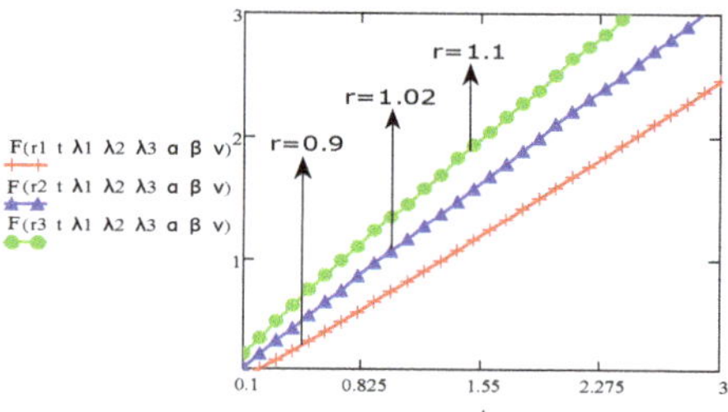

Figure 4. The aspect of velocity $F(\mathrm{r}, t)$ for various values of r and fixed values for $\Omega = 1, R = 1, p = 1, s = 1, \alpha = 0.3, \beta = 0.7, v = 0.053, n = 12, \lambda_1 = 5, \lambda_2 = 20$ and $\lambda_3 = 55$.

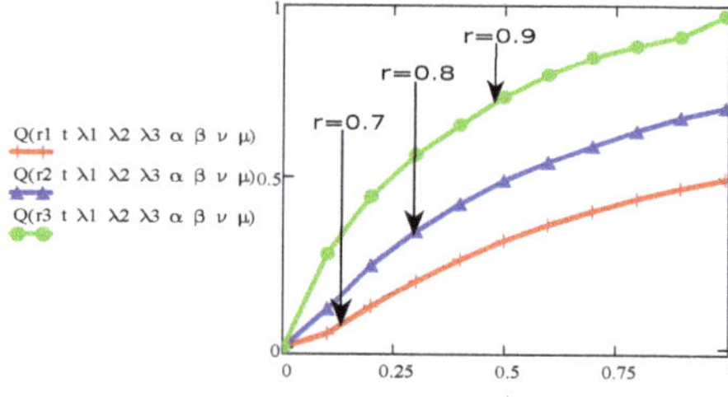

Figure 5. The aspect of stress $Q(\mathrm{r}, t)$ for various values of r and fixed values for $\mu = 0.09, \Omega = 1.5, R = 1, p = 1, s = 1, \alpha = 0.3, \beta = 0.7, v = 0.053, n = 12, \lambda_1 = 5, \lambda_2 = 20$ and $\lambda_3 = 55$.

Also, to uncover the aspects of other physical and fraction parameters on velocity and stress the graphs drawn and presented in Figures 6–17. All the graphs for velocity and stress in Figures 6–17 are plotted for concerned parameter against the change factor "time". One can say that the fractional parameter α, relaxation parameters λ_1 and λ_2 have decreasing behaviour for velocity and stress with respect to t. This fact can be observed in Figures 6–9, 12 and 13 respectively. Whereas, the velocity and stress are also increasing functions for the retardation parameter λ_3, fractional parameter β, viscosity μ and v with respect to time t as shown in Figures 10, 11, 14–17. Figures 2–17 make it possible to check the point to point variations, increment or decrement in two parameters (among which the graphs are made) for velocity and stress profile.

Finally, the comparison among the different generalized cases of Burgers' fluid are made in Figure 18 for velocity function against time t. The results of these generalized cases are derived by the implementation of the "Stehfest's algorithm" and "MATHCAD" to Equation (16). It is observed that the second grade fluids have more velocity as compared to other Newtonian and non-Newtonian fluid cases of this model while taking the same values of different dependent parameters and variation in time t. The Oldroyd-B fluid behaves as like the second grade fluid but have less velocity as compared to the second grade fluid's velocity. Similarly, Figure 18 also shows the behaviour of velocity for Newtonian and Maxwell fluids. The Maxwell fluids have minimum velocity for this flow model with prescribed conditions among all the Newtonian and non-Newtonian cases. As a description it is clearing that in all the Figures 2–18, the units of the material constants are SI units. The comparison of solutions that obtained from two different methods analytically [29,30] and numerically (derived from Equations (16) and (20)) are given in Tables 1 and 2 for two different cases Maxwell and Newtonian fluids. These tables showing clearly that two different methods for the same problem have the same results. Which shows the consistency of our numerical technique and results for fractional model of Burgers' fluid with already published literature [29,30].

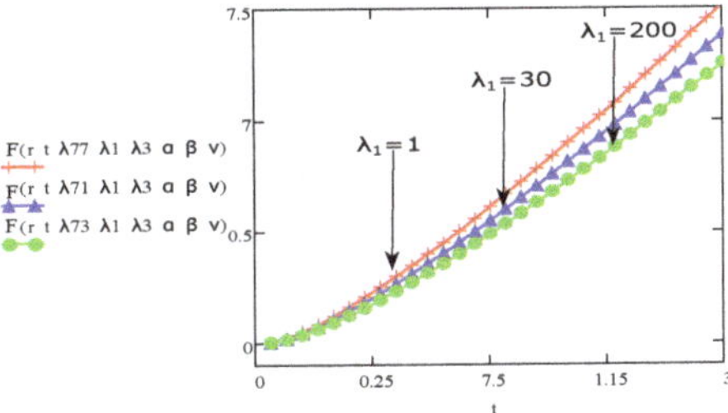

Figure 6. The aspect of velocity $F(r, t)$ for various values of λ_1 and fixed values for $\Omega = 1, R = 1, p = 1, s = 1, \alpha = 0.3, \beta = 0.8, v = 0.013, n = 12, r = 0.8, \lambda_2 = 20$ and $\lambda_3 = 55$.

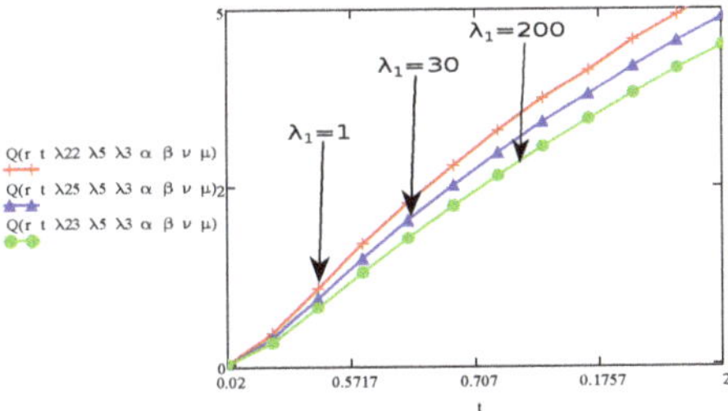

Figure 7. The aspect of stress $Q(r, t)$ for various values of λ_1 and fixed values for $\mu = 0.09, \Omega = 2.5, R = 1, p = 1, s = 1, \alpha = 0.3, \beta = 0.8, v = 0.013, n = 12, r = 0.8, \lambda_2 = 20$ and $\lambda_3 = 55$.

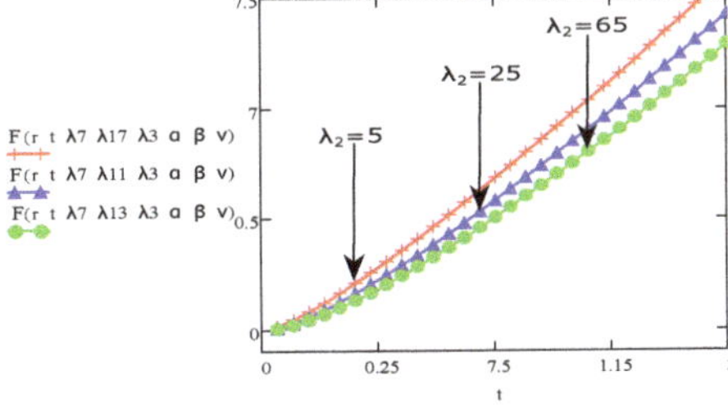

Figure 8. The aspect of velocity $F(r, t)$ for various values of λ_2 and fixed values for $\Omega = 1, R = 1, p = 1, s = 1, \alpha = 0.2, \beta = 0.7, v = 0.013, n = 12, r = 0.8, \lambda_1 = 5$ and $\lambda_3 = 55$.

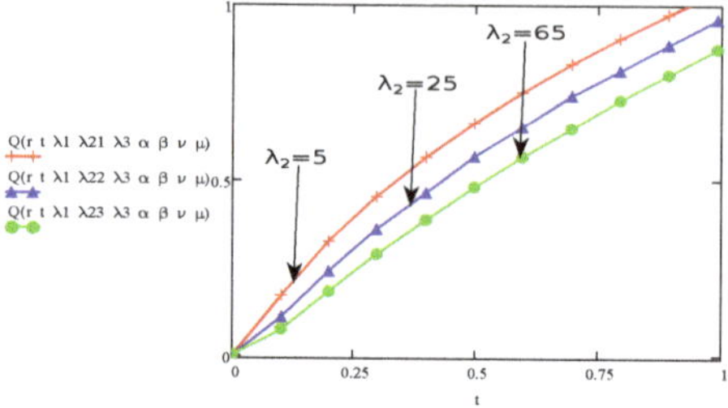

Figure 9. The aspect of stress $Q(r, t)$ for various values of λ_2 and fixed values for $\mu = 0.09, \Omega = 1.2, R = 1, p = 1, s = 1, \alpha = 0.2, \beta = 0.7, \nu = 0.013, n = 12, r = 0.8, \lambda_1 = 5$ and $\lambda_3 = 55$.

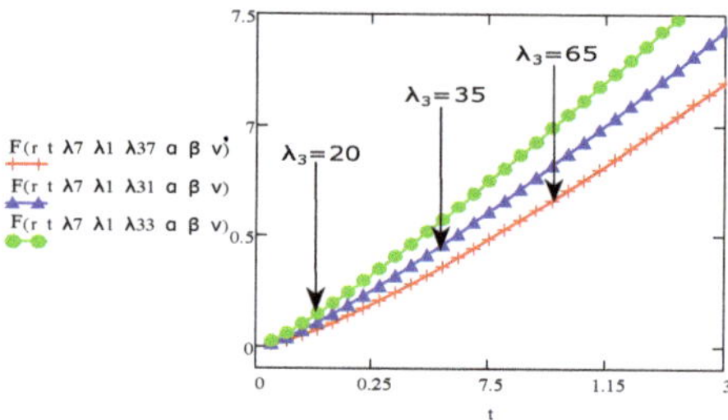

Figure 10. The aspect of velocity $F(r, t)$ for various values of λ_3 and fixed values for $\Omega = 1, R = 1, p = 1, s = 1, \alpha = 0.2, \beta = 0.8, \nu = 0.013, n = 12, r = 0.8, \lambda_1 = 5$ and $\lambda_2 = 20$.

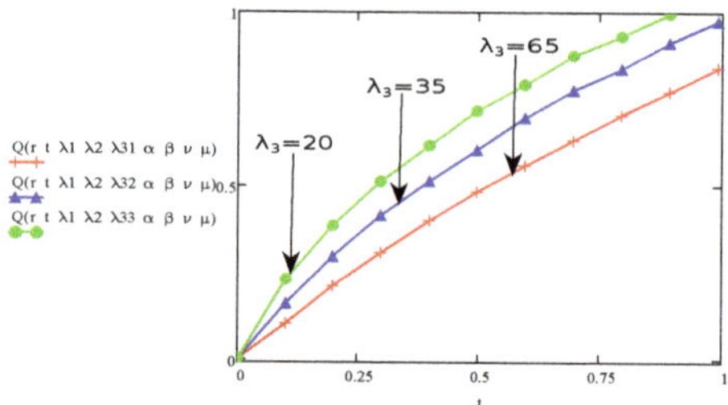

Figure 11. The aspect of stress $Q(r, t)$ for various values of λ_3 and fixed values for $\mu = 0.09, \Omega = 1.2, R = 1, p = 1, s = 1, \alpha = 0.2, \beta = 0.8, \nu = 0.013, n = 12, r = 0.8, \lambda_1 = 5$ and $\lambda_2 = 20$.

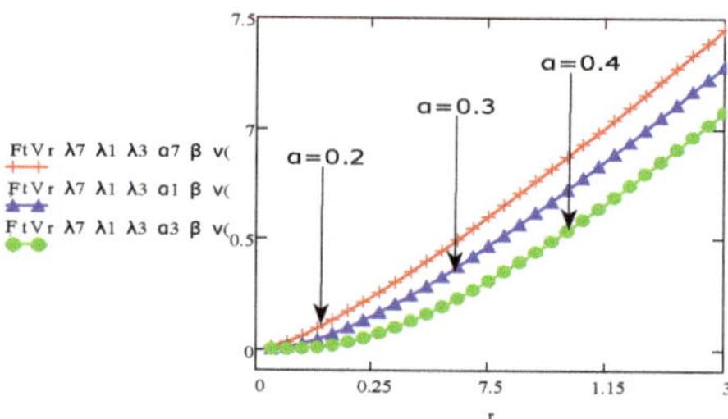

Figure 12. The aspect of velocity $F(r, t)$ for various values of α and fixed values for $\Omega = 1, R = 1, p = 1, s = 1, \beta = 0.7, \nu = 0.013, n = 12, r = 0.8, \lambda_1 = 5, \lambda_2 = 20$ and $\lambda_3 = 55$.

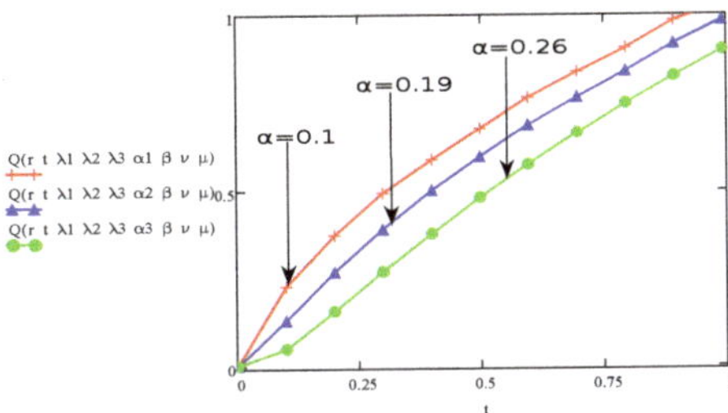

Figure 13. The aspect of stress $Q(r, t)$ for various values of α and fixed values for $\mu = 0.09, \Omega = 1.2, R = 1, p = 1, s = 1, \beta = 0.7, \nu = 0.013, n = 12, r = 0.8, \lambda_1 = 5, \lambda_2 = 20$ and $\lambda_3 = 55$.

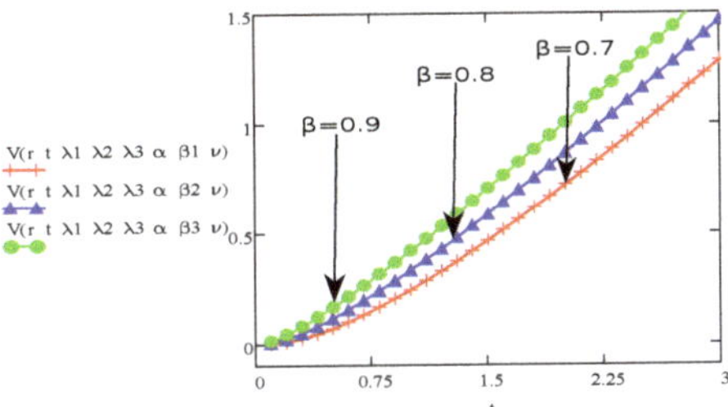

Figure 14. The aspect of velocity $F(r, t)$ for various values of β and fixed values for $\Omega = 1, R = 1, p = 1, s = 1, \alpha = 0.3, \nu = 0.013, n = 12, r = 0.8, \lambda_1 = 5, \lambda_2 = 20$ and $\lambda_3 = 55$.

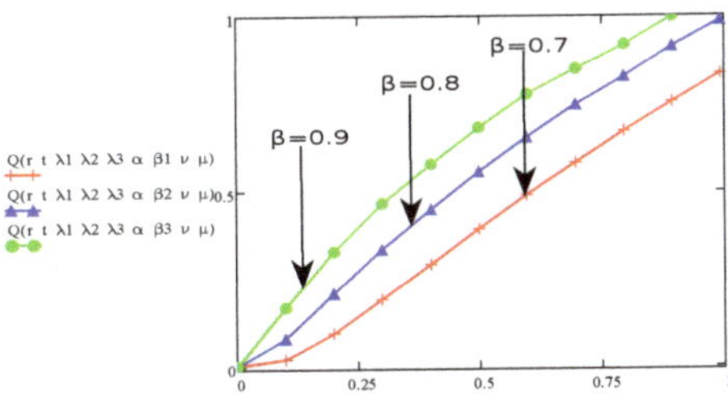

Figure 15. The aspect of stress $Q(r, t)$ for various values of β and fixed values for $\mu = 0.09, \Omega = 1.2, R = 1, p = 1, s = 1, \alpha = 0.3, \nu = 0.013, n = 12, r = 0.8, \lambda_1 = 5, \lambda_2 = 20$ and $\lambda_3 = 55$.

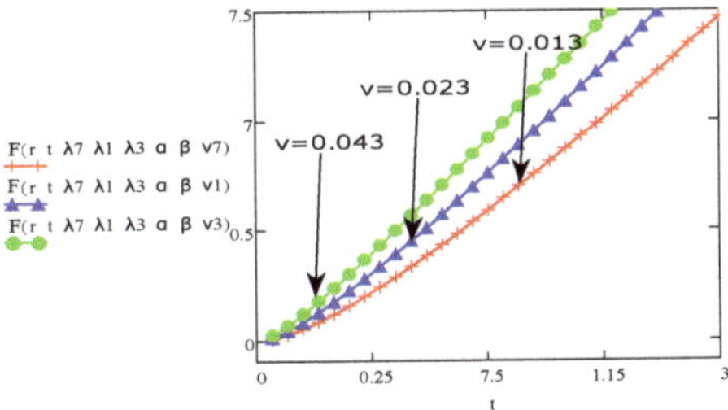

Figure 16. The aspect of velocity $F(r, t)$ for various values of ν and fixed values for $\Omega = 1, R = 1, p = 1, s = 1, \alpha = 0.3, \beta = 0.8, n = 12, r = 0.8, \lambda_1 = 5, \lambda_2 = 20$ and $\lambda_3 = 55$.

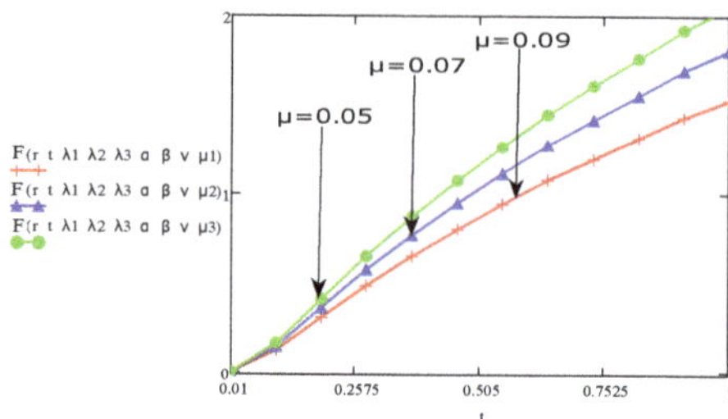

Figure 17. The aspect of stress $Q(r, t)$ for various values of μ and fixed values for $\Omega = 2.5, R = 1, p = 1, s = 1, \alpha = 0.3, \beta = 0.8, \nu = 0.013, n = 12, r = 0.8, \lambda_1 = 5, \lambda_2 = 20$ and $\lambda_3 = 55$.

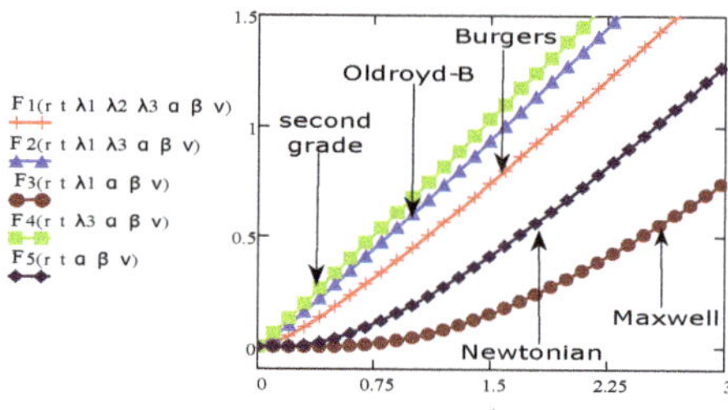

Figure 18. Comparisons of fluid velocity for different fluid models for same value of parameters.

Table 1. Comparison of exact solutions [30] and numerical solutions (obtained from "Stehfest's algorithm") for the fractional order derivative model of the Maxwell fluid.

r	Exact w(r,t) [30]	Numerical V(r,t)	Error
0	0.000	0.000	0.000
0.02	0.029	0.029	0.000
0.04	0.058	0.058	0.000
0.06	0.087	0.087	0.000
0.08	0.115	0.116	−0.001
0.10	0.145	0.145	0.000
0.12	0.174	0.174	0.000
0.14	0.203	0.203	0.000
0.16	0.233	0.233	0.000
0.18	0.262	0.263	−0.001
0.20	0.292	0.292	0.000
0.22	0.322	0.322	0.000
0.24	0.352	0.352	0.000
0.26	0.383	0.381	0.002
0.28	0.414	0.411	0.003
0.30	0.444	0.441	0.003
0.32	0.475	0.473	0.002
0.34	0.506	0.503	0.003
0.36	0.538	0.534	0.004
0.38	0.569	0.565	0.004
0.40	0.601	0.596	0.005
0.42	0.632	0.628	0.004
0.44	0.664	0.661	0.003
0.46	0.695	0.694	0.001
0.48	0.727	0.725	0.002
.....	.....	.....	.....

Table 2. Comparison of exact solutions [29] and numerical solutions (obtained from "Stehfest's algorithm") for the fractional order derivative model of the Newtonian fluid.

r	Exact w(r,t) [29]	Numerical V(r,t)	Error
0	0.000	0.000	0.000
0.02	0.023	0.023	0.000
0.04	0.046	0.046	0.000
0.06	0.069	0.069	0.000
0.08	0.093	0.092	0.001
0.10	0.116	0.115	0.001
0.12	0.139	0.138	0.001
0.14	0.162	0.161	0.001
0.16	0.185	0.185	0.000
0.18	0.208	0.208	0.000
0.20	0.232	0.232	0.000
0.22	0.255	0.256	−0.001
0.24	0.279	0.280	−0.001
0.26	0.304	0.304	0.000
0.28	0.329	0.329	0.000
0.30	0.354	0.354	0.000
0.32	0.379	0.379	0.000
0.34	0.405	0.404	0.001
0.36	0.431	0.430	0.001
0.38	0.458	0.456	0.002
0.40	0.484	0.482	0.002
0.42	0.511	0.509	0.002
0.44	0.538	0.536	0.002
0.46	0.564	0.564	0.000
0.48	0.591	0.592	−0.001

5. Conclusions

The flow of incompressible Burgers' fluid with leading equations in the terms of Caputo non-integer order derivatives is studied. The under consider flow is flowing through the circular channel of infinite length. The numerical solutions for velocity and stress has been drawn by utilizing the Laplace transformation along with the modified Bessel's function. After the analysis of result following fruitful remarks are observed:

- The ordinary fluid have less velocity as compared to fractional order derivative fluid models. This result can be verify from the graph of fractional parameter α available in Figure 12, which has decreasing altitude when the velocity is increasing.
- The velocity of the fluid increases for Burgers' fluid model as fluid becomes more thick in this model.
- The graph of the parameters β, λ_3, ν, t, r, and μ showed an increase/upward in behaviour with increase in velocity and stress function.
- The parameters λ_1, λ_2 and α are behaving opposite to the influence of velocity and shear stress.
- The fractional Burgers' fluid is flowing faster than the Maxwell and Newtonian fluids.
- Our obtained solutions given in Equations (16) and (20) derived from "Stehfest's inversion algorithm" and the exact solutions given in [29,30] are equivalent.
- In future, authors will try to study the fluid motion by considering the effects of temperature and magnetic field.

Author Contributions: R.S. and M.I. made the mathematical model and mathematical calculation of the paper. M.A.I. and M.I. made the numerical results and graphs of the paper. I.K. wrote the manuscript. M.I. and R.S. checked the calculation and revised the manuscript. All authors read and approved the final manuscript.

Funding: This research is supported by the Government College University, Faisalabad, Pakistan, and the Higher Education Commission, Pakistan.

Conflicts of Interest: The authors declare that they have no competing interests.

References

1. Taylor, G.I. Stability of a viscous liquid contained between two rotating cylinders. *Philos. Trans. R. Soc. Lond. Ser. A Contain. Pap. A Math. Phys. Character* **1923**, *223*, 289–343. [CrossRef]
2. Childress, S. *An Introduction to Theoretical Fluid Mechanics*; American Mathematical Soc.: New York, NY, USA, 2009, Volume 19.
3. Waters, N.; King, M. The unsteady flow of an elastico-viscous liquid in a straight pipe of circular cross section. *J. Phys. D Appl. Phys.* **1971**, *4*, 204–211. [CrossRef]
4. Rahaman, K.; Ramkissoon, H. Unsteady axial viscoelastic pipe flows. *J. Non-Newton. Fluid Mech.* **1995**, *57*, 27–38. [CrossRef]
5. Wood, W. Transient viscoelastic helical flows in pipes of circular and annular cross-section. *J. Non-Newton. Fluid Mech.* **2001**, *100*, 115–126. [CrossRef]
6. Fox, R.W.; McDonald, A.T. *Introduction to Fluid Mechanics*; John Wiley & Sons. Inc.: New York, NY, USA, 1994.
7. Fetecau, C. Analytical solutions for non-Newtonian fluid flows in pipe-like domains. *Int. J. Non-Linear Mech.* **2004**, *39*, 225–231. [CrossRef]
8. Ting, T.W. Certain non-steady flows of second-order fluids. *Arch. Ration. Mech. Anal.* **1963**, *14*, 1–26. [CrossRef]
9. Srivastava, P. Non-steady helical flow of a visco-elastic liquid(Nonsteady helical flow of viscoelastic liquid contained in circular cylinder, noting occurrence of oscillations in fluid decaying exponentially with time). *Arch. Mech. Stosow.* **1966**, *18*, 145–150.
10. Sherief, H.; Faltas, M.; El-Sapa, S. Pipe flow of magneto-micropolar fluids with slip. *Can. J. Phys.* **2017**, *95*, 885–893. [CrossRef]
11. Fetecău, C.; Fetecău, C. On the uniqueness of some helical flows of a second grade fluid. *Acta Mech.* **1985**, *57*, 247–252. [CrossRef]
12. Hayat, T.; Khan, M.; Ayub, M. Some analytical solutions for second grade fluid flows for cylindrical geometries. *Math. Comput. Model.* **2006**, *43*, 16–29. [CrossRef]
13. Fetecau, C.; Fetecau, C.; Vieru, D. On some helical flows of Oldroyd-B fluids. *Acta Mech.* **2007**, *189*, 53–63. [CrossRef]
14. Vieru, D.; Akhtar, W.; Fetecau, C.; Fetecau, C. Starting solutions for the oscillating motion of a Maxwell fluid in cylindrical domains. *Meccanica* **2007**, *42*, 573–583. [CrossRef]
15. Nadeem, S.; Asghar, S.; Hayat, T.; Hussain, M. The Rayleigh Stokes problem for rectangular pipe in Maxwell and second grade fluid. *Meccanica* **2008**, *43*, 495–504. [CrossRef]
16. Podlubny, I. *Fractional Differential Equations, vol. 198 of Mathematics in Science and Engineering*; Academic Press, San Diego, CA, USA, 1999.
17. Bagley, R.L.; Torvik, P.J. On the fractional calculus model of viscoelastic behavior. *J. Rheol.* **1986**, *30*, 133–155. [CrossRef]
18. Friedrich, C. Relaxation and retardation functions of the Maxwell model with fractional derivatives. *Rheol. Acta* **1991**, *30*, 151–158. [CrossRef]
19. Tan, W.; Xian, F.; Wei, L. An exact solution of unsteady Couette flow of generalized second grade fluid. *Chin. Sci. Bull.* **2002**, *47*, 1783–1785. [CrossRef]
20. Tong, D.; Shan, L. Exact solutions for generalized Burgers' fluid in an annular pipe. *Meccanica* **2009**, *44*, 427–431. [CrossRef]
21. Shah, S.H.A.M. Some helical flows of a Burgers' fluid with fractional derivative. *Meccanica* **2010**, *45*, 143–151. [CrossRef]

22. Xu, M.; Tan, W. Theoretical analysis of the velocity field, stress field and vortex sheet of generalized second order fluid with fractional anomalous diffusion. *Sci. China Ser. A Math.* **2001**, *44*, 1387–1399. [CrossRef]

23. Song, D.Y.; Jiang, T.Q. Study on the constitutive equation with fractional derivative for the viscoelastic fluids–modified Jeffreys model and its application. *Rheol. Acta* **1998**, *37*, 512–517. [CrossRef]

24. Wenchang, T.; Wenxiao, P.; Mingyu, X. A note on unsteady flows of a viscoelastic fluid with the fractional Maxwell model between two parallel plates. *Int. J. Non-Linear Mech.* **2003**, *38*, 645–650. [CrossRef]

25. Wenchang, T.; Mingyu, X. Unsteady flows of a generalized second grade fluid with the fractional derivative model between two parallel plates. *Acta Mech. Sin.* **2004**, *20*, 471–476. [CrossRef]

26. Abdullah, M.; Butt, A.R.; Raza, N.; Haque, E.U. Semi-analytical technique for the solution of fractional Maxwell fluid. *Can. J. Phys.* **2017**, *95*, 472–478. [CrossRef]

27. Stehfest, H. Algorithm 368: Numerical inversion of Laplace transforms [D5]. *Commun. ACM* **1970**, *13*, 47–49. [CrossRef]

28. Kamran, M.; Imran, M.; Athar, M. Exact solutions for the unsteady rotational flow of an Oldroyd-B fluid with fractional derivatives induced by a circular cylinder. *Meccanica* **2013**, *48*, 1215–1226. [CrossRef]

29. Safdar, R.; Imran, M.; Sadiq, N.; Ahmad, F. Unsteady Rotational Flow of a Burgers' Fluid through a Pipe with Non-Integer Order Fractional Derivatives. *J. Appl. Environ. Biol. Sci* **2018**, *8*, 144–156.

30. Imran, M.; Athar, M.; Kamran, M. On the unsteady rotational flow of a generalized Maxwell fluid through a circular cylinder. *Arch. Appl. Mech.* **2011**, *81*, 1659–1666. [CrossRef]

Article

Two-Dimensional Advection–Diffusion Process with Memory and Concentrated Source

Najma Ahmed [1], Nehad Ali Shah [2] and Dumitru Vieru [3,*]

[1] Abdus Salam School of Mathematical Sciences, Government College University, 5400 Lahore, Pakistan
[2] Department of Mathematics, Lahore Leads University, 5400 Lahore, Pakistan
[3] Department of Theoretical Mechanics, Technical University "Gheorghe Asachi" of Iasi, 6600 Iasi, Romania
* Correspondence: dumitru.vieru@tuiasi.ro

Received: 25 June 2019; Accepted: 3 July 2019; Published: 4 July 2019

Abstract: Two-dimensional advection–diffusion processes with memory and a source concentrated in the symmetry center of the domain have been investigated. The differential equation of the studied model is a fractional differential equation with short-tail memory (a differential equation with Caputo–Fabrizio time-fractional derivatives). An analytical solution of the initial-boundary value problem has been determined by employing the Laplace transform and double sine-Fourier transforms. A numerical solution of the studied problem has been determined using finite difference approximations. Numerical simulations for both solutions have been carried out using the software Mathcad.

Keywords: Advection–diffusion; fractional derivative; concentrated source; integral transform

1. Introduction

Partial differential equations with time-fractional derivatives are suitable to describe complex problems in chaotic dynamics, signal processing, wave propagation, classical and continuum mechanics [1–6]. These equations admit the non-local memory effects; therefore, the fractional models can eliminate the disadvantage of classical models that cannot satisfactorily describe some complex processes such as anomalous diffusion.

In recent years, fractional advection–diffusion equations were intensively studied due to their multiple applications in many practical problems such as transport of pollutants in air and water, contaminating streams in groundwater, migration of pollutants in rivers and seawater, and tracer dispersion in porous media.

Bhrawy and Baleanu [7] developed an efficient Legendre–Gauss–Lobatto method to solve the advection–diffusion equation with a space-fractional Caputo derivative and non-homogeneous initial-boundary conditions. The advection–dispersion equation with a fractional order dependent on time and the Coimbra derivative was numerically solved by Abdelkawy et al. [8]. Zhuang et al. [9] numerically solved the advection–diffusion equation with derivatives of variable fractional order and a nonlinear source. They presented the explicit and implicit Euler approximation, fractional method of line, matrix transfer technique and the extrapolation method. Fazio et al. [10] have analyzed the advection–diffusion equation with the time-fractional Caputo derivative by using a finite difference method with non-uniform grids. They obtained an accuracy method compared with the classical discrete fractional formula for the time-fractional Caputo derivative. Time–space fractional advection–diffusion equations with Riesz derivatives have been numerically investigated by Arshad et al. [11]. A fractional advection–diffusion equation with a nonlinear source has been studied by Jannelli et al. [12]. They obtained a numerical solution and an analytical solution based on the Mittag–Leffler functions. Using the finite volume element method, Badr et al. [13] have studied the fractional advection–diffusion equation with the time-fractional Caputo derivative. Stability of the method has been analyzed. Pimenov [14] has numerically investigated the fractional advection–diffusion equation with heredity. Singh et al. [15] have obtained analytical solutions for the fractional advection–dispersion equation by employing the q-fractional homotopy analysis transform method. Using the integral transform method, Avci and Yetim [16] have found the fundamental solutions of the fractional advection–diffusion with fractional derivatives of the Atangana–Baleanu type. Fundamental solutions for the Dirichlet problem of the two-dimensional fractional advection–diffusion equation with short-tail memory have been studied by Mirza and Vieru [17]. Using the Laplace transform and the sine–cosine Fourier transform, Mirza et al. [18] have determined analytical solutions for the fractional advection–diffusion equation in the one-dimensional case and with boundary conditions of Robin-type. Interesting fractional diffusive processes have been investigated by Hristov in [19–21]. Povstenko and Klekot [22] and Povstenko and Kyrylych [23] have studied the fractional advection diffusion equation with several boundary conditions. It is important to underline that the time–space fractional advection–diffusion equation could be solved with the Monte–Carlo method as a limit of continuous-time random walks [24].

In this work, the analytical and numerical study of two-dimensional advection–diffusion processes with short-tail memory and with the source concentrated in the symmetry center of the geometric domain is carried out.

The mathematical model is described by a fractional differential equation with the time-fractional Caputo–Fabrizio derivative. An analytical solution of the initial-boundary value problem has been determined by employing the Laplace and finite sine-Fourier transforms. A numerical solution of the studied problem has been determined using finite difference approximations. Numerical simulations for both solutions have been carried out using the software Mathcad and a good agreement is found.

2. The Statement of the Problem

Let us consider the two-dimensional advection–diffusion process with memory and the time-dependent concentrated-source in a finite domain:

$D = \{(x,y), a \leq x \leq b, c \leq y \leq d, a < b, c < d\}$ with the border ∂D.

The mathematical model is described by the following time-fractional advection–diffusion equation [7,9]:

$$^{CF}D_t^\alpha u(x,y,t) + k_1 \frac{\partial u(x,y,t)}{\partial x} + k_2 \frac{\partial u(x,y,t)}{\partial y} - \varepsilon_1 \frac{\partial^2 u(x,y,t)}{\partial x^2} - \varepsilon_2 \frac{\partial^2 u(x,y,t)}{\partial y^2} = f(x,y,t),$$
$$0 < \alpha \leq 1, \ k_1, \ k_2 \geq 0, \ \varepsilon_1, \ \varepsilon_2 > 0, \ (x,y) \in D. \tag{1}$$

where $u(x,y,t)$ is the field variable representing the solute concentration, $\vec{v} = k_1 \vec{e}_x + k_2 \vec{e}_y$ is the constant fluid velocity (the drift velocity), $\varepsilon_1, \varepsilon_2$ are the dispersion coefficients in the x-direction and y-direction, respectively and $f(x,y,t)$ represents the sources/sinks.

The operator $^{CF}D_t^\alpha(\cdot)$ denotes the time-fractional Caputo–Fabrizio derivative defined as [25]:

$$^{CF}D_t^\alpha u(x,y,t) = h(\alpha,t) * \dot{u}(x,y,t),$$
$$h(\alpha,t) = \begin{cases} \frac{M(\alpha)}{1-\alpha} \exp\left(\frac{-\alpha t}{1-\alpha}\right), & 0 \leq \alpha < 1, \\ \delta(t), & \alpha = 1, \end{cases} \tag{2}$$
$$\dot{u}(x,y,t) = \frac{\partial u(x,y,t)}{\partial t},$$

where "∗" denotes the convolution product and $\delta(t)$ is Dirac's distribution. In the above definition, the function $M(\alpha)$ is so called the normalization function, which should satisfy conditions $M(0) = M(1) = 1$. In the present paper, for simplicity, we will consider for the normalization function $M(\alpha) \equiv 1, \alpha \in [0,1]$.

The following properties of the Caputo–Fabrizio operator immediately result from Equation (2):

$$^{CF}D_t^0 u(x,y,t) = u(x,y,t) - u(x,y,0), \ ^{CF}D_t^1 u(x,y,t) = \frac{\partial u(x,y,t)}{\partial t}, \ ^{CF}D_t^\alpha C = 0, \ C = \text{const.} \tag{3}$$

The Laplace transform of the Caputo–Fabrizio derivative is given by:

$$L\left\{{}^{CF}D_t^\alpha u(x,y,t)\right\} = L\left\{h(\alpha,t) * \dot{u}(x,y,t)\right\} = L\{h(\alpha,t)\}L\left\{\dot{u}(x,y,t)\right\} = \frac{sL\{u(x,t,t)\} - u(x,y,0)}{(1-\alpha)s + \alpha}. \tag{4}$$

The fractional advection–diffusion equation (Equation (1)), is studied along with following initial and boundary conditions:

$$u(x,y,0) = \psi_0(x,y), \ (x,y) \in D, \tag{5}$$

$$u(x,y,t) = \psi_1(x,y,t), \ t > 0, \ (x,y) \in \partial D, \ t > 0, \tag{6}$$

Making the transformation:

$$u(x,y,t) = e^{f_0(x,y)}\varphi(x,y,t) + \psi_0(x,y), \ f_0(x,y) = \frac{k_1 x}{2\varepsilon_1} + \frac{k_2 y}{2\varepsilon_2}, \tag{7}$$

Equation (1) becomes:

$$^{CF}D_t^\alpha \varphi(x,y,t) + \frac{1}{2}\left(\frac{k_1^2}{\varepsilon_1} + \frac{k_2^2}{\varepsilon_2}\right)\varphi(x,y,t) - \varepsilon_1\frac{\partial^2\varphi(x,y,t)}{\partial x^2} - \varepsilon_2\frac{\partial^2\varphi(x,y,t)}{\partial y^2} = F(x,y,t) + G(x,y), \tag{8}$$

where

$$\begin{aligned} F(x,y,t) &= f(x,y,t)\exp(-f_0(x,y)) \\ G(x,y) &= \left[\varepsilon_1\frac{\partial^2\psi_0(x,y)}{\partial x^2} + \varepsilon_2\frac{\partial^2\psi_0(x,y)}{\partial y^2} - k_1\frac{\partial\psi_0(x,y)}{\partial x} - k_2\frac{\partial\psi_0(x,y)}{\partial y}\right]\exp(-f_0(x,y)). \end{aligned} \tag{9}$$

After transformation (7), the initial and boundary conditions become:

$$\varphi(x,y,0) = 0, \ (x,y) \in D, \tag{10}$$

$$\varphi(x,y,t) = F_1(x,y,t) + G_1(x,y), \ (x,y) \in \partial D, \ t > 0, \tag{11}$$

where

$$F_1(x,y,t) = \psi_1(x,y,t)\exp(-f_0(x,y)), \ G_1(x,y) = -\psi_0(x,y)\exp(-f_0(x,y)). \tag{12}$$

3. Analytical Solution of the Problem

The solution of the problem described by Equation (8) and conditions (10) and (11) will be determined by using the integral transform method.

Applying the Laplace transform to Equation (8), using Equation (4) and Equation (10), we obtain the transformed equation:

$$\frac{[1+\beta(1-\alpha)]s + \alpha\beta}{(1-\alpha)s + \alpha}\overline{\varphi}(x,y,s) - \varepsilon_1\frac{\partial^2\overline{\varphi}(x,y,s)}{\partial x^2} - \varepsilon_2\frac{\partial^2\overline{\varphi}(x,y,s)}{\partial y^2} = \overline{F}(x,y,s) + \frac{1}{s}G(x,y), \tag{13}$$

where $\beta = \frac{1}{2}\left(\frac{k_1^2}{\varepsilon_1} + \frac{k_2^2}{\varepsilon_2}\right)$ and $\overline{\varphi}(x,y,s) = \int\limits_0^\infty \varphi(x,y,t)e^{-st}dt, \ \overline{F}(x,y,s) = \int\limits_0^\infty F(x,y,t)e^{-st}dt$ are the Laplace transforms of functions $\varphi(x,y,t), \ F(x,y,t)$, respectively.

Equation (13) has to satisfy the boundary condition:

$$\begin{aligned} \overline{\varphi}(x,y,s) &= \overline{\varphi}_1(x,y,s), \ (x,y) \in \partial D, \\ \overline{\varphi}_1(x,y,s) &= \overline{F}_1(x,y,s) + \frac{1}{s}G_1(x,y). \end{aligned} \tag{14}$$

Now, we apply this to Equation (13), the double sine-Fourier transform with respect to variables x and y defined as [26]:

$$\widetilde{\widetilde{\varphi}}(n,k,s) = \int\limits_c^d \int\limits_a^b \sin\left[\frac{n(x-a)\pi}{b-a}\right] \sin\left[\frac{k(y-c)\pi}{d-c}\right] \overline{\varphi}(x,y,s)dxdy, \quad n,\ k = 1,2,\dots \tag{15}$$

A direct calculation leads to the following relationships:

$$\int\limits_c^d \int\limits_a^b \frac{\partial^2 \overline{\varphi}(x,y,s)}{\partial x^2} \sin\left[\frac{n(x-a)\pi}{b-a}\right] \sin\left[\frac{k(y-c)\pi}{d-c}\right] dxdy = \overline{\widetilde{H}}_1(n,k,s) - \left(\frac{n\pi}{b-a}\right)^2 \widetilde{\widetilde{\varphi}}(n,k,s), \tag{16}$$

where

$$\overline{\widetilde{H}}_1(n,k,s) = \frac{n\pi}{b-a} \int\limits_c^d \left[\overline{\varphi}_1(a,y,s) - (-1)^n \overline{\varphi}_1(b,y,s)\right] \sin\left[\frac{k(y-c)\pi}{d-c}\right] dy \tag{17}$$

$$\int\limits_a^b \int\limits_c^d \frac{\partial^2 \overline{\varphi}(x,y,s)}{\partial y^2} \sin\left[\frac{n(x-a)\pi}{b-a}\right] \sin\left[\frac{k(y-c)\pi}{d-c}\right] dydx = \overline{\widetilde{H}}_2(n,k,s) - \left(\frac{k\pi}{d-c}\right)^2 \widetilde{\widetilde{\varphi}}(n,k,s) \tag{18}$$

where

$$\overline{\widetilde{H}}_2(n,k,s) = \frac{k\pi}{d-c} \int\limits_a^b \left[\overline{\varphi}_1(x,c,s) - (-1)^k \overline{\varphi}_1(x,d,s)\right] \sin\left[\frac{n(x-a)\pi}{b-a}\right] dx. \tag{19}$$

Applying the double sine-Fourier transform to Equation (13), using the relationships (16)–(19) and notations

$$\begin{aligned}
a_{nk} &= 1 + (1-\alpha)\left[\beta + \varepsilon_1\left(\tfrac{n\pi}{b-a}\right)^2 + \varepsilon_2\left(\tfrac{k\pi}{d-c}\right)^2\right] \\
b_{nk} &= \alpha\left[\beta + \varepsilon_1\left(\tfrac{n\pi}{b-a}\right)^2 + \varepsilon_2\left(\tfrac{k\pi}{d-c}\right)^2\right], n,k = 1,2,\dots, \\
\overline{\widetilde{\Psi}}(n,k,s) &= \overline{\widetilde{F}}(n,k,s) + \tfrac{1}{s}\widetilde{\widetilde{G}}(n,k) + \varepsilon_1 \overline{\widetilde{H}}_1(n,k,s) + \varepsilon_2 \overline{\widetilde{H}}_2(n,k,s),
\end{aligned} \tag{20}$$

the transform $\widetilde{\widetilde{\varphi}}(n,k,s)$ is written in the following form:

$$\widetilde{\widetilde{\varphi}}(n,k,s) = \frac{(1-\alpha)s + \alpha}{a_{nk}s + b_{nk}} \overline{\widetilde{\Psi}}(n,k,s). \tag{21}$$

Now, we will write Equation (21) in a suitable form which, after inversion, satisfies the initial and boundary conditions (10) and (11). To do this, we consider the following auxiliary functions:

$$\begin{aligned}
v_i(y,t) &= [g_i(y,t) - g_i(d,t)]H(y-c) + [g_i(y,t) - g_i(c,t)]H(d-y) - \\
&\quad - g_i(y,t)H(y-c)H(d-y); \quad i = 1,2, \ y \in [c,d],
\end{aligned} \tag{22}$$

$$\begin{aligned}
w_i(x,t) &= [h_i(x,t) - h_i(b,t)]H(x-a) + [h_i(x,t) - h_i(a,t)]H(b-x) - \\
&\quad - h_i(x,t)H(x-a)H(b-x); \quad i = 1,2, \ x \in [a,b],
\end{aligned} \tag{23}$$

where

$$\begin{aligned}
g_1(y,t) &= F_1(a,y,t) + G_1(a,y), \quad g_2(y,t) = F_1(b,y,t) + G_1(b,y), \\
h_1(x,t) &= F_1(x,c,t) + G_1(x,c), \quad h_2(x,t) = F_1(x,d,t) + G_1(x,d),
\end{aligned} \tag{24}$$

and

$$H(z) = \frac{\text{sign}(z)[1 + \text{sign}(z)]}{2} = \begin{cases} 0, & z \le 0 \\ 1, & z > 0 \end{cases}, \tag{25}$$

is the Heaviside step unit function.

With the above functions, we define:

$$\theta_1(x,y,t) = \frac{b-x}{b-a}v_1(y,t) + \frac{x-a}{b-a}v_2(y,t) = \frac{x[v_2(y,t)-v_1(y,t)]}{b-a} + \frac{bv_1(y,t)-av_2(y,t)}{b-a}, \tag{26}$$

$$\theta_2(x,y,t) = \frac{d-y}{d-c}w_1(x,t) + \frac{y-c}{d-c}w_2(x,t) = \frac{y[w_2(x,t)-w_1(x,t)]}{d-c} + \frac{dw_1(x,t)-cw_2(x,t)}{d-c}, \tag{27}$$

$$\begin{aligned}\theta_3(x,y,t) = \ & h_1(a,t)H(b-x)H(d-y) + h_1(b,t)H(x-a)H(d-y) + \\ & h_2(a,t)H(b-x)H(y-c) + h_2(b,t)H(x-a)H(y-c),\end{aligned} \tag{28}$$

The function $\theta(x,y,t)$ defined by

$$\theta(x,y,t) = \theta_1(x,y,t) + \theta_2(x,y,t) + \theta_3(x,y,t), \tag{29}$$

satisfies the following properties:

$$\theta(a,y,t) = g_1(y,t), \quad \theta(b,y,t) = g_2(y,t), \quad \theta(x,c,t) = h_1(x,t), \quad \theta(x,d,t) = h_2(x,t) \tag{30}$$

Applying the Laplace transform and double sine-Fourier transform to functions (26)–(28), we obtain:

$$\begin{aligned}\overline{\widetilde{\theta}}_1(n,k,s) &= \frac{a-(-1)^n b}{n\pi}\overline{P}_{1k}(s) + \frac{1-(-1)^n}{n\pi}\overline{P}_{2k}(s),\\[4pt] \overline{P}_{1k}(s) &= \int\limits_c^d [\overline{v}_2(y,s)-\overline{v}_1(y,s)]\sin\left[\frac{k(y-c)\pi}{d-c}\right]dy,\\[4pt] \overline{P}_{2k} &= \int\limits_c^d [b\overline{v}_1(y.s)-a\overline{v}_2(y,s)]\sin\left[\frac{k(y-c)\pi}{d-c}\right]dy,\end{aligned} \tag{31}$$

$$\begin{aligned}\overline{\widetilde{\theta}}_2(n,k,s) &= \frac{c-(-1)^k d}{k\pi}\overline{Q}_{1n}(s) + \frac{1-(-1)^k}{k\pi}\overline{Q}_{2n}(s),\\[4pt] \overline{Q}_{1n}(s) &= \int\limits_a^b [\overline{w}_2(x,s)-\overline{w}_1(x,s)]\sin\left[\frac{n(x-a)\pi}{b-a}\right]dx,\\[4pt] \overline{Q}_{2n}(s) &= \int\limits_a^b [d\overline{w}_1(x,s)-c\overline{w}_2(x,s)]\sin\left[\frac{n(x-a)\pi}{b-a}\right]dx,\end{aligned} \tag{32}$$

$$\overline{\widetilde{\theta}}_3(n,k,s) = \left[\overline{h}_1(a,s)+\overline{h}_1(b,s)+\overline{h}_2(a,s)+\overline{h}_2(b,s)\right]\left(\frac{1-(-1)^n}{n\pi}\right)\left(\frac{1-(-1)^k}{k\pi}\right)(b-a)(d-c) \tag{33}$$

respectively,

$$\begin{aligned}\overline{\widetilde{\theta}}(n,k,s) = \ & \frac{a-(-1)^n b}{n\pi}\overline{P}_{1k}(s) + \frac{1-(-1)^n}{n\pi}\overline{P}_{2k}(s) + \frac{c-(-1)^k d}{k\pi}\overline{Q}_{1n}(s) + \frac{1-(-1)^k}{k\pi}\overline{Q}_{2n}(s) + \\ & \left[\overline{h}_1(a,s)+\overline{h}_1(b,s)+\overline{h}_2(a,s)+\overline{h}_2(b,s)\right]\left(\frac{1-(-1)^n}{n\pi}\right)\left(\frac{1-(-1)^k}{k\pi}\right)(b-a)(d-c).\end{aligned} \tag{34}$$

The function $\overline{\widetilde{\varphi}}$, given by Equation (21), can be written as:

$$\overline{\widetilde{\varphi}}(n,k,s) = \overline{\widetilde{\theta}}(n,k,s) + \frac{1}{a_{nk}s+b_{nk}}\overline{\widetilde{\Psi}}_1(n,k,s), \tag{35}$$

where

$$\overline{\widetilde{\Psi}}_1(n,k,s) = [(1-\alpha)s+\alpha]\overline{\widetilde{\Psi}}(n,k,s) - (a_{nk}s+b_{nk})\overline{\widetilde{\theta}}(n,k,s). \tag{36}$$

Applying the inverse Laplace and Fourier transforms, we obtain the solution

$$\begin{aligned}\varphi(x,y,t) = \ & H(t)\theta(x,y,t) + \\ & \frac{4}{(b-a)(d-c)}\sum_{n=1}^{\infty}\sum_{k=1}^{\infty}\sin\left[\frac{n(x-a)\pi}{b-a}\right]\sin\left[\frac{k(y-c)\pi}{d-c}\right]\int\limits_0^t \frac{\exp(-(t-\tau)b_{nk}/a_{nk})}{a_{nk}}\widetilde{\Psi}_1(n,k,\tau)d\tau.\end{aligned} \tag{37}$$

Using the property (30), it is easy to see that function (37) satisfies boundary condition (11). The solution for the classical advection equation is obtained by making $\alpha = 1$ into Equations (20) and (35)–(37).

4. Particular Case: A Time-Dependent Concentrated Source in the Center of Domain D

In this section, for the source $f(x, y, t)$ and for the initial and boundary conditions, we consider:

$$
\begin{aligned}
f(x, y, t) &= \gamma_0 \delta\left(x - \tfrac{a+b}{2}\right)\delta\left(y - \tfrac{c+d}{2}\right)\exp(-pt), \; p \geq 0, \\
u(x, y, 0) &= \psi_0(x, y) = \gamma_0 \exp(f_0(x, y)), \; (x, y) \in D, \; \gamma_0 = const., \\
u(x, y, t) &= \psi_1(x, y, t) = 2\gamma_0 \exp(f_0(x, y)), \; (x, y) \in \partial D, \; t > 0.
\end{aligned}
\tag{38}
$$

The physical significance of the above source $f(x, y, t)$ is that of a source of intensity $\gamma_0 \exp(-pt)$ concentrated in the center $\left(\tfrac{a+b}{2}, \tfrac{c+d}{2}\right)$ of the rectangular domain D.

Using (38), Equations (9), (11), (12) and (14) lead to:

$$
\begin{aligned}
F(x, y, t) &= \gamma_0 \delta\left(x - \tfrac{a+b}{2}\right)\delta\left(y - \tfrac{c+d}{2}\right)\exp(-pt - f_0(x, y)), \; G(x, y) = -\tfrac{\beta\gamma_0}{2}, \\
F_1(x, y, t) &= 2\gamma_0, \; G_1(x, y) = -\gamma_0, \; \varphi_1(x, y, t) = \gamma_0.
\end{aligned}
\tag{39}
$$

By using $\overline{\varphi}_1(x, y, s) = \tfrac{\gamma_0}{s}$ in Equations (17) and (19), we have:

$$
\begin{aligned}
\widetilde{\overline{H}}_1(n, k, s) &= \tfrac{\gamma_0}{s}\tfrac{d-c}{b-a}\tfrac{n[1-(-1)^n][1-(-1)^k]}{k}, \\
\widetilde{\overline{H}}_2(n, k, s) &= \tfrac{\gamma_0}{s}\tfrac{b-a}{d-c}\tfrac{k[1-(-1)^n][1-(-1)^k]}{n}.
\end{aligned}
\tag{40}
$$

Using the "sifting" property of the Dirac distribution [27]

$$
\int_a^b f(x)\delta(x - c)dx =
\begin{cases}
\tfrac{1}{2}f(c^+), & c = a, \\
\tfrac{1}{2}[f(c^-) + f(c^+)], & c \in (a, b), \\
\tfrac{1}{2}f(c^-), & c = b, \\
0, & c \in (a, b).
\end{cases}
\tag{41}
$$

we find that the double sine-Fourier transform of the functions

$$
\begin{aligned}
\overline{F}(x, y, s) &= \tfrac{\gamma_0}{s+p}\delta\left(x - \tfrac{a+b}{2}\right)\delta\left(y - \tfrac{c+d}{2}\right)\exp(-f_0(x, y)), \\
G(x, y) &= \tfrac{-\beta\gamma_0}{2},
\end{aligned}
\tag{42}
$$

are given by

$$
\begin{aligned}
\widetilde{\overline{F}}(n, k, s) &= \tfrac{\gamma_0}{s+p}\sin\left(\tfrac{n\pi}{2}\right)\sin\left(\tfrac{k\pi}{2}\right)\exp\left(-\tfrac{k_1(a+b)}{2\varepsilon_1^2} - \tfrac{k_2(c+d)}{2\varepsilon_2^2}\right), \\
\widetilde{G}(n, k) &= \tfrac{-\beta\gamma_0}{2}\tfrac{(b-a)(d-c)}{kn\pi^2}[1 - (-1)^n][1 - (-1)^k].
\end{aligned}
\tag{43}
$$

Since $\sin\left(\tfrac{m\pi}{2}\right) = \tfrac{(-1)^m - 1}{2}i^{m+1}$, $i^2 = -1$, $m \in \mathbb{N}$, the function $\widetilde{\overline{F}}(n, k, s)$ is written as

$$
\widetilde{\overline{F}}(n, k, s) = \tfrac{-\gamma_0}{s+p}\exp\left(-\tfrac{k_1(a+b)}{2\varepsilon_1^2} - \tfrac{k_2(c+d)}{2\varepsilon_2^2}\right)\tfrac{[1-(-1)^n][1-(-1)^k]}{4}i^{n+k}.
\tag{44}
$$

By using Equations (40) and (43) in the third equation (20), we obtain:

$$
\begin{aligned}
\widetilde{\overline{\Psi}}(n, k, s) = {} & \tfrac{[1-(-1)^n][1-(-1)^k]}{4} \times \\
& \left[\tfrac{2\gamma_0}{s}\tfrac{2\varepsilon_1(n\pi)^2(d-c)^2 + 2\varepsilon_2(k\pi)^2(b-a)^2 - \beta(b-a)^2(d-c)^2}{(k\pi)(n\pi)(b-a)(d-c)} - \tfrac{\gamma_0\gamma_1 i^{n+k}}{s+p}\right],
\end{aligned}
\tag{45}
$$

where

$$\gamma_1 = \exp\left(-\frac{k_1(a+b)}{4\varepsilon_1} - \frac{k_2(c+d)}{4\varepsilon_2}\right)$$

In the considered particular case, Equations (22)–(24) and (31)–(34) become:

$$\begin{aligned}
g_1(y,t) &= g_2(y,t) = \gamma_0, v_1(y,t) = v_2(y,t) = -\gamma_0 H(y-c)H(d-y),\\
h_1(x,t) &= h_2(x,t) = \gamma_0, w_1(x,t) = w_2(x,t) = -\gamma_0 H(x-a)H(b-x),
\end{aligned}$$

(46)

$$\begin{aligned}
\theta_1(x,y,t) &= -\gamma_0 H(y-c)H(d-y),\ \theta_2(x,y,t) = -\gamma_0 H(x-a)H(b-x),\\
\theta_3(x,y,t) &= \gamma_0 H(b-x)H(d-y) + \gamma_0 H(x-a)H(d-y)+\\
&\quad \gamma_0 H(b-x)H(y-c) + \gamma_0 H(x-a)H(y-c).
\end{aligned}$$

(47)

The transformed Laplace–Fourier of the function (47) are given by:

$$\begin{aligned}
\widetilde{\overline{\theta}}_1(n,k,s) &= \widetilde{\overline{\theta}}_2(n,k,s) = -\tfrac{1}{s}\gamma_0(b-a)(d-c)\frac{1-(-1)^n}{n\pi}\frac{1-(-1)^k}{k\pi},\\
\widetilde{\overline{\theta}}_3(n,k,s) &= \tfrac{4}{s}\gamma_0(b-a)(d-c)\frac{1-(-1)^n}{n\pi}\frac{1-(-1)^k}{k\pi},\\
\widetilde{\overline{\theta}}(n,k,s) &= \tfrac{2}{s}\gamma_0(b-a)(d-c)\frac{1-(-1)^n}{n\pi}\frac{1-(-1)^k}{k\pi}.
\end{aligned}$$

(48)

Function $\widetilde{\overline{\Psi}}_1(n,k,s)$ given by Equation (36) can be written as:

$$\widetilde{\overline{\Psi}}_1(n,k,s) = C_{nk}\frac{1}{s+p} + D_{nk}\frac{1}{s} + E_{nk},$$

(49)

where

$$\begin{aligned}
A_{nk} &= [1-(-1)^n][1-(-1)^k]\\
B_{nk} &= \frac{1}{(b-a)(d-c)}\left[2\varepsilon_1(n\pi)^2(d-c)^2 + 2\varepsilon_2(k\pi)^2(b-a)^2 - \beta(b-a)^2(d-c)^2\right],\\
C_{nk} &= \tfrac{1}{4}\gamma_0\gamma_1[(1-\alpha)p - \alpha]i^{n+k}A_{nk},\\
D_{nk} &= \frac{\gamma_0 A_{nk}}{2(n\pi)(k\pi)}[\alpha B_{nk} - 4(b-a)(d-c)b_{nk}],\\
E_{nk} &= \frac{A_{nk}}{4(n\pi)(k\pi)}\left[2(1-\alpha)\gamma_0 B_{nk} - 8\gamma_{00}(b-a)(d-c)a_{nk} - (1-\alpha)\gamma_0\gamma_1(n\pi)(k\pi)i^{n+k}\right],
\end{aligned}$$

(50)

and transform (35) becomes:

$$\widetilde{\overline{\varphi}}(n,k,s) = \widetilde{\overline{\theta}}(n,k,s) + \frac{1}{a_{nk}}\frac{1}{s+b_{nk}/a_{nk}}\left[\frac{C_{nk}}{s+p} + \frac{D_{nk}}{s} + E_{nk}\right]$$

(51)

Applying the inverse Laplace transform and the inverse Fourier transform to function (51), we obtain:

$$\begin{aligned}
\varphi(x,y,t) &= H(t)\theta(x,y,t) + \frac{4}{(b-a)(d-c)}\sum_{n=1}^{\infty}\sum_{k=1}^{\infty}\frac{1}{a_{nk}}\Bigg\{E_{nk}\exp\left(\frac{-b_{nk}t}{a_{nk}}\right)+\\
&\quad \int_0^t \exp\left(\frac{-b_{nk}(t-\tau)}{a_{nk}}\right)(D_{nk} + C_{nk}e^{-p\tau})d\tau\Bigg\}\sin\left(\frac{n(x-a)\pi}{b-a}\right)\sin\left(\frac{k(y-c)\pi}{d-c}\right),
\end{aligned}$$

(52)

where the function $\theta(x,y,t)$ is given by

$$\begin{aligned}
\theta(x,y,t) &= \gamma_0[H(x-a) + H(b-x)][H(y-c) + H(d-y)]-\\
&\quad \gamma_0 H(x-a)H(b-x) - \gamma_0 H(y-c)H(d-y).
\end{aligned}$$

(53)

Using the property

$$\vartheta(t) = H(t-\alpha_1) + H(\alpha_2-t) - H(t-\alpha_1)H(\alpha_2-t) \equiv 1,\ t \in [\alpha_1,\alpha_2],$$

(54)

it is found that $\theta(a, y, t) = \theta(b, y, t) = \gamma_0$ for $y \in [c, d]$, respectively $\theta(x, c, t) = \theta(b, d, t) = \gamma_0$ for $x \in [a, b]$; therefore, function (53) satisfied boundary condition (14). Now, to return to the concentration $u(x, y, t)$, the function $\varphi(x, y, t)$ given by Equation (52) will be replaced into transformation (7).

Figures 1–4 have been plotted to highlight the variation of the non-dimensional concentration $\varphi(x, y, t)$ versus the fractional parameter α.

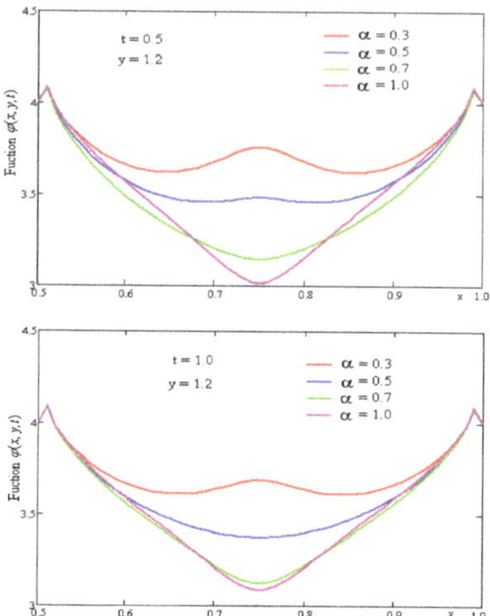

Figure 1. Profiles of function $\varphi(x, 1.2, t)$ for different values of the fractional parameter α, y = 1.2, t = 0.5 and t = 1.0.

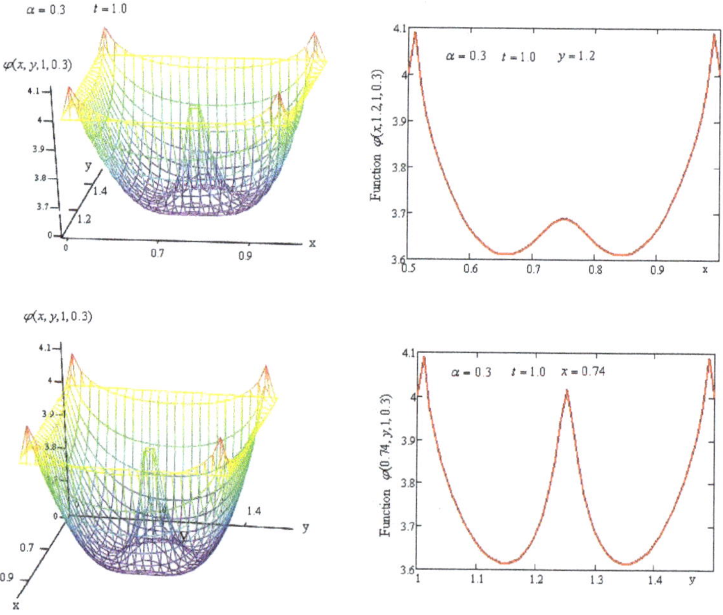

Figure 2. Profiles of surface $\varphi(x, y, 1)$ for $\alpha = 0.3$ and their sections with the planes $y = 1.2$ and $x = 0.74$, respectively.

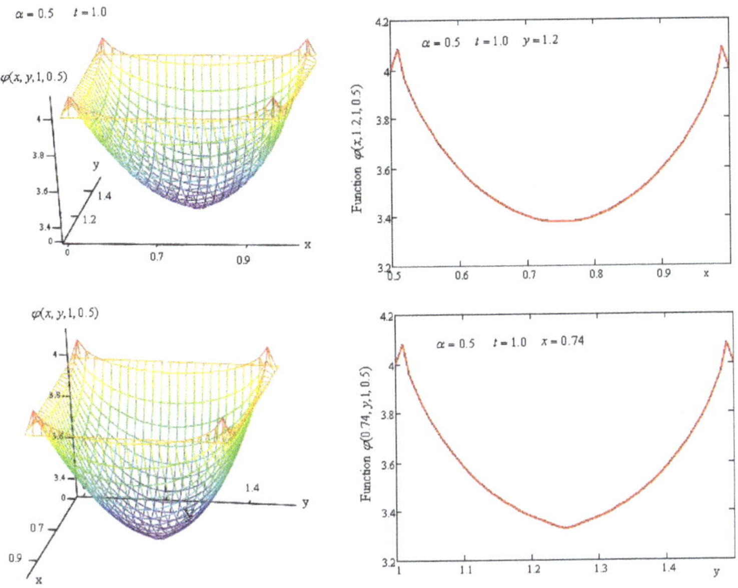

Figure 3. Profiles of surface $\varphi(x, y, 1)$ for $\alpha = 0.5$ and their sections with the planes $y = 1.2$ and $x = 0.74$, respectively.

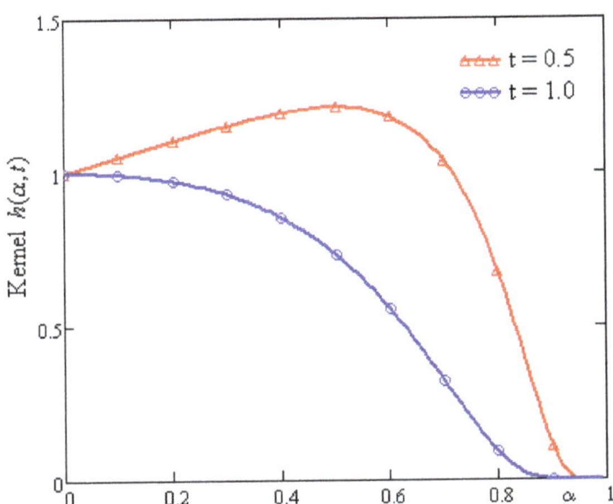

Figure 4. Profile of kernel $h(\alpha, t)$ for two values of the time t.

To draw these graphs we have used the following values of the parameters: $a = 0.5$, $b = 1.0$, $c = 1.0$, $d = 1.5$, $p = 0.25$, $\gamma_0 = 4$, $\varepsilon_1 = \varepsilon_2 = 0.2$, $k_1 = 0.25$, $k_2 = 0.5$.

In Figure 1, we show the plotted curves $\varphi(x, 1.2, 0.5)$ and $\varphi(x, 1.2, 1.0)$ for different values of the fractional parameter $\alpha \in \{0.3, 0.5, 0.7, 1.0\}$. It is observed from Figure 1 that the values of concentration φ decrease with the fractional parameter and have a peak (maximum/minimum) in the central area of the geometrical domain $D = [a, b] \times [c, d]$.

In Figures 2 and 3, we show the plotted surfaces $\varphi(x, y, 1)$ for the fractional parameters $\alpha = 0.3$ and $\alpha = 0.5$, respectively, along with the sections through planes $y = 1.2$, $x = 0.74$.

For small values of the fractional parameter, the concentration φ has maximum values in the central area of the rectangular $D = [a, b] \times [c, d]$. If the fractional parameter increases, the concentration has a minimum in the same area. This behavior is due to the kernel of the time-fractional Caputo–Fabrizio derivative (see Figure 4) which, for $t = 0.5$, increases for $\alpha \leq 0.55$ and decreases for $\alpha > 0.55$. Therefore, at small values of the fractional parameter, the weight function in the fractional derivative has bigger values and the influence on the concentration is stronger.

5. Numerical Solution

In this section, based on the finite difference approximation of the derivatives, we develop a numerical scheme to the problem formulated in Section 2 of the paper.

We consider Equation (8) along with the initial and boundary conditions (10) and (11), with $(x, y, t) \in [a, b] \times [c, d] \times [0, T]$, $T > 0$. The discrete sets of nodes are defined as:

$$
\begin{aligned}
x_i &= a + ih_x, \ i = 0, 1, \ldots, N_1, \ h_x = (b-a)/N_1, \\
y_j &= c + jh_y, \ j = 0, 1, \ldots, N_2, \ h_y = (d-c)/N_2, \\
t_n &= nh_t, \ n = 0, 1, \ldots, N, \ h_t = T/N.
\end{aligned} \tag{55}
$$

The numerical estimation of the solution $\varphi(x, y, t)$ of Equation (8) in the point (x_i, y_j, t_n) is denoted by $\varphi_{i,j}^n = \varphi(x_i, y_j, t_n)$. The initial and boundary conditions (10) and (11) are written as:

$$
\begin{aligned}
\varphi_{i,j}^0 &= \varphi(x_i, y_j, t_0) = 0, \ i = 0, 1, \ldots, N_1, \ j = 0, 1, \ldots, N_2, \\
\varphi_{0,j}^n &= \varphi(a, y_j, t_n) = F_1(a, y_j, t_n) + G_1(a, y_j) = P_{0,j}^n, \ j = 0, 1, \ldots, N_2, \ n = 1, 2, \ldots, N, \\
\varphi_{N_1,j}^n &= \varphi(b, y_j, t_n) = F_1(b, y_j, t_n) + G_1(b, y_j) = Q_{N_1,j}^n, \ j = 0, 1, \ldots, N_2, \ n = 1, 2, \ldots, N, \\
\varphi_{i,0}^n &= \varphi(x_i, c, t_n) = F_1(x_i, c, t_n) + G_1(x_i, c) = R_{i,0}^n, \ i = 0, 1, \ldots, N_1, \ n = 1, 2, \ldots, N, \\
\varphi_{i,N_2}^n &= \varphi(x_i, d, t_n) = F_1(x_i, d, t_n) + G_1(x_i, d) = S_{i,N_2}^n, \ i = 0, 1, \ldots, N_1, \ n = 1, 2, \ldots, N,
\end{aligned} \tag{56}
$$

Using the following approximation for the first time-derivative:

$$
\left. \frac{\partial \varphi(x_i, y_j, t)}{\partial t} \right|_{t=t_k} = \frac{1}{h_t} \left[\varphi(x_i, y_j, t_{k+1}) - \varphi(x_i, y_j, t_k) \right] = \frac{1}{h_t} \left(\varphi_{i,j}^{k+1} - \varphi_{i,j}^k \right), \ k = 0, 1, \ldots, N-1, \tag{57}
$$

the time-fractional Caputo–Fabrizio derivative, in the point $t = t_n$, can be approximated by:

$$
\begin{aligned}
{}^{CF}D_t^\alpha \varphi(x, y, t) \Big|_{(x_i, y_j, t_n)} &= \frac{1}{1-\alpha} \int_0^{t_n} \left. \frac{\partial \varphi(x_i, y_j, t)}{\partial t} \right|_{t=\tau} \exp\left(\frac{-\alpha(t_n-\tau)}{1-\alpha} \right) d\tau = \\
\sum_{k=0}^{n-1} \frac{1}{1-\alpha} \int_{t_k}^{t_{k+1}} & \left. \frac{\partial \varphi(x_i, y_j, t)}{\partial t} \right|_{t=\tau} \exp\left(\frac{-\alpha(t_n-\tau)}{1-\alpha} \right) d\tau = \\
\sum_{k=0}^{n-1} \frac{1}{(1-\alpha)h_t} & \left(\varphi_{i,j}^{k+1} - \varphi_{i,j}^k \right) \int_{t_k}^{t_{k+1}} \exp\left(\frac{-\alpha(t_n-\tau)}{1-\alpha} \right) d\tau = \sum_{k=0}^{n-1} a_{n,k} \left(\varphi_{i,j}^{k+1} - \varphi_{i,j}^k \right), \ n = 1, 2, \ldots N,
\end{aligned} \tag{58}
$$

where

$$
a_{n,k} = \frac{1}{\alpha h_t} \left[\exp\left(\frac{-\alpha(t_n-t_{k+1})}{1-\alpha} \right) - \exp\left(\frac{-\alpha(t_n-t_k)}{1-\alpha} \right) \right], \ \alpha \in (0, 1), \ n = 1, 2, \ldots, N, \ k = 0, 1, \ldots, (n-1) \tag{59}
$$

It is known [28,29] that Equation (58) approximates the time-fractional Caputo–Fabrizio derivative with an error of order $O(h_t^2)$.

The space derivatives of the second order are approximated by:

$$
\left.\frac{\partial^2 \varphi(x,y,t)}{\partial x^2}\right|_{(x_i,y_j,t_n)} = \frac{1}{h_x^2}\left(\varphi_{i+1,j}^n - 2\varphi_{i,j}^n + \varphi_{i-1,j}^n\right), i = 1,2,\ldots,N_1-1, j = 0,1,\ldots,N_2, n = 1,2,\ldots N,
$$
$$
\left.\frac{\partial^2 \varphi(x,y,t)}{\partial y^2}\right|_{(x_i,y_j,t_n)} = \frac{1}{h_y^2}\left(\varphi_{i,j+1}^n - 2\varphi_{i,j}^n + \varphi_{i,j-1}^n\right), i = 0,1,\ldots,N_1, j = 1,2,\ldots,N_2-1, n = 1,2,\ldots N.
$$
(60)

Using the approximation formulae (58) and (60), Equation (8) is written as:

$$
\sum_{k=0}^{n-1} a_{n,k}\left(\varphi_{i,j}^{k+1} - \varphi_{i,j}^k\right) + \beta\varphi_{i,j}^n + \beta_1\left(\varphi_{i+1,j}^n - 2\varphi_{i,j}^n + \varphi_{i-1,j}^n\right) +
$$
$$
\beta_2\left(\varphi_{i,j+1}^n - 2\varphi_{i,j}^n + \varphi_{i,j-1}^n\right) = H_{i,j}^n, \, i = 1,2,\ldots,N_1-1, \cdots j = 1,2,\ldots,N_2-1, \, n = 1,2,\ldots N,
$$
(61)

where

$$
\beta_1 = \frac{-\varepsilon_1}{h_x^2}, \, \beta_2 = \frac{-\varepsilon_2}{h_y^2}, \, H_{i,j}^n = F\left(x_i, y_j, t_n\right) + G\left(x_i, y_j\right)
$$
(62)

For $n = 1$, Equation (61) becomes:

$$
d_{1,0}\varphi_{i,j}^1 + \beta_1\left(\varphi_{i+1,j}^1 + \varphi_{i-1,j}^1\right) + \beta_2\left(\varphi_{i,j+1}^1 + \varphi_{i,j-1}^1\right) = H_{i,j}^1,
$$
(63)

where $d_{1,0} = a_{1,0} + \beta - 2(\beta_1 + \beta_2)$.

Making $j = 1$ and $i = 1,2,\ldots,N_1-1$ into Equation (63), we obtain the following algebraic system:

$$
MX_1 + M_0X_2 = D_1^1,
$$
(64)

where M is a square matrix of order (N_1-1) whose elements are given by:

$$
M_{0,k} = d_{1,0}\delta_{0,k} + \beta_1\delta_{1,k}, k = 0,1,\ldots,N_1-2,
$$
$$
M_{i,k} = \beta_1\delta_{i,k+1} + d_{1,0}\delta_{i+1,k+1} + \beta_1\delta_{i+2,k+1}, i = 1,2,\ldots,N_1-2, k = k = 0,1,\ldots,N_1-2,
$$
(65)

so,

$$
M = \begin{pmatrix}
d_{1,0} & \beta_1 & 0 & 0 & 0 & \cdots & 0 & 0 & 0 \\
\beta_1 & d_{1,0} & \beta_1 & 0 & 0 & \cdots & 0 & 0 & 0 \\
0 & \beta_1 & d_{1,0} & \beta_1 & 0 & \cdots & 0 & 0 & 0 \\
\cdots & \cdots & \cdots & \cdots & \cdots & \cdots & \cdots & \cdots & \cdots \\
0 & 0 & 0 & 0 & 0 & \cdots & \beta_1 & d_{1,0} & \beta_1 \\
0 & 0 & 0 & 0 & 0 & 0 & \cdots & \beta_1 & d_{1,0}
\end{pmatrix}
$$
(66)

the matrix M_2 is the diagonal square matrix of order (N_1-1) with elements $M_{1i,j} = \beta_2\delta_{i,j}, i, j = 0,1,\ldots,N_1-2, \delta_{i,j} = \begin{cases} 0, i \neq j, \\ 1, i = j, \end{cases}$ is the Kronecker delta and $X_1, X_2 D_1^1$ are the column-matrices.

$$
X_1 = \begin{pmatrix} \varphi_{1,1}^1 \\ \varphi_{2,1}^1 \\ \cdots \\ \varphi_{N_1-1,1}^1 \end{pmatrix}, \, X_2 = \begin{pmatrix} \varphi_{1,2}^1 \\ \varphi_{2,2}^1 \\ \cdots \\ \varphi_{N_1-1,2}^1 \end{pmatrix}, \, D_1^1 = \begin{pmatrix} D_{11}^1 \\ D_{12}^1 \\ \cdots \\ D_{1N_1-1}^1 \end{pmatrix}, \, \begin{aligned} & D_{11}^1 = H_{1,1}^1 - \beta_1P_{0,1}^1 - \beta_2R_{1,0}^1, \\ & D_{1i}^1 = H_{i,1}^1 - \beta_2R_{i,0}^1, \, i = 2,3,\ldots,N_1-2, \\ & D_{1N_1-1}^1 = H_{N_1-1,1}^1 - \beta_1Q_{N_1,1}^1 - \beta_2R_{N_1-1,0}^1. \end{aligned}
$$
(67)

The algebraic system (64) has (N_1-1) equations with $2(N_1-1)$ unknown functions.

Making $j = 2,3,\ldots,N_2-2$ and $i = 1,2,\ldots,N_1-1$ into Equation (63), we obtain the algebraic system:

$$
M_0X_{j-1} + MX_j + M_0X_{j+1} = D_j^1,
$$
(68)

where

$$D^1_{j1} = H^1_{1,j} - \beta_1 P^1_{0,j},$$
$$D^1_{ji} = H^1_{i,j}, \; i = 2,3,\ldots,N_1 - 2,$$
$$D^1_{jN_1-1} = H^1_{N_1-1,j} - \beta_1 Q^1_{N_1,j}.$$

(69)

Making $j = N_2 - 1$ and $i = 1,2,\ldots,N_1 - 1$ into Equation (63), we obtain the algebraic system:

$$M_0 X_{N_2-2} + M X_{N_2-1} = D^1_{N_2-1},$$

(70)

where

$$D^1_{N_2-11} = H^1_{1,N_2-1} - \beta_1 P^1_{0,N_2-1} - \beta_2 S^1_{1,N_2},$$
$$D^1_{N_2-1i} = H^1_{i,N_2-1} - \beta_2 S^1_{i,N_2}, \; i = 2,3,\ldots,N_1 - 2,$$
$$D^1_{N_2-1N_1-1} = H^1_{N_1-1.N_2-1} - \beta_1 Q^1_{N_1,N_2-1} - \beta_2 S^1_{N_1-1,N_2}.$$

(71)

The algebraic systems (64), (68) and (70) can be written as a unique algebraic system of $(N_1 - 1)(N_2 - 1)$ equations with $(N_1 - 1)(N_2 - 1)$ unknown functions $\varphi^1_{i,j}$, $i = 1,2,\ldots,N_1 - 1$, $j = 1,2,\ldots,N_2 - 1$, namely

$$\begin{pmatrix} M & M_0 & 0 & 0 & 0 & \ldots & 0 & 0 & 0 \\ M_0 & M & M_0 & 0 & 0 & \ldots & 0 & 0 & 0 \\ 0 & M_0 & M & M_0 & 0 & \ldots & 0 & 0 & 0 \\ \ldots & \ldots & \ldots & \ldots & \ldots & \ldots & \ldots & \ldots & \ldots \\ 0 & 0 & 0 & 0 & 0 & \ldots & M_0 & M & M_0 \\ 0 & 0 & 0 & 0 & 0 & 0 & \ldots & M_0 & M \end{pmatrix} \begin{pmatrix} X_1 \\ X_2 \\ X_3 \\ \ldots \\ X_{N_2-2} \\ X_{N_2-1} \end{pmatrix} = \begin{pmatrix} D^1_1 \\ D^1_2 \\ D^1_3 \\ \ldots \\ D^1_{N_2-2} \\ D^1_{N_2-1} \end{pmatrix}$$

(72)

The solution of system (72) gives the values of function $\varphi(x,y,t)$ in points (x_i, y_j, t_1) for $i = 1,2,\ldots,N_1 - 1$, $j = 1,2,\ldots,N_2 - 1$.

The numerical procedure is continued in a similar way for $n = 2,3,\ldots,N$ when Equation (61) is written in the equivalent form:

$$d_{n,n-1}\varphi^n_{i,j} + \beta_1\left(\varphi^n_{i+1,j} + \varphi^n_{i-1,j}\right) + \beta_2\left(\varphi^n_{i,j+1} + \varphi^n_{i,j-1}\right) = G^n_{i,j}, \; i = 1,2,\ldots,N_1 - 1, \; j = 1,2,\ldots,N_2 - 1, \quad (73)$$

where

$$d_{n,n-1} = a_{n,n-1} + \beta - 2(\beta_1 + \beta_2), \; n = 2,3,\ldots,N,$$
$$G^n_{i,j} = H^n_{i,j} + a_{n,n-1}\varphi^{n-1}_{i,j} - \sum_{k=0}^{n-2} a_{n,k}\left(\varphi^{k+1}_{i,j} - \varphi^k_{i,j}\right), \; i = 1,2,\ldots,N_1 - 1, \; j = 1,2,\ldots,N_2 - 1.$$

(74)

Finally, the proposed numerical procedure gives the values of the solution $\varphi(x,y,t)$ in grid points (x_i, y_j, t_n), $i = 1,2,\ldots,N_1 - 1$, $j = 1,2,\ldots,N_2 - 1$, $n = 1,2,\ldots,N$.

To compare the numerical values obtained with the analytical solution, respectively, by the numerical method, we have considered the following particular problem:

$$F(x,y,t) = 5000\exp\left(-\frac{k_1 x}{2\varepsilon_1} - \frac{k_2 y}{2\varepsilon_2}\right), \; G(x,y) = 0, \; (x,y,t) \in [0,0.05] \times [0,0.05] \times [0,0.1],$$
$$F_1(x,y,t) = 0, \; G_1(x,y) = 0, \varphi(x,y,0) = 0,$$
$$k_1 = k_2 = 0.5, \; \varepsilon_1 = \varepsilon_2 = 2,$$

(75)

In this case, the double Fourier transform of the function $F(x,y,t)$ is:

$$\widetilde{F}(n,k) = \frac{n\pi(b-a)\left[\exp\left(\frac{-ak_1}{2\varepsilon_1}\right) - (-1)^n \exp\left(\frac{-bk_1}{2\varepsilon_1}\right)\right]}{\left(\frac{k_1}{2\varepsilon_1}\right)^2 (b-a)^2 + (n\pi)^2} \times \frac{k\pi(b-a)\left[\exp\left(\frac{-ck_2}{2\varepsilon_2}\right) - (-1)^k \exp\left(\frac{-dk_2}{2\varepsilon_2}\right)\right]}{\left(\frac{k_2}{2\varepsilon_2}\right)^2 (d-c)^2 + (k\pi)^2},$$

(76)

and the analytical solution is given by:

$$\varphi(x,y,t) = \frac{4\alpha}{(b-a)(d-c)} \sum_{n=1}^{\infty} \sum_{k=1}^{\infty} \frac{\widetilde{F}(n,k)}{b_{nk}} \sin\left[\frac{n(x-a)\pi}{b-a}\right] \sin\left[\frac{k(y-c)\pi}{d-c}\right] +$$
$$\frac{4}{(b-a)(d-c)} \sum_{n=1}^{\infty} \sum_{k=1}^{\infty} \frac{\widetilde{F}(n,k)[(1-\alpha)b_{nk}-\alpha a_{nk}]}{a_{nk}b_{nk}} \sin\left[\frac{n(x-a)\pi}{b-a}\right] \sin\left[\frac{k(y-c)\pi}{d-c}\right] e^{-\frac{b_{nk}t}{a_{nk}}} . \tag{77}$$

The values of function $\varphi(x,y,t)$ at the instant $t = 0.01$, obtained with the numerical method and with expression (77) given in the Table 1, are in good agreement.

Table 1. Values of function $\varphi(x,y,0.01)$.

x	$y_1 = 0.005$		$y_2 = 0.010$		$y_3 = 0.015$	
	Numerical Solution	Equation (77)	Numerical Solution	Equation (77)	Numerical Solution	Equation (77)
0.005	0.080	0.081	0.129	0.130	0.158	0.159
0.010	0.129	0.130	0.214	0.216	0.267	0.270
0.015	0.158	0.159	0.267	0.270	0.338	0.340
0.020	0.174	0.175	0.297	0.299	0.379	0.381
0.025	0.179	0.181	0.307	0.309	0.391	0.394
0.030	0.174	0.175	0.297	0.299	0.378	0.381
0.035	0.158	0.159	0.268	0.269	0.338	0.340
0.040	0.125	0.120	0.213	0.215	0.267	0.270
0.045	0.080	0.081	0.129	0.130	0.158	0.159

x	$y_4 = 0.020$		$y_5 = 0.025$		$y_6 = 0.030$	
	Numerical Solution	Equation (77)	Numerical Solution	Equation (77)	Numerical Solution	Equation (77)
0.005	0.174	0.175	0.179	0.181	0.174	0.175
0.010	0.297	0.299	0.306	0.309	0.297	0.299
0.015	0.378	0.380	0.391	0.394	0.379	0.381
0.020	0.424	0.427	0.439	0.441	0.424	0.427
0.025	0.439	0.441	0.455	0.457	0.438	0.442
0.030	0.423	0.426	0.438	0.441	0.423	0.426
0.035	0.377	0.380	0.391	0.393	0.377	0.380
0.040	0.296	0.298	0.306	0.308	0.296	0.299
0.045	0.174	0.175	0.179	0.181	0.174	0.175

x	$y_7 = 0.035$		$y_8 = 0.040$		$y_9 = 0.045$	
	Numerical Solution	Equation (77)	Numerical Solution	Equation (77)	Numerical Solution	Equation (77)
0.005	0.158	0.159	0.128	0.130	0.080	0.081
0.010	0.267	0.269	0.213	0.215	0.128	0.130
0.015	0.338	0.340	0.267	0.269	0.158	0.159
0.020	0.377	0.380	0.296	0.298	0.174	0.175
0.025	0.390	0.393	0.306	0.308	0.179	0.180
0.030	0.377	0.380	0.296	0.299	0.174	0.175
0.035	0.337	0.340	0.266	0.269	0.157	0.159
0.040	0.266	0.269	0.212	0.214	0.128	0.129
0.045	0.157	0.159	0.128	0.129	0.079	0.081

6. Discussion and Conclusions

A two-dimensional advection–diffusion process with short-tail memory and concentrated source situated in the symmetry center of the geometrical domain has been investigated.

The studied mathematical model is described by a fractional differential equation with a time-fractional Caputo–Fabrizio derivative with exponential kernel.

Analytical solutions of the proposed initial-boundary value problem were obtained using the integral transform method (Laplace and finite Fourier transforms are employed).

Numerical solutions of the studied problem have been determined using finite difference approximations.

Numerical simulations for both analytical and numerical solutions have been carried out using the software Mathcad.

The fractional order of the Caputo–Fabrizio derivative has a significant influence on the pollutant concentration. For small values of the fractional parameter, the concentration has maximum values in the central area of the geometrical domain. If the fractional parameter increases, the concentration has a minimum in the same area. This behavior is due to the memory kernel of the time-fractional Caputo–Fabrizio derivative.

Author Contributions: The authors have equally contributed to this paper.

Funding: This research received no external funding.

Conflicts of Interest: On behalf of all authors, the corresponding author states that there is no conflict of interest.

References

1. Mainardi, F. *Fractals and Fractional Calculus in Continuum Mechanics*; Springer: Wien, Austria, 1997.
2. Laskin, N. *Fractional Quantum Mechanics*; World Scientific: Hackensack, NJ, USA, 2018.
3. Laskin, N. Generalized classical mechanics. *Eur. Phys. J. Spec. Top.* **2013**, *222*, 1929–1938. [CrossRef]
4. Hristov, J. Fractional derivative with non-singular kernels: From the Caputo-Fabrizio definition and beyond: Appraising analysis with emphasis on diffusion models. In *Frontiers in Fractional Calculus*; Bhalekar, S., Ed.; Bentham Publication: Kolhapur, India, 2017; Chapter 10; pp. 243–269.
5. Baleanu, D.; Golmankhaneh, A.K.; Nigmatullin, R.; Golmankhaneh, A.K. Fractional Newtonian mechanics. *Cent. Eur. J. Phys.* **2010**, *8*, 120–125. [CrossRef]
6. Povstenko, Y. *Fractional Thermoelasticity, Solid Mechanics and Its Applications*; Springer International Publishing: Cham, Switzerland, 2015; p. 219.
7. Bhrawy, A.H.; Baleanu, D. A spectral Legendre-Gauss-Lobatto collocation method for a space-fractional advection diffusion equation with variable coefficients. *Rep. Math. Phys.* **2013**, *72*, 219–233. [CrossRef]
8. Abdelkawy, M.A.; Zaky, M.A.; Bhrawy, A.H.; Baleanu, D. Numerical simulation of time variable fractional order mobile-immobile advection-dispersion model. *Rom. Rep. Phys.* **2015**, *67*, 773–791.
9. Zhuang, P.; Liu, F.; Anh, V.; Turner, I. Numerical methods for the variable-order fractional advection-diffusion equation with a nonlinear source term. *SIAM J. Num. Anal.* **2009**, *47*, 1760–1781. [CrossRef]
10. Fazio, R.; Jannelli, A.; Agreste, S. A finit difference method on non-uniform meshes for time-fractional advection-diffusion equations with a source term. *Appl. Sci.* **2018**, *8*, 960. [CrossRef]
11. Arshad, S.; Baleanu, D.; Huang, J.; Al Qurashi, M.M.; Tang, Y.; Zhao, Y. Finite difference method for time-space fractional advection-diffusion equations with Riesz derivative. *Entropy* **2018**, *20*, 321. [CrossRef]
12. Jannelli, A.; Ruggieri, M.; Speciale, M.P. Exact and numerical solutions of time-fractional advection-diffusion equation with a nonlinear source term by means of the Lie symmetries. *Nonlinear Dyn.* **2018**, *92*, 543–555. [CrossRef]
13. Badr, M.; Yazdani, A.; Jafari, H. Stability of a finite volume element method for the time-fractional advection-diffusion equation. *Numer. Methods Partial Differ. Eqs.* **2018**, *34*, 1459–1471. [CrossRef]
14. Pimenov, V.G. Numerical method for advection-diffusion equation with heredity. *Itogi Naukii Tekniki Ser. Sovrem. Mat. Pril. Temat. Obz.* **2017**, *132*, 86–90. [CrossRef]
15. Singh, J.; Secer, A.; Swroop, R.; Kumar, D. A reliable analytical approach for a fractional model of advection-dispersion equation. *Nonlinear Eng.* **2018**, in press. [CrossRef]
16. Avci, D.; Yetim, A. Analytical solution to the advection-diffusion equation with the Atangana-Baleanu derivative over a finite domain. *J. BAUN Inst. Sci. Technol.* **2018**, *20*, 382–395. [CrossRef]
17. Mirza, I.A.; Vieru, D. Fundamental solutions to advection-diffusion equation with time-fractional Caputo-Fabrizio derivative. *Comput. Math. Appl.* **2017**, *73*, 1–10. [CrossRef]
18. Mirza, I.A.; Vieru, D.; Ahmed, N. Fractional advection-diffusion equation with memory and Robin-type boundary condition. *Math. Model. Nat. Phenom.* **2019**, *14*, 306. [CrossRef]

19. Hristov, J. Integral balance approach to 1-D space-fractional diffusion models. In *Mathematical Methods in Engineering Applications in Dynamics of Complex Systems*; Tas, K., Baleanu, D., Machado, J.A.T., Eds.; Springer: Cham, Switzerland, 2019; pp. 1–21.

20. Hristov, J. The heat radiation diffusion equation with memory: Constitutive approach and approximate integral-balance solutions. In *Heat Conduction, Methods, Applications and Research*; Hristov, J., Bennacer, R., Eds.; NOVA Science Publishers: New York, NY, USA, 2019.

21. Hristov, J. Response functions in linear viscoelastic constitutive equations and related fractional operators. *Math. Model. Nat. Phenom.* **2018**, *14*, 305. [CrossRef]

22. Povstenko, Y.; Klekot, J. The Dirichlet problem for the time-fractional advection-diffusion equation in a line segment. *Bound. Value Probl.* **2016**, *2016*, 89. [CrossRef]

23. Povstenko, Y.; Kyrylych, T. Time-fractional diffusion with mass absorption in a half-space line domain due to boundary value of concentration varying harmonically in time. *Entropy* **2018**, *19*, 346. [CrossRef]

24. Fulger, D.; Scalas, E.; Germano, G. Monte Carlo simulation of uncoupled continuous-time random walks yielding a stochastic solution of the space-time fractional diffusion equation. *Phys. Rev. E* **2008**, *77*, 021122. [CrossRef]

25. Caputo, M.; Fabrizio, M. A new definition of fractional derivative without singular kernel. *Progr. Fract. Differ. Appl.* **2015**, *1*, 73–85.

26. Andrews, L.C.; Shivamoggi, B.K. *Integral Transforms for Engineers*; SPIE Press: Washington, DC, USA, 1999.

27. Bracewell, R. The Shifting Property. In *The Fourier Transform and Its Applications*, 3rd ed.; McGraw-Hill Science/Engineering/Math.: New York, NY, USA, 1999; pp. 74–77.

28. Atangana, A.; Alqahtani, R.T. Numerical approximation of the space-time Caputo-Fabrizio fractional derivative and application to groundwater pollution equation. *Adv. Diff. Eqs.* **2016**, *2016*, 156. [CrossRef]

29. Rangaig, N.A. Finite difference approximation for Caputo-Fabrizio time fractional derivative on non-uniform mesh and some applications. *Phys. J.* **2018**, *3*, 255–263.

MDPI
St. Alban-Anlage 66
4052 Basel
Switzerland
Tel. +41 61 683 77 34
Fax +41 61 302 89 18
www.mdpi.com

Symmetry Editorial Office
E-mail: symmetry@mdpi.com
www.mdpi.com/journal/symmetry